Y. Tateno T. Iinuma M. Takano (Eds.)

Computed Radiography

With 181 Figures

Springer-Verlag
Tokyo Berlin Heidelberg New York London Paris

Yukio Tateno
Division of Clinical Research
National Institute of Radiological Science
9-1, Anagawa 4-chome, Chiba, 260 Japan

Takeshi Iinuma
Division of Clinical Research
National Institute of Radiological Science
9-1, Anagawa 4-chome, Chiba, 260 Japan

Masao Takano
Fuji Photo Film Co., Ltd.
Miyanodai Development Center
798, Miyanodai Kaisei-cho,
Ashigarakami-gun, Kanagawa, 258 Japan

Revised translation of the original Japanese edition:
Journal of Medical Imagings Vol. 4 Suppl. 1 1984
Originally published by SHUJUNSHA Co., Ltd.

ISBN 978-4-431-66886-2 **ISBN 978-4-431-66884-8 (eBook)**
DOI 10.1007/978-4-431-66884-8

Softcover reprint of the hardcover 1st edition 1987

Typesetting: Asco Trade Typesetting Ltd., Hong Kong

Preface

Recent years have witnessed many technological developments in the field of medical imaging with the introduction of such systems as computed tomography, ultrasonography, and nuclear magnetic resonance. Significant though these technologies are they have failed to affect the dominance of X-ray imaging in medicine owing to its high degree of diagnostic information. Since X-rays were first introduced, however, these have been no major technological innovations and the system has remained essentially the same as that developed by Dr. Roentgen. This position has now changed with the introduction of computed radiography (CR), which represents a transformation of the conventional screen-film system. By virtue of its superior image quality, sensitivity, and image-processing capacity, CR provides greater diagnostic information and seeks to replace conventional X-rays. Since its development in 1981, CR has rapidly established itself as an important new imaging system and its impact in the field of radiology has been remarkable.

Computed Radiography presents both the technical basis of this technology as well as a critical appraisal of its use in a variety of clinical fields. Part of the book appeared in a Japanese monograph in 1984, but over half of the contributions were newly written for the present volume. It is hoped that *Computed Radiography* will facilitate a thorough understanding of this system and its applications.

August 1987 Yukio Tateno

Contents

Part I. Technical aspects

Chapter 1

Introduction

Masao Takano[1]

Almost 90 years have passed since Roentgen took the first X-ray photograph in 1896. The dominant events in technological progress since that time include the development of the screen/film radiography system using a fluorescent screen and the fluorography system using an image intensifier (II) tube (Fig. 1.1). In addition, a number of imaging systems for visualizing the internal parts of the human body have been developed. Since the 1970s, computed imaging systems, which produce cross-sectional images of internal parts of the body by means of various kinds of signal, have appeared with the rapid progress in technology; these systems include X-ray CT, ultrasonic (US), nuclear magnetic resonance (MRI), and positron CT. X-ray CT and US, in particular, are extensively distributed and in common use today.

How are such new imaging systems being practically applied in medicine? The results of a survey carried out in the USA in 1980 are shown in Table 1.1 [1]. According to the survey, 90% of medical images used in examinations are in principle the same as the X-ray images of Roentgen. X-ray CT started to appear in the early 1970s and it was originally thought that it would replace conventional radiography. In reality, however, it accounts for only 8% of medical images. Today, other new technologies such as MRI and positron CT are reaching the level of practical application, and X-ray CT and US will probably be increasingly used. It is very unlikely, however, that such new technologies will exert any significant impact upon the dominant position of conventional radiography. In addition, digital fluorography (DF) made an appearance in about 1980. This new technology has been applied to digital substraction angiography (DSA) but it accounts for only 1% of all X-ray examinations.

It may be said in short that although a wave of technological innovation has made advances in the form of new digital imaging systems in medicine, there have been practically no changes in conventional radiography, which still maintains a dominant position. Why is it that conventional radiography has failed to be affected by digital imaging? It is because the amount of information contained in an analog X-ray image is considerably greater than that contained in an image of any digital imaging system. To be more specific, whereas a radiograph is a high-quality image of at least 4–6 megabytes, X-ray CT images contain only 0.5 megabytes (a tenth of that contained in the former), MRI 0.3 megabytes, and US only 0.07 megabytes [2]. A technological difficulty which had to be overcome to eliminate this big difference in the amount of information yielded lay in the development of a practical digital radiography system to replace the analog X-ray imaging system, i.e., a computed radiography (CR) system.

The CR system aims to replace the present screen/film system of analog X-rays. The system has to be

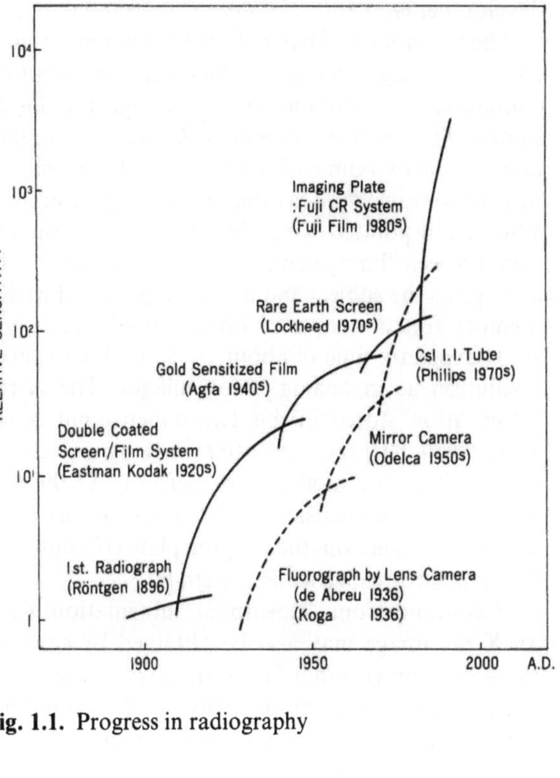

Fig. 1.1. Progress in radiography

[1] Fuji Photo Film Co., Ltd., Miyanodai Development Center, 798, Miyanodai Kaisei-cho, Ashigarakami-gun, Kanagawa, 258 Japan

Table 1.1. Projected number of radiological procedures for 1980

Procedure	Hospitals	Physician's offices	Total
Plain-film radiography[a]	120 400 000	19 900 000	140 300 000
Contrast studies	20 400 000	2 500 000	22 900 000
Ultrasound	6 000 000	1 300 000	7 300 000
Nuclear medicine	5 800 000	SV	5 800 000
Computed tomography	2 900 000	400 000	3 300 000
Special vascular procedures	1 000 000	SV	1 000 000
Diagnostic cardiac catheterization	500 000	0	500 000
Total	157 000 000	24 100 000	181 100 000

SV very small volume
[a] Includes portables

such that its image quality in diagnosis (wide latitude, certainty of visual diagnosis), imaging sensitivity (exposure dose), image-processing capacity, imaging costs, and the degree of imaging freedom (anatomical region, collimation, tube voltage, etc.) have to exceed or equal those of conventional screen/film systems, that it is possible to use existing X-ray imaging systems and techniques, and that it will not bring about drastic changes in the framework of X-ray diagnosis. The CR system also has to be such that its use of digital technology will allow a higher diagnostic efficiency and accuracy through the effective application of image processing and display, allow the long-term storing and efficient retrieving of X-ray images, and lend itself to new systems and developments.

The technological basis of the CR system is that it converts a large amount of diagnostic information contained in a conventional X-ray image into digital signals. The key to a successful CR system is a means of detecting two-dimensional positional information through ultraweak X-rays passing through a subject. With a pin-pointed X-ray beam used in a conventional screen/film system, the sensor itself has to be as large as the subject and has to be provided with a memory function so as to obtain simultaneously a short exposure time of about $\frac{1}{100}$ s and high spatial resolution as in analog X-ray images. The X-ray information stored in this two-dimensional X-ray sensor will be processed after imaging as electric signals of each digital picture element. As the first practical two-dimensional X-ray sensor provided with such a function, the imaging plate (IP) has been developed to realize a CR system [3].

Two-dimensional positional information from an X-ray image may also be obtained by combining a one-dimensional fan-beam X-ray and a one-dimensional X-ray sensor array, or by scanning a subject with pencil-beam X-rays. A number of systems have been proposed, and they are presented in Table 1.2. The basis of a CR system is what type of system is used to produce the X-rays, what type of sensor system is used to convert X-ray image information to digital information, and how X-ray image information is used in the diagnosis as a digital radiograph.

The new medical imaging systems of X-ray CT, US, and MRI developed in recent years have been making a contribution to medical diagnosis by adding new diagnostic information of a different quality to that obtained with a conventional X-ray image. The CR systems, aim at transforming the X-ray image itself, which is of prime importance in current medical image diagnosis, and at developing a new system of image diagnosis for the future. In that sense, it may be said that the CR systems have a possibility of exerting a great impact upon medicine.

Concept of CR system

The basic concept of the CR system may be summarized as follows: (1) To reduce the exposure dose to a minimum; (2) to provide rich diagnostic information, produce consistently sharp images, and offer a wide latitude; (3) to allow a new diagnostic technique to be developed through image processing; (4) to convert X-ray image information into digital signals, store the signals thus converted, and allow retrieval and communication.

A system was envisaged which would improve the quality of images used in medical diagnosis. Since diagnosis in radiography is the result of information processing based on observation with the naked eye, pattern recognition, and diagnostic experience, it was thought sufficient to obtain an accurate image by passing X-rays through a subject and processing

Table 1.2. Digital radiography

System	Maker (announcement)	Exposure system	X-ray beam profile	Minimum exposure time (s)	Imaging system							Display system		
					Principle	Detector profile	Imaging process	Pixel/ line number	Spatial resolution (pixel/mm)	Gray level (bit)	Scan time (s)	Output	Matrix size (pixel)	Gray level (bit)
FCR	Fuji Film (1981)	Conventional	Cone	>1/100	PSL	Area	PSL plate (BaFBr:Eu) ↓ Laser scan ↓ PMT	2510-1760	10-5	10	35–55	Film (10 pixels/mm)	max. 2510 × 2000	10
System One	Digirad (1983)	Conventional	Cone	>1/100	PSL	Area	PSL plate (−) ↓ Laser scan ↓ ( )	2048	6	12	30	CRT	512 × 512	(6)
PDX-4800	ADAC (1982)	Conventional	Cone	>1/100	PC	Area	PC plate (α-Se) ↓ Electro meter scan	1024	5-3	12	90	CRT	1024 × 1024	(6)
Digitalchest	Picker (1981)	Scanning	Fan	4	PL	Line	PL belt (Gd_2O_2S:Tb) ↓ PD array	1024	2	12	4	CRT	512 × 512	(6)
DR-2000	Xonics (1983)	Scanning	Fan	1.5	PL	Line	SC array (C_3I) ↓ II ↓ PD array	2048	4	12	1.5	CRT	1024 × 1024	(6)
Micro-Dose	AS&E (1979)	Scanning	Point	5	PL	Line	SC (MaI) ↓ PMT	1024	6-2.5	10	5	CRT	640 × 512	(6)

PSL photostimulated luminescence, *PC* photoconduction, *PL* photoluminescence, *SC* scintillator, *II* image intensifier, *PD* photodiode, *PMT* photomultiplier tube

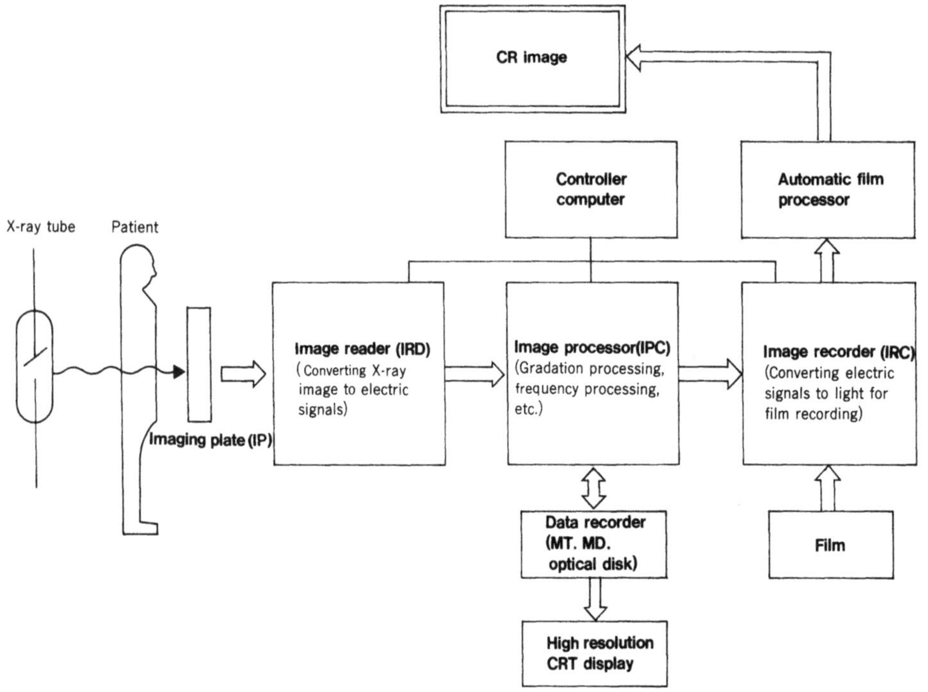

Fig. 1.2. Generalized block diagram

the image obtained to allow detection with a visual system. Therefore, no semiconductor element arrays or II tube/television cameras were used in the detector for X-ray images. Also, no CRT (cathode ray tube) cameras were used to produce the X-ray images, which are final images. Instead, a high-precision laser printer system was used.

In organizing the system, the following points were of importance. In detecting and recording X-ray image information of high quality and sensitivity, a new integrating area detector, the imaging plate, was developed.

For signal reading and conversion of the signals thus read to time-serial signals, no television cameras were used as they would cause deterioration of image quality; instead, a high-precision optical scanning type reading mechanism was used.

To improve the precision of the diagnosis, an algorithm for digital image processing was devised and from this hardware was developed, taking advantage of the results of studies on the quality of X-ray images and characteristics of the visual system.

For X-ray images which are outputs (to be called CR images here as they are different from conventional radiographs), a high-precision laser optical scanning-type recording mechanism was used to record completely the X-ray image information.

In Fig. 1.2, a generalized block diagram of the CR system is shown. The basic X-ray image detector is called the imaging plate (IP) and is made by densely applying crystals of phosphors on a support layer of a polymer material. When X-rays are irradiated onto it, the energy is temporarily stored within the crystals; when an He-Ne laser beam scanning light is irradiated onto it, the X-ray energy stored in the crystals is emitted as fluorescent blue light. The optical signals are then converted to time-serial electric signals, and after A/D (Analog to Digital) conversion, image processing, D/A (Digital to Analog) conversion, and conversion to optical signals, an X-ray image is recorded by a laser beam on a film. The remaining X-ray afterimage on the IP can be erased by applying light for repeated use of the plate. The X-ray image recorded on the film is made as small as possible to allow diagnosis with the naked eye in order to reduce material and energy consumption.

In the following chapters, the functions and performance of the individual elements comprising the CR system are described.

References

1. Johnson JL, Abenathy DL (1983) Radiology 146:851–853
2. Takeshi Iinuma (1984) Jpn Radiation Technology Soc J 40:1004–1015
3. Sonoda M, Takano M, Miyahara J, Kato H (1983) Radiology, 148:833–838

Chapter 2

Imaging plate

Junji Miyahara[1]

Introduction

The imaging plate (IP) is a new and flexible X-ray sensor for the CR system which uses the conventional medical X-ray imaging system and can be employed as a substitute for the screen/film system. The IP is made by densely applying particles of inorganic crystals called photostimulable phosphor onto a polyester film. The phosphor layer has the function of recording an X-ray image. The photostimulable phosphor is a special luminescent material which stores X-ray energy and emits light proportional to the stored X-ray energy when stimulation energy such as visible light is irradiated to it.

The X-ray image is stored to be read out at a later time.

Photostimulable phosphor

Knowledge of the phenomenon of a substance emitting light upon receiving an external energy reportedly dates back to before 1500 BC in China. Such substances are called phosphor. After the discovery of the fluorescent Boronia Stone in Italy in 1603, academic interest in this phenomenon began to grow. In the second half of the nineteenth century, the phenomenon of a substance emitting when excited by ultraviolet (UV) light and emitting again when stimulated by infrared light after exposure to UV light was discovered. This phenomenon is called photostimulated luminescence (PSL); zinc sulfide, alkalihalide, silicon, diamond, oxidized compounds, organics are known to be PSL substances. A large number of substances are associated with this phenomenon. The PSL phenomenon occurs as the information of the initial stimulus (primary excitation) is stored in the substance; this initial information is read out when light (secondary stimulation) is applied to it.

[1] Fuji Photo Film Co., Ltd., Miyanodai Development Center, 798, Miyanodai Kaisei-cho, Ashigarakami-gun, Kanagawa, 258 Japan

A crystal of barium fluorohalide ($BaFX:Eu^{2+}$, X = Cl, Br, I) containing a very small amount of bivalent europium ions was adopted as a photostimulable phosphor showing PSL for the IP after extensive compounding of and testing a large number of substances. This substance shows the strongest PSL among all substances known to control the compounding process. A BaFX compound is an ion crystal having the structure shown in Fig. 2.1. This compound becomes colored when a color center called the F center is formed under irradiation by X-rays or ultraviolet rays for a long time [1]. The color center is a defect in the crystal which absorbs light of specific wavelength within the visible radiation; the F center is where an electron has been caught at the point where a specific atom forming the crystal has been removed (called the vacancy). The europium ion crystallized in a very small amount is called the activator. It replaces the barium ion in the BaFX crystal and is crystallized when a phosphor forms, creating a so-called luminescence-center [2]. It has gradually become appreciated that these F and luminescence-centers together play an important role in storing X-ray information [3].

Figure 2.2 shows energy levels expressing the PSL mechanism of the photostimulable phosphor [4].

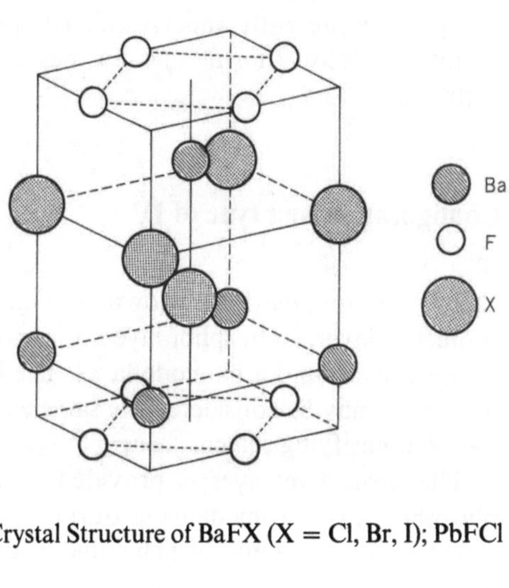

Fig. 2.1. Crystal Structure of BaFX (X = Cl, Br, I); PbFCl type

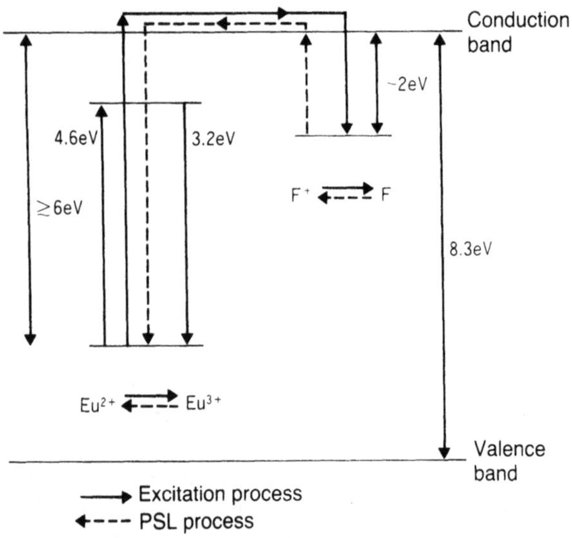

Fig. 2.2. Photostimulated luminescence (PSL). Mechanism of—BaFBr:Eu^{2+}. It is the same mechanism for BaFCl: Eu^{2+} as for BaFI:Eu^{2+}

The europium ions in the crystal are ionized by the primary excitation (imaging) of the X-rays from bivalent to trivalent ions and electrons are released to the conduction band. The electrons thus released are captured in the vacancy of the halogen ions (previously formed) by Coulomb's force and a semistable F center state is produced. It is in this state that the X-ray image information is stored. If visible light (secondary stimulation light, reading light) to be absorbed to the F center is then irradiated, the electrons captured in the F center will be released again to the conduction band; they will be captured by the trivalent europium ions to generate excited bivalent ions, and energy will be discharged in the form of luminescence.

The above is the PSL mechanism of the BaFX:Eu phosphor as presently understood. The mechanism is not yet fully known, and further studies are required.

Configuration and type of IP

The IP is composed, as shown in Fig. 2.3, of a protective layer, a phosphor layer, a support layer, a backing layer, and a bar code label. The basic configuration may be considered the same as that of an X-ray intensifying screen.

The protective layer is provided to guard the phosphor layer from damage during handling or transfer within the unit, which could lead to defects on the image. The protective layer is basically present to prevent scratching, elongation, contraction, or cracking as a result of temperature and/or humidity changes, stop deterioration due to light or X-rays, and provide sufficient strength against bending or wear during transfer within the unit. With respect to image characteristics, the protective layer should be as thin and as transmissive to light as possible. A material satisfying these opposing requirements is a polyethylene *terephtalate* (PET) film.

The phosphor layer is made by applying a PSL dispersed in an organic polymer (binder) onto a support and drying it. The organic polymer is added to bind the phosphor particles. The properties are such that the phosphor has excellent dispersion characteristics; it is possible to form a uniform membrance on the support layer; no property changes are caused by temperature, humidity, X-rays, laser light, etc., and proper flexibility and mechanical strength are provided. Various highly polymerized materials such as nitrocellulose, polyester, acryl, and uretan are used as materials for the binder.

As a photostimulable phosphor, particles of BaFX:Eu^{2+} (X = Cl, Br, I), which have an average diameter of 5–10 μm, are used. In general, the larger the particles the greater their inherent light-emitting efficiency, reducing the absorption of light scattered in the phosphor layer, and thus the apparent sensitivity of the IP itself will be improved. On the other hand, the sharpness of the image will be reduced, and the inherent noise of the IP, associated with its granularity, will be disadvantageous. In other words, the structure noise will be inexpedient. It is not only the particle size, however, that determines the sensitivity and image characteristics; various factors other than the laser beam in the phosphor layer and the particle size, which affects the scattering and absorption of the light emitted, will also have an important influence. To improve sharpness, for instance, a system for coloring the inside of the phosphor layer is used for the conventional screen. In this case, the light emitted instantaneously when the X-rays are irradiated shows multiple scattering and spreads to reach the X-ray film where the light emission is recorded. The key to improving the sharpness is, therefore, suppression of the spreading of the light emitted, which is achieved by coloring. In the case of the IP, on the other hand, the light emission is detected by a focusing system with an opening larger than a single picture element; the light emission from a picture element thus contains all the information of that picture element, even if the light emitted is scattered in the phosphor layer. To improve sharpness, therefore, the phosphor layer has to be colored such that no emitted light is absorbed; instead, the scattered laser light is absorbed. For the phosphor layer,

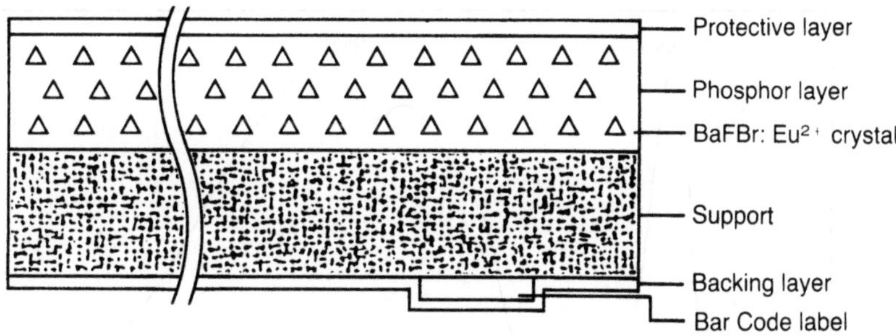

Fig. 2.3. Structure of imaging plate

various additional measures are taken to improve the image quality.

The supporting layer has to be such that it protects the phsophor layer from external forces and shocks to the IP; it has to be flat and have appropriate flexibility with mechanical strength. A black PET film containing carbon is used as the material for this layer. It prevents reflection of the laser light on the boundary of the phosphor and supporting layers and thus improves sharpness by giving light absorption properties to the supporting layer itself.

The backing layer protects the IP surface from scratches during transfer and in storage. A soft polymer film is laminated for this layer.

On the bar code label is recorded a number assigned to the IP. This provides verification that the IP has been used as well as ID and imaging information.

The type and size of the IP are presented in Table 2.1. The sensitivity of the IP cannot be individually determined as in the case of the conventional screen/film system. With the IP, standard type (ST) is generally used for a wide range of medical applications. The HR type for higher resolution uses a photostimulable phosphor of finer particles than the ST and is made thinner to improve the sharpness. The SB for subtraction is used in combination with the IP Rapid Changer. The MS type for simultaneous multisection tomography uses three each of three different types of IP with different sensitivities—a total of nine IPs—and is so organized that the amount of X-ray absorption is roughly the same from the top to the bottom layers.

Intrinsic characteristics of IP

Emission and stimulation spectrum

The photostimulable phosphor emits a blue-purple light, as shown in Fig. 2.4. This emitting light is caused by a very small amount of bivalent europium ions as the luminescence center contained in the phosphor. The luminescence intensity changes according to the wavelength of the light irradiated on the IP. This wavelength dependency of the PSL intensity is expressed in the photostimulation spectrum, as shown in Fig. 2.5. PSL can be most effectively obtained with red light at about 600 nm. The photostimulation spectrum is in good agreement with the absorption spectrum of F center (color

Table 2.1. IP types and sizes

Type	Application	Nominal size designation in.	cm	Plate size (mm)
ST[a]	General radiography	14 × 17	35 × 43	354 × 430
		14 × 14	35 × 35	354 × 354
		10 × 12		252 × 303
		—	24 × 30	238 × 298
		8 × 10	—	201 × 252
		—	18 × 24	178 × 238
HR[b]	Mammography (high resolution)	8 × 10	—	201 × 252
		—	18 × 24	178 × 238
SB[c]	Subtraction	14 × 14	35 × 35	354 × 354
		10 × 12	—	252 × 303
		—	24 × 30	238 × 298
MS[d]	Multisection tomography	14 × 14	35 × 35	354 × 354
		10 × 12	—	252 × 303
		—	24 × 30	238 × 298

[a] Standard type designed to cover a wide range of medical subjects and exposure conditions
[b] Designed for mammography and similar precise radiography, retaining a higher resolution than ST
[c] Designed exclusively for subtraction. Sensitivity variations within the IP and changes in optical density within a radiograph are kept to a minimum when subtraction is performed. The sensitivity is the same as with ST. An IP rapid changer for angiography is recommended
[d] Designed for simultaneous multisection tomography such that X-rays are absorbed as evenly as possible from the upper to lower plates. This type is available as a set of nine plates to enable multisection tomography of nine cross-sections, and plates are numbered from 1 to 9. A special cassette is required for MS

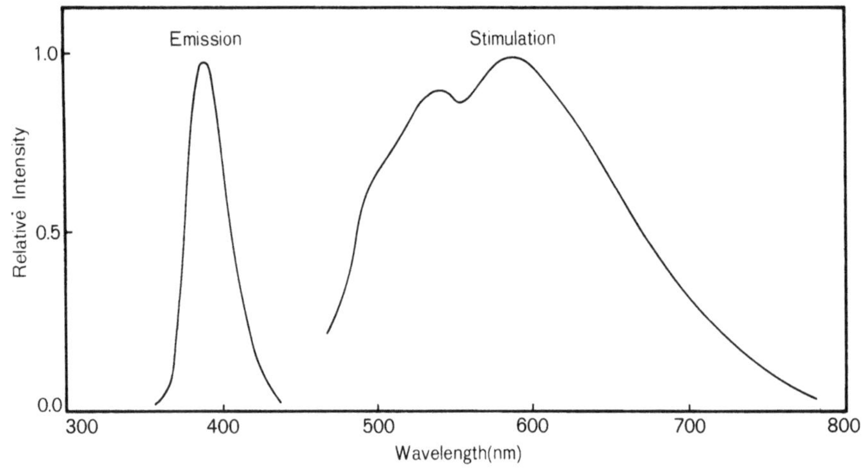

Fig. 2.4. Emission and stimulation spectra for photostimulable phosphor, BaFBr: Eu^{2+}, used in the IP. The peak of the emission spectrum is at 390–400 nm which is close to the spectral sensitivity wavelength peak of the emission detector (photomultiplier tube). The stimulation spectrum matches the wavelength of the He-Ne laser.

center, as mentioned above) in the phosphor generated as a result of X-irradiation; this is why the F center, related to the vacancy of halogen ions in the crystal, causes PSL.

With a CR system using the IP, it is necessary to improve the S/N ratio of the image such that the wavelength of the photostimulation light and that of the emitting photostimulated light with the X-ray information are sufficiently different to allow optical separation and that the peak of the PSL spectrum is at 400 nm; at this point, the photomultiplier tube (PMT) has a high detection efficiency. The two spectra shown in Fig. 2.4 have favorable characteristics to meet these requirements.

Time-response characteristics of emission

When He-Ne laser light stimulates the IP after X-irradiation, photostimulated light will be promptly emitted. When stimulation is stopped, emitting light will also stop. Light emission will not be totally stopped immediately but gradually with an attenuation characteristic of the light-emitting process of the photostimulable phosphor. This attenuation is especially significant for a CR system which reads the X-ray image information from the IP at high speed. When the attenuation is slow, even if scanning is made while the laser light irradiates a picture element, both the light emission (X-ray image information) of that picture element and light emission from previous picture elements will be detected, leading to deterioration of the image quality.

Figure 2.5 shows the attenuation characteristics of the photostimulated light emission in the initial stage immediately after the He-Ne laser light is turned off. The light emission attenuates exponentially. The time when the light emission intensity is equal to $1/e$ ($e = 2.718$) of the initial value is termed the light emission life. For the IP, the light emission life is 0.8 μs, which is determined by the bivalent europium ions, which form the luminescence center, and the crystal composition of the phosphor. This short light emission life makes it possible to read an X-ray image of a large area with a high density in a short period of time.

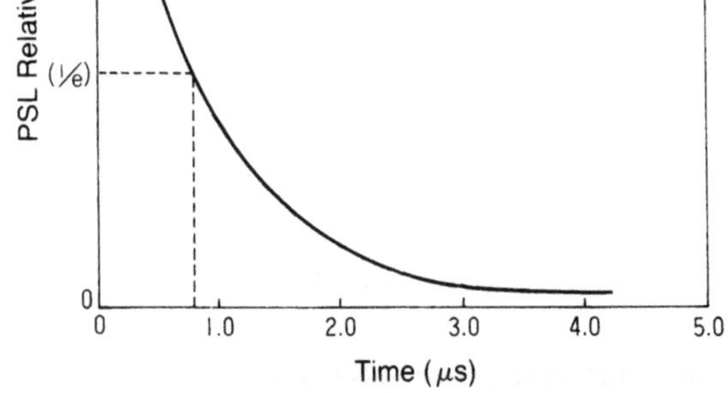

Fig. 2.5. Response characteristics of the light emission of photostimulable phosphor BaFBr: Eu^{2+} used for the IP. It has a light emission life (time to attenuate down to $1/e$) of 0.8 μs

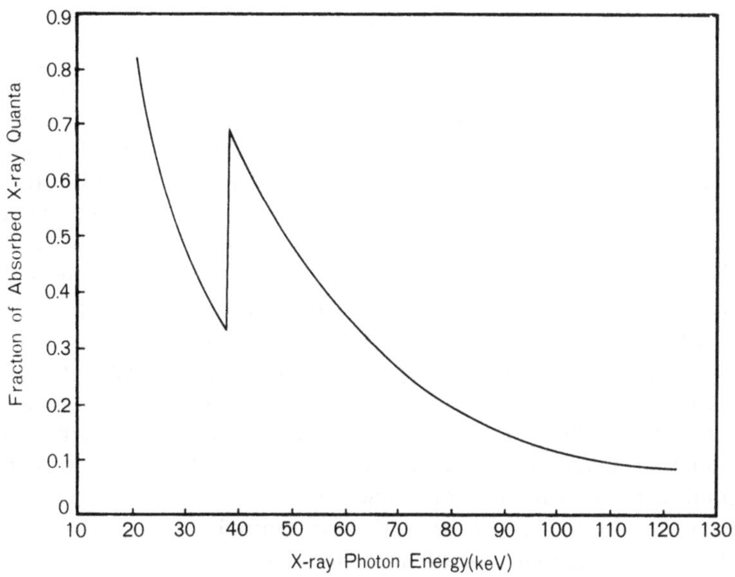

Fig. 2.6. Example of X-ray absorption spectrum of the IP. A sudden change is observed near 37 keV showing the K-edge for Ba

Dynamic range

The X-ray absorption spectrum of the IP is shown in Fig. 2.6. The horizontal axis shows the energy of the X-ray quanta irradiated to the IP, and the vertical axis indicates the X-ray absorption factor of the IP corresponding to the X-ray quanta. The rapid absorption at about 37 keV is caused by the K absorption edge of a barium atom, which is the major element composing the phosphor, and is not dependent on the light-emitting characteristics of the phosphor.

The dependence of the amount of light emission from the IP on the irradiation dose of X-rays when they are irradiated to the IP is shown in Fig. 2.7. As seen from the figure, a good linearity is shown for an X-ray dose of over $1:10^4$ range. Such a wide dynamic range is not seen with the conventional two-dimensional X-ray detector and is one of the major characteristics of the CR system using the IP. Even with an ordinary fluorescent screen, the light instantaneously emitted from the phosphor upon X-irradiation bears a linear relationship to the X-ray dose over a wide range. An X-ray film together with a fluorescent screen is required as a detector to record the light emission and at the same time have the function of a display, allowing the reader to see the results obtained. The X-ray film, therefore, has a narrow dynamic range of high photographic gradation and does not effectively take advantage of the wide-range light emission linearity of the fluorescent screen itself. With the CR system, on the other hand, as the photostimulated light emission of the IP is converted to electric signals with the PMT, the over $1:10^4$- dynamic range of the IP is fully utilized to provide useful information for the diagnosis.

This wide dynamic range makes it possible to detect precisely a small difference in the X-ray absorption characteristics of each tissue of the subject; at the same time, it permits a fully automatic image-processing system, allowing digital radiographs to be obtained which are always stable under any X-ray imaging conditions. The IP is capable of reading the X-ray information in two stages within a system. In the first step, an image is roughly read at *ultra-high speed* with a laser light, and a histogram of the

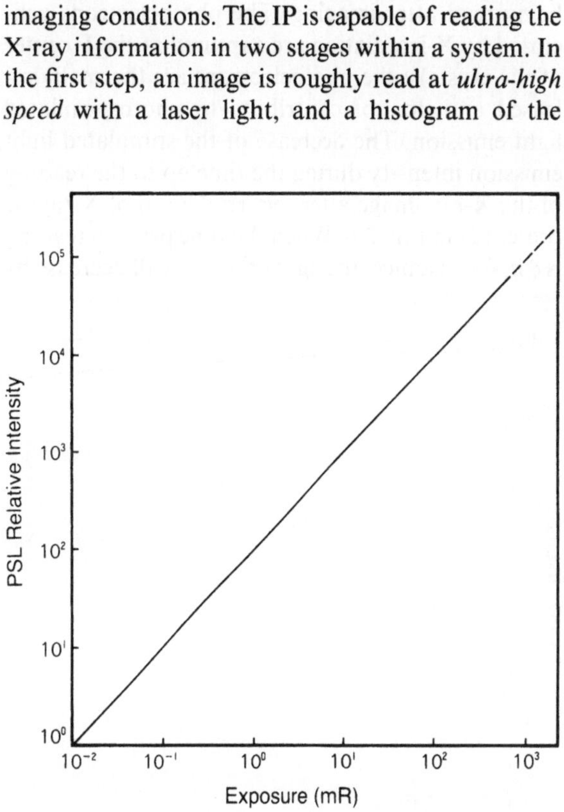

Fig. 2.7. Dynamic range of the IP. Excellent linearity is shown over a wide range of over $1:10^4$ from a low to a high dose

photostimulated light emission dose of the X-ray image is instantaneously calculated. In the second step, the sensitivity of the PMT and the gain of the amplifier are automatically set on the basis of the information thus obtained, and the X-ray image is read with an ultra-strong laser light at high precision. In the image-procession section, image processing is performed under conditions suitable for the intended diagnosis on the basis of the total information, including the image itself, the purpose of imaging and imaging conditions input from the console, the histogram information obtained in the first step, and the system's internal menu. With this system, it is possible to obtain digital X-ray images with optimum image processing consistently at an optimum density with any subject and with any X-ray exposure dose. This can be seen as an example of the concrete introduction of the concept of quality control to radiography.

Fading

Fading refers to the phenomenon whereby X-ray image information stored in the memory by X-irradiation decreases with the passage of time before being read after imaging. The photoelectrons generated by X-irradiation are captured at the F center of the crystals; the photoelectrons are thermally released and do not contribute to photostimulated light emission. The decrease of the stimulated light emission intensity during the time up to the reading of the X-ray image after the irradiation of X-rays is presented in Fig. 2.8. When the time prior to reading is 8 h, for instance, the light emission will decrease by about 25%. Fading will increase with time and with increasing storing temperatures. This cannot be avoided when photostimulated light emission is used. It is possible, however, to reduce the amount of fading, and it is one of the important keys to the modification of the photostimulable phosphor.

In practice, however, the reduction of light emission due to fading is electrically and automatically compensated for by the PMT gain within the CR system, and there are hardly any problems in the diagnosis. With an IP imaged at a less than normal exposure and subsequently left for a long time prior to reading, granular deterioration will occur due to an increase in the fluctuation of the detected X-ray quantums and the natural radiation. It is thus desirable to obtain an image within 8 h.

Influences of natural radiation

The IP is sensitive not only to X-rays but also to other forms of electromagnetic waves, such as ultraviolet and gamma rays, and corpuscular rays, such as alpha, beta, and electron rays, and accumulates their energy so as to be capable of detecting it in the form of an image. It is a sensor that is unusual in being so highly sensitive. This characteristic of the IP, however, indicates that it will be influenced by radiation from the walls or fixtures of the building in which it is located, natural radioactive elements, cosmic rays, or the very small amount of natural radioactive elements contained in the IP itself, which is inevitable with the present technology. In fact, if a well-erased IP is left for a long time and an image is

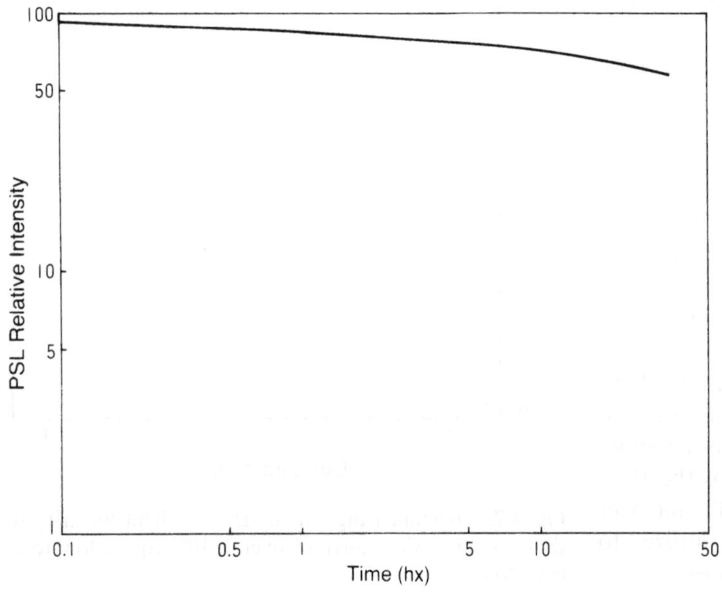

Fig. 2.8. IP fading characteristics (25°C)

constructed from it with a CR system under highly sensitive conditions, a number of small black spots will randomly appear depending on the length of time the IP was left. These black spots can be classified into those originating in the IP itself and those originating in the forms of natural radiation listed above. Figure 2.9 shows that the amount of black spots is dependent on time, with the natural radiation shielded by a 10-cm-thick lead plate and without such shielding. From the data, it can be seen that when reading the image on the IP, especially under high-sensitivity conditions, the IP should not be left for a long time and that the IP should be erased by irradiating light to remove the effects of radiation prior to imaging if it does have to be left for a long period. It should be noted that concrete, marble, and potassium-containing materials constantly emit natural radiation.

Image quality characteristics of the IP

With the conventional screen/film system, once the combination of screen and film has been chosen, the sensitivity and image quality (gradation, sharpness, granularity) will be automatically determined. With the IP, both the sensitivity and image quality can be freely selected according to the conditions of the laser optics, reading mechanism, and signal-processing of the CR unit. Thus, the sensitivity and image quality of the IP can be individually selected unlike the screen/film system.

In this section, to examine the inherent problems of the IP, the sensitivity and image quality under standard conditions will be discussed. For details about the relationship between image processing and image quality, the reader is referred to Chap. 4.

Sensitivity

With the screen/film system, the sensitivity is automatically determined when the combination is selected. With the IP, the amount of photostimulated light emission is linear with the X-ray dose for over $1:10^4$ range. Below this limit, therefore, the IP is available for any range of X-rays and the sensitivity can be freely set according to the image and purpose of diagnosis.

As stated above, the sensitivity can be determined for individual cases. For the sake of convenience, however, it may be expressed with a relative value of the amount of photostimulated light emission when laser light of a particular intensity is irradiated following a certain dose of X-irradiation. It does not mean, however, that this relative value will be immediately reflected in the imaging sensitivity. With the CR system, the information stored in the IP is retrieved in the form of electric signals, and it may be necessary, therefore, to define a new concept of sensitivity to relate the signal-to-noise (SN) ratio of these signals or detection quantum efficiency (DQE) to a factor as an image. This aspect remains to be studied.

Granularity

For the evaluation of granularity, the root mean square (RMS) granularity is generally used. Normally, this value indicates a standard deviation of fluctuation of film density. With the CR system, the fluctuations of the electric signals from the IP can be directly computer-processed to obtain the RMS value.

With a digital radiograph obtained with the CR system, the factors contributing to granularity may be roughly classified into the quantum noise of X-rays, the stimulated light emission quantum noise of the IP, the structure noise of the IP, optical noise, electric noise, and the inherent noise of the CR sys-

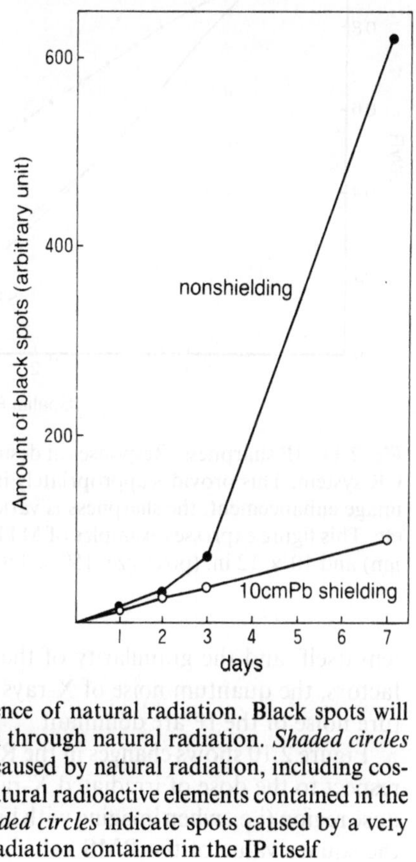

Fig. 2.9. Influence of natural radiation. Black spots will randomly form through natural radiation. *Shaded circles* indicate spots caused by natural radiation, including cosmic rays and natural radioactive elements contained in the IP itself; *unshaded circles* indicate spots caused by a very small dose of radiation contained in the IP itself

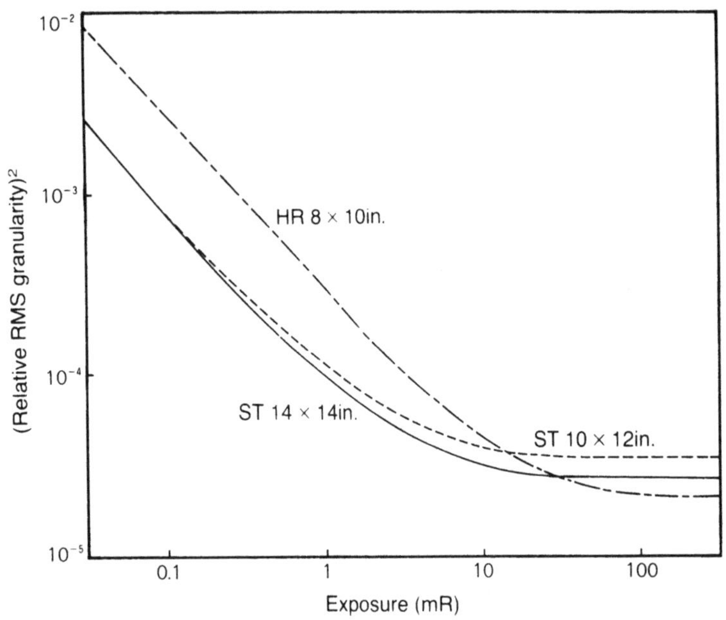

Fig. 2.10. IP granularity. The IP has a wide exposure range latitude in the CR system. It is necessary to conceptualize granularity for each dose region. In the low-dose region (high-speed exposure), granularity is mainly determined by the quantum noise of the X-ray exposure; in the high-dose region (low-speed exposure), it is determined by the structure noise of the IP. This figure expresses the granularity curve (RMS value) as a function of dosage

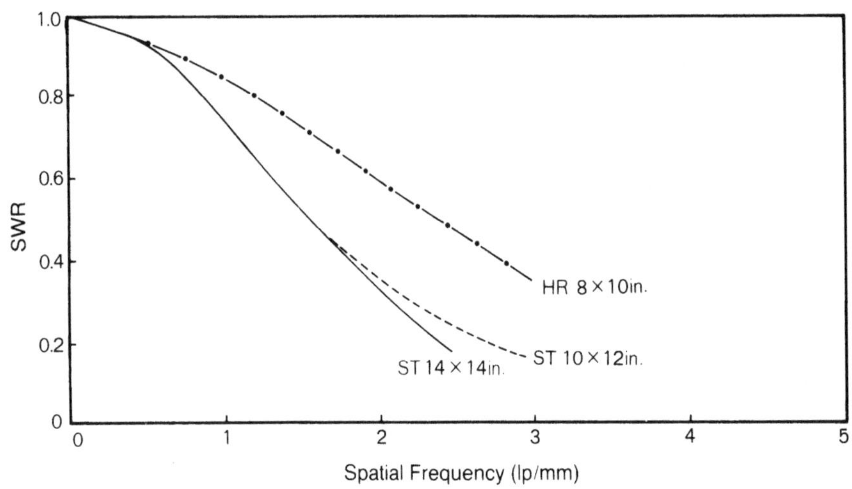

Fig. 2.11. IP sharpness. Responses at desired spatial frequency regions can be accentuated by image enhancement in the CR system. This provides appropriately intensified diagnostic images for every exposure region and method. Without image enhancement, the sharpness is varied by the pixel size, IP type laser power, laser beam diameter, reading speed, etc. This figure expresses examples of MTF curves without image enhancement for ST 14 × 14 in. (pixel size: 200 × 200 μm) and 10 × 12 in. (pixel size: 150 × 150 μm) and for HR 8 × 10 in. (pixel size: 100 × 100 μm)

tem itself, and the granularity of thc film. Of these factors, the quantum noise of X-rays and the structure noise of the IP are dominant.

Figure 2.10 shows changes in the RMS value with respect to the dose of irradiated X-rays. In the low-dose region (i.e., when imaging with high sensitivity), the square value of the RMS value approaches the straight line of the gradient of minus 1. This indicates that the granularity in this region is dominated by the granularity mainly caused by the fluctuation in X-rays (quantum noise). In short, the number of X-ray quanta descreases with the decrease in dose, leading to deterioration of granularity.

In the high-dose region (i.e., when imaging with

low sensitivity), on the other hand, the square value of the RMS value roughly approaches a constant value, and the dependency on the amount of X-ray irradiation almost disappears. This is because the number of X-ray quanta increases and its fluctuations do not contribute to the granularity, and the structure noise, which is almost inherent to the IP, is dominant.

Sharpness

Sharpness is generally expressed as the square wave response (SWR). To determine the SWR with the conventional screen/film system, a sharpness chart is imaged by X-ray and the image obtained is measured with a microdensitometer. With the IP, on the other hand, the electric signals from the IP obtained with the CR system are directly computer-processed to give the SWR.

Figure 2.11 shows the SWR of the IP. In the HR type, the phosphor and the IP have been especially designed to improve sharpness, and the reading resolution is increased.

References

1. Yuste M, Taurel L, Rahmami M, Lemoye D (1976) Optical absorption and ESR study of F-centers in BaFCl and SrFCl crystals. J. Phys. Chem. Solids 37:961–966
2. Stevels ALN, Pingault F. (1975) BaFCl; Eu^{2+}, A new phosphor for X-ray intensifing screens Philips Res. Rept. 30:277
3. Miyahara J (1986) Imaging Plate—A New area detector and its application. In Japanese Solid State Phys. 21:171–179
4. Takahashi K, Kohda K, Miyahara J, Kanemitsu Y, Amitani K, Shionya S (1984) Mechanism of photostimulated luminescence in BaFX;Eu^{2+} (X = Cl, Br) Phosphors. J. Lumi. 31 & 32:266

Chapter 3

Image reading: Conversion to digital signals

Hisatoyo Kato[1]

Principle of reading

An X-ray image stored on the IP is spatially continuous analog information, and the laser scanner, shown in Fig. 3.1, is used to read the information and convert it into digital signals. By scanning the converged laser light with a vibrating or polyhedral rotating mirror in a direction perpendicular to that of the scanning light, while precisely transferring the IP, it is possible to scan two-dimensionally all side of the IP one after another with a laser light spot. The photostimulated light excited by the laser light spots is directed to the photomultiplier tube by the high-efficiency light guides located along the scanning line of the laser light to be converted to electric signals. The electric signals thus converted are analog signals proportional to the amount of photostimulated light emission. They are properly amplified and logarithmically converted to pass through the A/D converter to become digital signals.

By repeating this process while the laser light spot scans all sides of the IP, it is possible to convert the X-ray image recorded with spatial continuation (analog) on the IP to time-serial digital signal arrays.

Factors determining image quality

The image quality of the CR image-reading system is determined by a number of factors, as shown in Fig. 3.2. Although there are a number of parameters describing image quality, they may as a rule be classified into two groups: (1) factors associated with whether the signals are good or bad, i.e., those which are photographically termed sharpness of image and those which are electrically and optically termed frequency response or component; (2) factors concerning the degree of noise in obstructing signals, i.e., those which are photographically termed granularity and those which are electrically and optically termed noise. These factors may also be classified into two other groups: (1) those arising mainly from the characteristics of the IP; (2) those arising from the optical and electrical characteristics of the reading system. The former is described in Chap. 2 of Part I; the latter will be taken up here. Another factor determining the image quality is quantization. Quantization is basically done to convert an X-ray pattern (analog signals) spatially and in terms of intensity into a discrete digital signal pattern. Quantization involves problems of image sharpness (frequency component) which can be transmitted, noise, and artifacts. The concept of "contrast detecting capability" as a basic physical characteristic of the image-detecting system is used in analyzing the image characteristics of the CR image-reading system.

Sharpness/frequency component

The sharpness is basically determined by the characteristics of the IP itself and the optical and electrical characteristics of the reading system; specifically, it is determined by how far the spot diameter and the excited laser light itself are dispersed and spread in

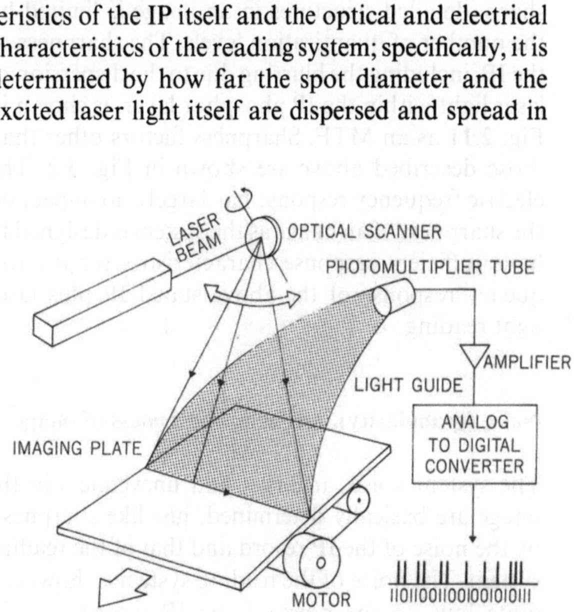

Fig. 3.1. Construction of the image-reading system

[1] Fuji Photo Film Co., Ltd., Miyanodai Development Center, 798, Miyanodai Kaisei-cho, Ashigarakami-gun, Kanagawa, 258 Japan

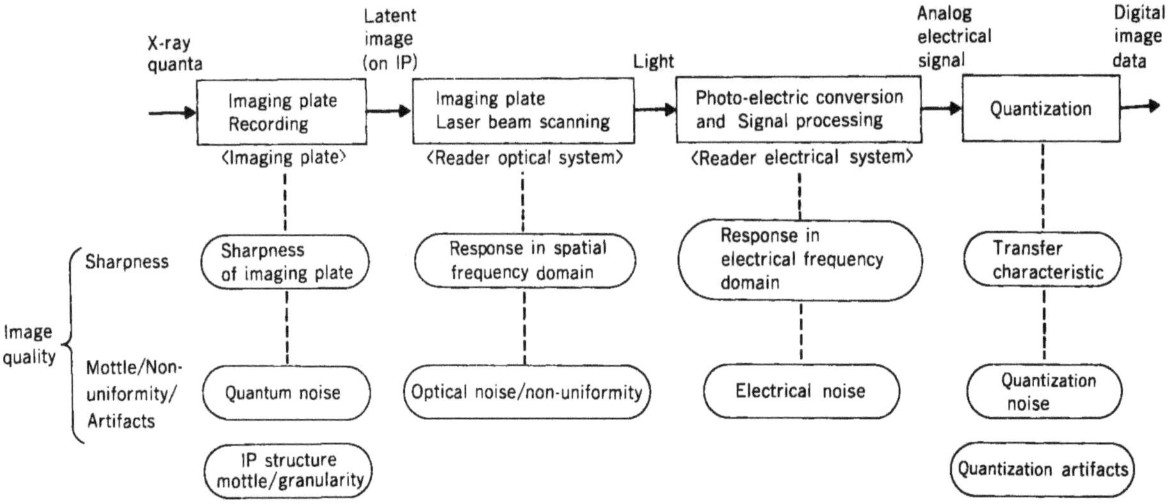

Fig. 3.2. Factors governing image quality in CR reading system

the fluorescent body layer of the IP. The emitted light is also dispersed in the same fluorescent body layer, but it has no effect upon the sharpness; there is a problem, however, of how to detect the emitted light efficiently. In any case, in order to improve sharpness, it is necessary to use a laser light with a sufficiently small spot diameter and an IP with a fluorescent body layer of low laser light dispersion. In addition, it should be noted that precisely transmissible frequency components are determined from the number of spatial quantization levels in converting the image into digital signals, as described in the sampling theory (see p. 19); thus, the delineation of sharp, detailed structures in an image is limited by the number of quantization levels. The sharpness of the IP, including the blurring due to the dispersion of laser light within the IP phosphor layer, is shown in Fig. 2.11 as an MTF. Sharpness factors other than those described above are shown in Fig. 3.2. The electric frequency response has largely no impact on the sharpness of an image as the system is designed to have sufficient response characteristics for the frequency response of the above-stated IP plus laser light reading.

Noise (granularity), artifacts, unevenness of image

The system noise, artifacts, and unevenness in the image are basically determined, just like sharpness, by the noise of the IP record and that of the reading system. The noise of the reading system is, however, quite low, so the noise of the IP record may be considered to be that of the entire reading system. The noise of the IP record may be classified roughly into X-ray quantum noise and the structure noise of the IP itself. As shown in Fig. 2.10, the X-ray quantum noise (which increases as the X-ray dose decreases) is dominant in the low X-ray dose region, and the structure noise (constant regardless of the X-ray dose) is dominant in the high X-ray dose region.

Artifacts and density irregularity (unevenness of image) are both considered types of noise. The former is generated from digital processing and is well known (see below). In addition, the moire pattern of the laser light scanning lines with a grid image recorded on the IP and dust images on the IP are also considered artifacts. Density irregularity includes scanning line irregularity and IP sensitivity irregularity; these could influence the image quality in diagnosis, but they are difficult to quantify.

Reading quantization

The question of how much detail, both spatially and in terms of intensity, of what is read should be converted to digital signals can be answered theoretically by setting the detailed structure of the image in sufficient detail to allow satisfactory transmission by following the sampling theory. In practice, however, it is extremely difficult to determine the extent of the required structural details to be transmitted. It may sound satisfactory to quantize as far as possible, but this would involve excessive costs and would result in an enormous amount of data and processing. It is, therefore, desirable to make quantization with as small a number of quantization levels as possible, and to limit the data to a minimum. On the other hand, if the level of quantization is too 'rough' during the conversion into digital signals, picture elements with mosaic (when spatial quantization is rough) or contour patterns (when quantization of strength is too rough) will appear as artifacts and obstruct the diagnosis. The reading quantization (conversion to digital signals) level of CR images was determined by producing a number of X-ray images of various quantization levels by simulation and subjectively and statistically evaluating them from the standpoint of X-ray diagnosis (Fig. 3.3.) The simulated images were constructed by converting an ordinary X-ray photograph to digital signals in sufficient detail (to be specific, 20 pixels/mm, 12 bits/pixel with a film digitizer of high resolution, reducing the resolution gradually by means of computer processing, and reproducing an image on a film with a laser printer. Evaluation was performed by a number of radiologists, who compared the original X-rays with the simulated photographs. The evaluation results of the chest X-rays are presented in Table 3.1. It can be seen in the table that if the range indicated by *No diff.* or *No subst.* is considered a range without deterioration from quantization, with a minimum quantization level of 5 pixels/mm and 8 bits/pixel or 6.7 pixels/mm and 7 bits/pixel, details of the X-rays can be sufficiently maintained for diagnosis. In terms of the amount of information, the former combination (5 pixels/mm, 8 bits/pixel) would be better as the amount of digital data is lower. The same evaluation was repeated with different kinds of imaging objects and subject sizes (film sizes) to obtain details about the appropriate levels for CR. Intensity refers to the

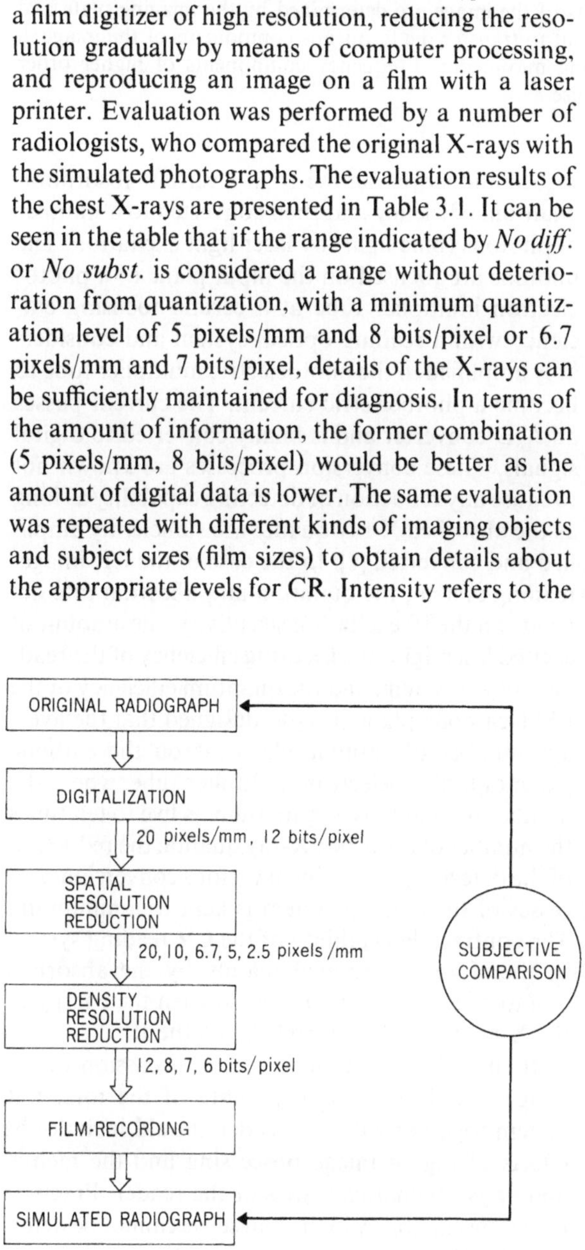

Fig. 3.3. Experimental procedure for determining *digitalization* conditions of X-ray images

Table 3.1. Result of visual evaluation of 20 chest radiographs by four radiologists

Spatial resolution (pixels/mm)	Density resolution (bits/pixel)		
	8	7	6
20	No diff.	No diff.	No diff.
10	No diff.	No diff.	Sev.
6.7	No diff.	No subst.	Sev.
5	No subst.	Det.	Sev.
2.5	Art.	Art.	Sev.

No diff. no difference observed, *No subst.* no substantial difference observed, *Art.* artifacts observed, *Det.* deterioration observed, *Sev.* severe artifacts observed

amount of light emission when reading the IP, which is proportional to the absorbed X-ray energy at the IP. It is necessary to make the quantization excessive in detail. This is because the increase in intensity due to quantization in a low-intensity area with a small amount of light emission is conspicuous compared with a high-intensity area with a large amount of light emission, if quantization is made linearly with intensity. Intensity is logarithmically converted to expand the low-intensity area, and the high-intensity area is compacted; quantization is then done to make the intensity increase uniformly less conspicuous. The value obtained by logarithmically converting strength is proportional to the "photography density" on a film. This is comparable with linearly quantizing the "image density" in photography.

The relationship between quantization and the sharpness/frequency component of the image may be explained by the sampling theory. For spatial quantization, there will be no problem if quantization is made with a sampling frequency indicated by the sampling theory to allow sufficient transmission of the frequency component of an image. The spatial quantization level shown in Fig. 3.4a is thought to be sufficient to meet the requirements. If quantization is made with a low sampling frequency (rough quantization level) as shown in Fig. 3.4b, on the other hand, so-called undersampling will result, causing a mixture of the high- and low-frequency components (Aliasing phenomenon) due to high harmonic modulation; image noise will result. In this case, as seen in Fig. 3.4 the frequency component of the image has been largely transmitted. It is a common misunderstanding that when the quantization level is too low undersampling occurs and the sharpness/frequency component of the image is reduced; noise will only increase on an image as a result of the Aliasing phenomenon. In any case, quantization

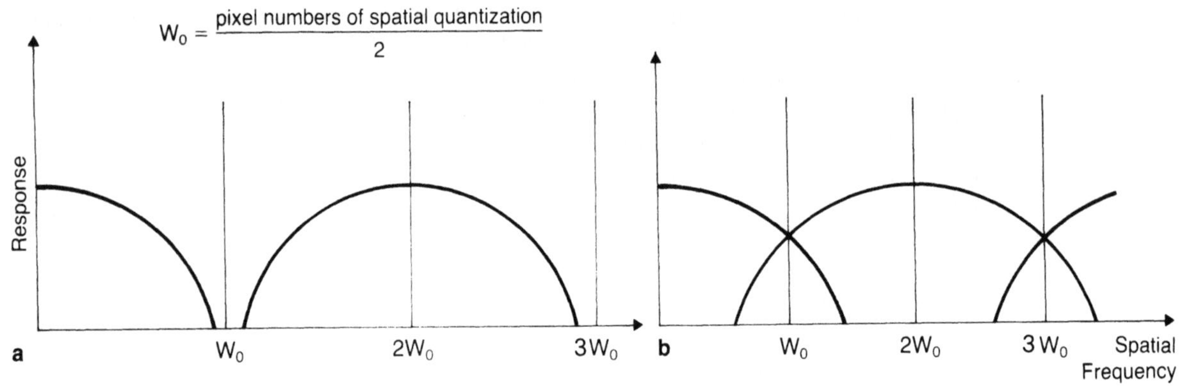

Fig. 3.4a, b. Sampling theory. Spatial frequency components of the image are determined by the imaging plate and reader system. **a** Quantization which has a resolution sufficient to transfer the frequency components of the image. **b** Quantization which has a coarse resolution. Undersampling may occur. Frequency components of higher order harmonics overlap zero order components and generate a noise

should be done with a sampling frequency (quantization level) that is appropriate according to the sampling theory. With respect to quantization of intensity, however, if the quantization is narrow compared with the intensity of the image noise, the image signals can be sufficiently transmitted; if the quantization is wide, some of the image signals will be lost, resulting in false contouring (contour artifact). The intensity quantization level shown in Table 3.2 is set to allow sufficient transmission of the signals.

Contrast detecting capability

Contrast detecting capability expresses the basic signal detection characteristics of an imaging system like photography in terms of to what extent weak signals or contrast can be detected. In the case of the CR image-reading system, this detecting capability is determined by how effectively the X-ray quanta entering the IP can be converted to electric signals. Figure 3.5 shows a flow of incoming X-ray quanta being converted to electric signals after going through various conversion processes with statistical fluctuations. First, the incoming X-rays are absorbed and stored on the IP at a certain absorption quantum efficiency and then converted to a number of photons (blue light) by laser light excitation. The photons are focused on the input plane of a photoelectron multiplier tube at a certain focusing efficiency with a reading optical system and considerably amplified in the photoelectric multiplier tube to become a photoelectric current. The current passes an A/D converter and is finally converted to digital signals. These conversion processes are known not to cause any reduction of detecting capability as long as the absorbed X-ray quanta are sufficiently amplified and converted [1]. In the case of the CR image-reading system, the quantum amplification rate depends on the IP excitation sensitivity, the amount of excited laser light, the focusing efficiency of the reading optical system, and the quantum efficiency of the PMT cathode plane. It is so designed that the average number of cathode electrons on the cathode plane of a photoelectron multiplier tube, where the number of quanta is at minimum, is five to ten times the number of absorbed X-ray quanta; the reduction of the detecting capability in various conversion processes of the reading system is kept to a minimum. The contrast detectability of the CR reading system is, therefore, determined mainly by the absorbed quantum efficiency (about 50% when the energy of the X-rays amounts to 80 kV_p) on the IP.

Table 3.2. Quantization levels necessary and sufficient for the CR system

Size of imaging plate (cm)	Spatial resolution (pixels/mm)	Density resolution (bits/pixel)
43 × 35	5	8
35 × 35	5	8
30 × 25	6.7	8
20 × 25	10	8

It should be noted here that a discussion of the image signal detecting capability of the total CR system together with the reader should include the effects of digital image processing and the mental and physical characteristics of the reader. Practical detecting capability in the real sense cannot be determined only with the pure physical contrast capability stated here.

Structure and specifications of reading unit

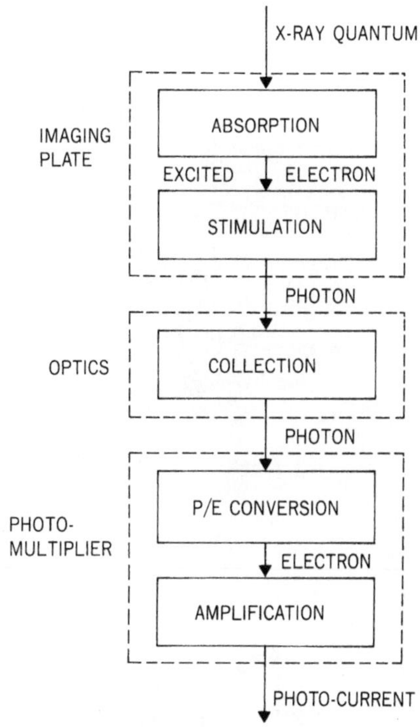

Fig. 3.5. Statistical analysis of the CR system

In clinical situations, a reading unit alone is not very practical unless it is provided with additional peripheral reading devices to meet the various demands of radiograhy technicians in different situations. Figure 3.6 shows the opertions and structure of the reading unit when a cassette type imaging unit is used. The exposed cassette is put into the cassette-inserting unit, called the cassette feeder, which can be operated under normal lighting conditions; here, the IP is removed from the cassette to the reading section. Four different sizes of cassette can be handled: 43 × 35 cm, 35 × 35 cm, 25 × 30 cm, and 20 × 25 cm. Reading takes some time, and so it may not be complete if the IP insertion interval is too short. To keep the IP on standby if necessary, therefore, an IP stacker is provided between the cassette feeder and reading section. The images stored in the IP are read in the reading section by laser light scanning, the remaining images are erased in the erasing section, and the IP is then sent to the IP sorter to be sorted and stored according to size, awaiting reuse. Since the IP can be loaded onto a cassette in a bright room it may be manually loaded, but the IP loader enables the loading to be carried out automatically. The IP thus circulates between the imaging unit and the CR reading unit; meanwhile, the image information read in the reading section is converted to digital signals and sent together with the signals from the patient ID information and imaging information entering section to the computer/image-processing section, where it is processed and printed on photographic film as the CR image. The imaging operation with the cassette-type CR system is so designed that it is equivalent to that with the screen/film

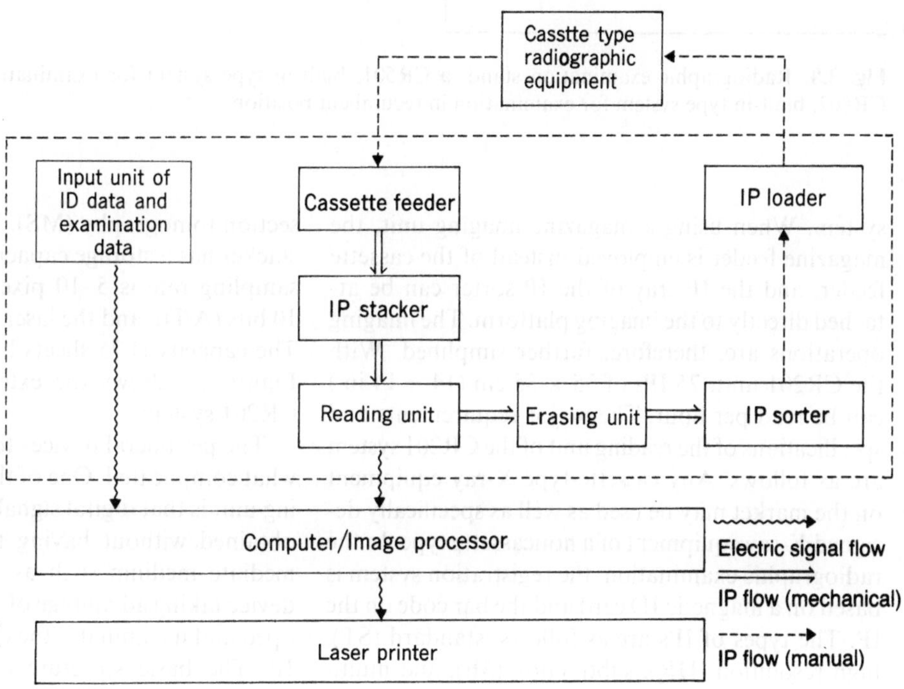

Fig. 3.6. Basic block diagram of image reader (IRD)

Fig. 3.7. CR201 image reader (IRD)

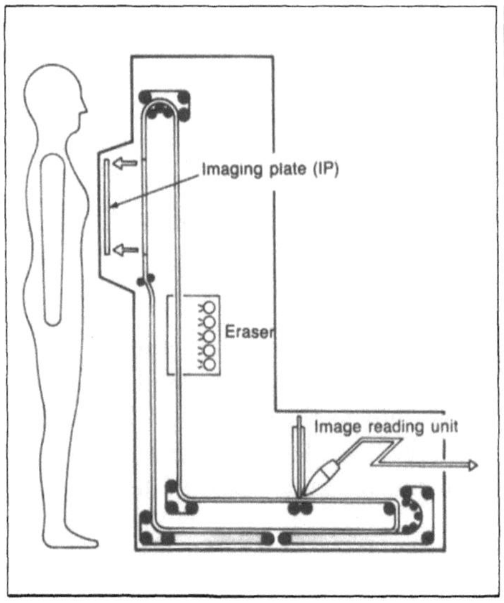

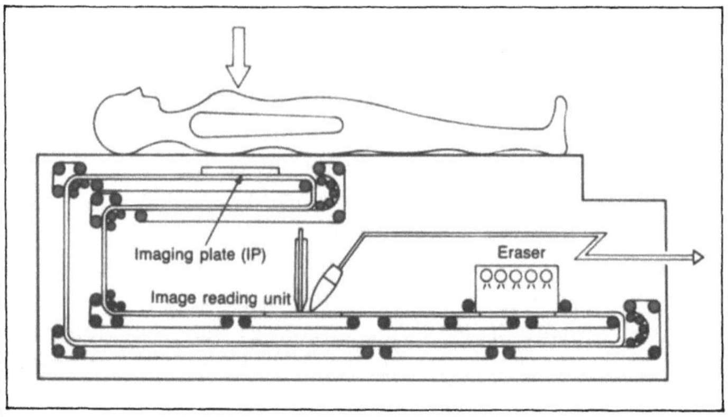

Fig. 3.8. Radiographic examination stand. **a** CR501, built-in type system for examination in standing position; **b** CR502, built-in type system for examination in recumbent position

system. When using a magazine imaging unit, the magazine feeder is employed instead of the cassette feeder, and the IP tray of the IP sorter can be attached directly to the imaging platform. The imaging operations are, therefore, further simplified. With the CR201 unit, 75 IPs of 35 × 35 cm (14 × 14 in.) can be read per hour. The major requirements and specifications of the reading unit of the CR201 system are as follows. Any cassette-type X-ray equipment on the market may be used as well as specifically designed X-ray equipment of a noncassette type. In the radiographic examination, the registration system is based on a magnetic ID card and the bar code on the IP. The types of IPs are as follows: standard (ST), high resolution (HR), subtraction (SB), and multisection tomography (MS). In image reading, the IP stacker has a storage capacity of up to 60 sheets, the sampling rate is 5–10 pixels/mm, the gray level is 10 bits (A/D), and the laser spot diameter is 100 μm. The capacity is 75 sheets/h for the 14 × 14 in. size. Figure 3.7 shows the external appearance of the CR201 system.

The peripheral devices for IP handling are somewhat complicated. One of the merits of the CR reading unit is that digital signals of X-ray images can be obtained without having to go through an intermediate medium such as a film image; a reading device taking advantage of this merit has been developed and it eliminates the difficulties in handling the IP. The basic structure of this reading device is

Table 3.3. Specifications of built-in type system

	Fuji CR501	Fuji CR502
Radiographic examination		
ID/setting of exposure conditions	Magnetic ID card registration system	Magnetic ID card registration system
Size of exposure (in.)	14 × 17, 14 × 14, 10 × 12	14 × 17, 14 × 14, 10 × 12, 8 × 10
Waiting time for exposure (s)		
Minimum	12	12
Maximum (for 14 × 14 in. exposure)	80	55
Image reading		
Sampling rate (pixels/mm)	5–7	5–7
Gray level [bits (A/D)]	10	10
Control		
Processing time (s)[a]	170	170
Processing capacity (plates/h for 14 × 14 in. exposure)	65	65

[a] For a period starting from the first exposure at the radiographic examination stand to completion of processing on the exposed CR film

shown in Fig. 3.8; it is called an on-line type reading device. It incorporates a reading section in the imaging platform for the standing or lying positions. As the cycle of imaging, reading, and erasing is repeated and a number of IPs are circulating within the imaging system, the technician is not required to replace the IPs; he can thus devote himself to the positioning of the patient or the entering of patient ID and imaging data. The time saved may, therefore, be used in patient care. Additionally, since a computer can control the IP quality and the conditions of the unit, a considerable saving in manpower is possible. Widespread use of this type of reading device would be desirable. The specifications of the on-line type image reading device are given in Table 3.3.

References

1. Dainth JC, Chaw R (1974) Image science. Academic Press, London, p 172

Chapter 4

Image processing

Masamitsu Ishida[1]

System operating principle

The operating principle of the CR system is best clarified by reference to the four-quadrant display shown in Fig. 4.1.

First quadrant. The characteristics peculiar to the IP, i.e., the relationship between the irradiation dose of X-rays and the intensity of photostimulated luminescence (PSL) stimulated by the laser, are shown. The relationship between the two is linear over a range of more than $1:10^4$; this linear relationship contributes to the high sensitivity and wide dynamic range of the CR system.

Second quadrant. The relationship between the input and output signals to and from the image reader (IRD) is shown. The IRD features a built-in sensitivity/image range automatic setting mechanism, which determines the reading conditions according to the image information recorded on the IP (the X-ray dose and dynamic range of image). In Fig. 4.1, the reading conditions are the line indicated by *A* for example 1, with a higher X-ray dose and a narrower image range, and the line indicated by *B* for example 2 with a lower X-ray dose and a wider image range. This automatic setting mechanism is discussed in detail below. With the characteristics peculiar to the IP shown in the first quadrant and the automatic setting mechanism shown in the second quadrant, the imaging and observation characteristics are independently controlled, and the read image information is sent to the image processor (IPC) shown in the third quadrant.

Third quadrant. With the IPC, the image is processed to display an image suitable for diagnosis, i.e., the display characteristics are independently controlled; specifically, gradation, frequency, and subtraction processing are performed.

[1] Fuji Photo Film Co., Ltd., Miyanodai Development Center, 798, Miyanodai Kaisei-cho, Ashigarakami-gun, Kanagawa, 258 Japan

Fourth quadrant. The image signals sent to the image recorder (IRC) are reconverted to optical signals to obtain a radiograph. The characteristic curve of the film used with the CR system is automatically compensated within the IRC so that the density vs exposure curve is linear (this conversion curve is not shown in Fig. 4.1). The fourth quadrant, therefore, shows the characteristic curve of the output radiograph in the CR system. Unlike the characteristic curve of conventional radiographs, that of the CR system automatically changes according to the X-ray dose and the image range.

Sensitivity/image range automatic setting mechanism

To control automatically the imaging characteristics so as to realize image density stability, i.e., modifying the image density instability due to over- or under-exposure during X-ray imaging, the sensitivity/image range automatic setting mechanism is built into the IRD. This mechanism utilizes the fact that the histogram of an X-ray image has a particular pattern according to the anatomical region (e.g., chest, abdomen, and bone) and technique (plain or tomographic or contrast imaging). Before the IRD reads the image information (that finally displayed on a film), the information recorded on the IP is preread by a weak laser light. A histogram of the preread image information is then prepared, the characteristics of the histogram extracted using the information concerning the anatomical region and/or technique, and the sensitivity/image range automatically determined. To be more specific, the maximum and minimum doses (*S1*, *S2* in Fig. 4.1) of the image signals effective for diagnosis are detected from the histogram of the image signals in the first quadrant of Fig. 4.1; reading conditions *A* and *B* are so determined that *S1* and *S2* are mapped to *Q1* and *Q2*, respectively, on the IRD output. In other words, the straight line shown by *A* is the optimum reading condition for example 1 with a higher X-ray dose and a narrower image range, and *B* is that for example 2 with a lower X-ray dose and a wider image range.

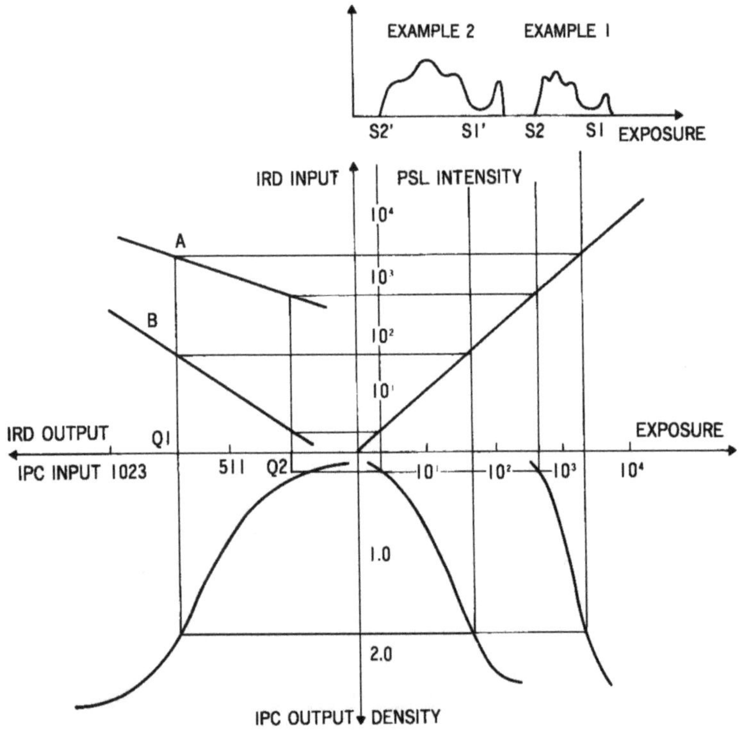

Fig. 4.1. Operating characteristics of CR system

Gradation processing

With the automatic sensitivity/image range setting mechanism described above, image signals can be read within a certain setting and range even if the X-ray dose and/or the X-ray energy change. Unlike the gradation characteristics (H&D curve of the conventional screen/film system), the gradation porcessing in the CR system controls the density of the image signals obtained, i.e., it independently controls the display characteristics.

Outline

The gradation processing is shown in the third quadrant of Fig. 4.1. It converts the 10-bit input image signals QL, of the IPC to IPC output image signals, i.e., the IRC input signals QV with the nonlinear conversion curve of the following equation:

$$QV = f(QL) \tag{1}$$

For radiographs, there are a number of anatomical region and imaging techniques, and it is not desirable to reproduce all images with one type of gradation processing. The CR system is, therefore, so designed that proper gradation processing is performed according to the imaging menu, i.e., the objective of imaging. In examination of the chest or breast, for instance, gradation processing is performed as shown in Fig. 4.2. In order to reproduce an image over a wide image signal range of the chest, gradation processing for the chest features, high contrast and low contrast in the regions corresponding to the lung and mediastinum, respectively, is carried out to prevent the density of the mediastinum being too close to that of bone in the output film. In the case of mammography, the contrast in the low-density region (low QL) is increased and the contrast in the high-density region (high QL) is suppressed so as to reveal any calcification up to the edge region.

Gradation-processing parameters

The nonlinear conversion curve $f(*)$ to determine the gradation-processing conditions is controlled by four parameters—gradation type (GT), rotation center (GC), rotation amount (GA), and gradation shifting amount (GS).

Gradation type. This is the most basic parameter of the four. It indicates the fundamental form of the nonlinear conversion curve. There are currently 15 types of basic form, and the gradation type is indicated by a letter of the alphabet (A to P) on a CR film.

Rotation center, rotation amount. These are parameters controlling the gradient of the nonlinear conversion curve. With these two parameters, the gradation type can be changed as shown in Fig. 4.3.

Gradation shift. This parameter indicates the amount to shift the nonlinear conversion curve horizontally (on the IPC input image signals), as shown in Fig. 4.4. As a result, the output radiograph density is totally changed.

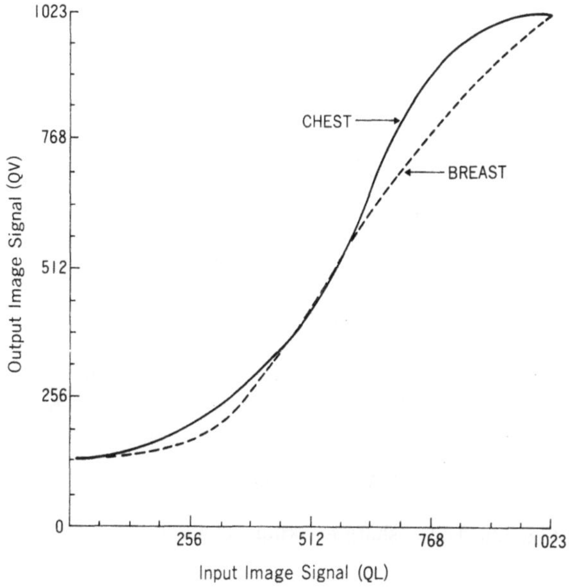

Fig. 4.2. Gradation processing (tonal conversion)

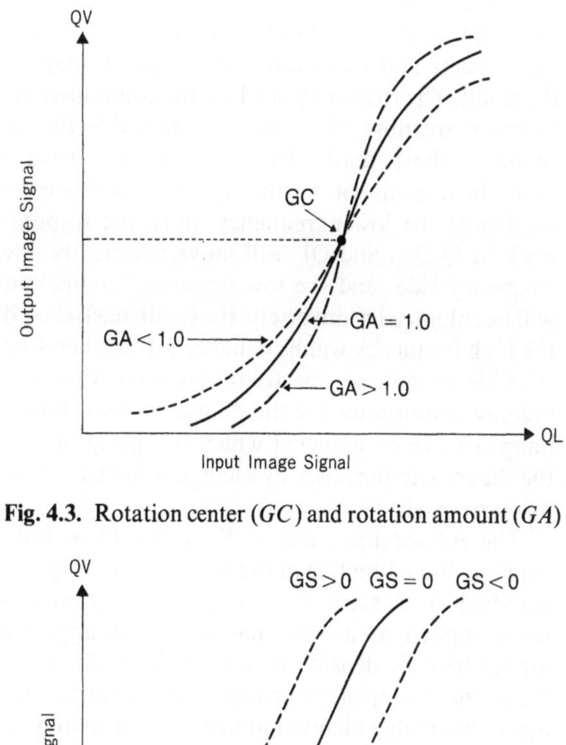

Fig. 4.3. Rotation center (*GC*) and rotation amount (*GA*)

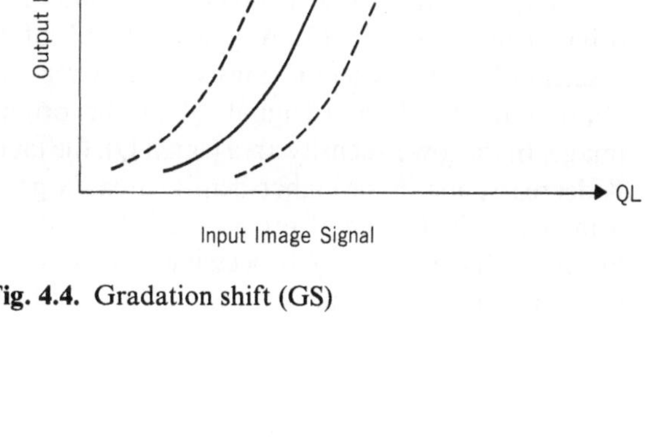

Fig. 4.4. Gradation shift (GS)

Frequency processing

Control of the frequency response (image sharpness) of an imaging system is called frequency processing. The frequency response of a screen/film system becomes smaller as the frequency increases. In other words, the contrast becomes smaller for a high-frequency component of an image. With the CR system, however, the frequency response can be increased to improve the image contrast by frequency processing.

Principle

This technique is generally called 'unsharp masking' because an unsharp (blurred) image Qus is used for the mask image to enhance the spatial frequency response. The principle is shown in Fig. 4.5 (one-dimensionally) and by the following equation:

$$\begin{aligned} QL(x, y) &= g(Q(x, y)) \\ &= Q(x, y) + K(Q(x, y)) \times [Q(x, y) - Qus(x, y)] \end{aligned} \quad (2)$$

Q, Qus, and QL represent the original, unsharp, and processed images, respectively; K indicates a weight factor for determining the degree of enhancement. With the CR system, gradation processing is performed after frequency processing, and the QL of Eq. (2) and that of Eq. (1) represent the same image signals.

The dotted line in Fig. 4.5 indicates the frequency response of the unsharp image, Qus, which is obtained by calculating the moving average as follows:

$$Qus(x, y) = \frac{\sum_{i=x-[(M-1)/2]}^{x+[(M-1)/2]} \sum_{j=y-[(N-1)/2]}^{y+[(N-1)/2]} Q(i, j)}{M \times N} \quad (3)$$

Here, M × N represents the unsharp mask size, and generally M = N. In Fig. 4.5, the broken line indicates the difference between the original and the unsharp images, i.e., the frequency response of Q-

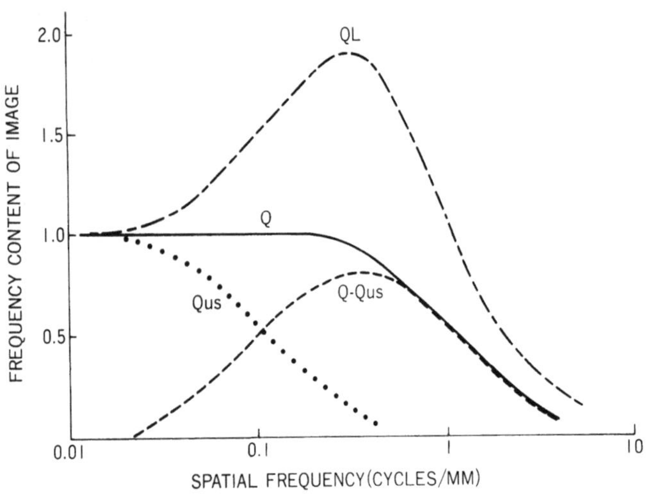

Fig. 4.5. The unsharp masking technique

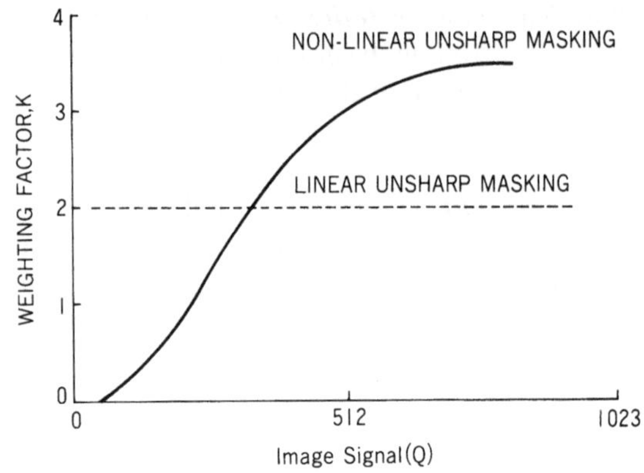

Fig. 4.6. Weighting factors for linear and nonlinear unsharp masking techniques

Qus; the single-dot chain line indicates the frequency response of the final processed image, which is obtained by multiplying Q-Qus by a certain enhancement factor and then adding the original image to the result. The frequency at which the components of an image are most enhanced is determined by the size of the unsharp mask. That is, if a large mask is used, the response of the unsharp image will become smaller at the lower frequency; thus, the response peak of Q-Qus and QL will move toward the low frequency side, and the low-frequency component will be enhanced. Conversely, if a small mask is used, the high frequency will be enhanced. It is, therefore, possible to enhance the frequency band that is of greatest importance for the diagnosis, and thus an image can be constructed which is appropriate for the diagnostic purposes by changing the size of the unsharp mask.

The enhancement factor, K, in Eq. (2) is not a constant but a function of the original image, Q. If K is used as a constant and strong frequency processing is applied, an artifact may sometimes appear in areas where the density changes suddenly. Such artifacts tend to appear in images made using contrast agent. Strongly enhanced images appear to noisy in the low-density part of the image (the part with a small dose of X-ray quanta absorbed by the IP). To eliminate such defects, K is made a function of the original image as in Eq. (2). A typical example of this is seen in Fig. 4.6, where K consistently increases in relation to the density (signal Q) of the original image. In the lower-density area (small Q), the factor K decrease, and in the higher-density area (large Q), it increases. This kind of processing is termed nonlinear unsharp masking processing; when K is a constant, the processing is linear. With nonlinear processing, the above-stated artifact is removed, and the noise in a low-density area is considerably reduced compared with linear processing. On the other hand, the shadow contrast in a high-density area is enhanced almost to the same extent as in linear processing.

Although unsharp masking processing is effective even when applied to the conventional radiograph, it is more effective when applied to a CR system with a wide dynamic range like the IP. An image with a wide dynamic range has an expanded visible area for diagnosis, but it is not suitable for diagnosis if it is directly displayed on a film, since it results in an image with a generally low contrast. Frequency-enhancement processing is a process which improves the response of the frequency component, i.e., a process which increases local contrast of the image without changing the entire density range of the image. By combining the wide dynamic range of the IP and frequency processing, therefore, it is possible to construct an image that improves the diagnostic accuracy.

Frequency-processing parameters

The manner of frequency response determining the frequency-processing conditions is controlled by three parameters—frequency rank (RN), frequency type (RT), and degree of enhancement (RE).

Frequency rank. RN refers to a frequency band in which the frequency contents of the image are enhanced by frequency processing. As stated above, a frequency band is determined by the size of the unsharp mask. With the current CR system, it is de-

Table 4.1. Relationship between frequency rank and spatial frequency at which response is maximum in frequency processing

RN	0	1	2	3	4	5	6	7	8	9
fo (c/mm)	0.09	0.13	0.18	0.25	0.35	0.50	0.71	1.00	1.40	2.00

RN frequency rank, *fo* frequency on IP

scribed by a frequency value, *fo*, at which the response is maximum in frequency processing (*fo* indicates a spatial frequency on the IP), and is divided into ten frequency ranks. The relationship between the frequency rank and *fo* is shown in Table 4.1.

Frequency type, degree of enhancement. These two parameters specify the enhancement factor function, K(Q), in Eq. (2) and Fig. 4.6. RT specifies the shape of the function when the maximum degree of enhancement is normalized to 1.0; currently there are eight types of RT. RE indicates the maximum value of the degree of enhancement. The function K(Q) is, therefore, calculated by the following equation:

$$K(Q) = RE \times RT(Q) \tag{4}$$

Effects of image processing

Accurate detection of abnormalities is one of the most fundamental tasks required of an X-ray imaging system. Since the X-ray imaging system can be considered to involve two processes, image acquisition and image display, it is possible to adopt two approaches in attempting to increase the detection of abnormalities. One is to employ an image-acquisition system that provides superior physical image quality, with a large signal-to-noise ratio (SNR), a high spatial resolution, and a wide dynamic range. The other is to optimize the image display so that the radiologist can extract as much diagnostic information as possible from the acquired X-ray image. The second approach alone may increase the detectability of abnormalities because the detection performance of human observers depends not only on the physical properties of the images, but also on the display conditions.

To optimize control of the display characteristics, image processing—gradation processing and frequency processing, i.e., unsharp masking, are employed as stated above. The detectability of low-contrast radiographic patterns can be improved through these processings, as shown in Fig. 4.7. For evaluation, the well-known receiver-operating characteristic (ROC) curve is used and the pattern size is 5 mm in diameter. The long-dashed line shows the detectability when the characteristic curve (H and D curve) for display is made the same as the screen/film combination of the Hi-STD/RX. The chained line indicates the case where the gamma of the characteristic curve is doubled (doubling the contrast); the solid line shows the case where the frequency processing of RN = 1 and RE = 3.0 is performed.

The detectability of low-contrast radiographic patterns is improved through image processing, as discussed above, and this may be explained as follows [1]. The physical SNR_R of an X-ray image in which the X-ray exposed IP was read by the IRD is shown by the following equation (hereinafter, the noise of the IRC will be ignored as it is small):

$$SNR_R = \frac{S_R}{N_R} \tag{4}$$

If the read image is subjected to image processing, e.g., processing to double the contrast, both the signal and noise will be doubled and the SNR will not change. This is contradictory to the results given in Fig. 4.7. Thus, we assume the existence of an ad-

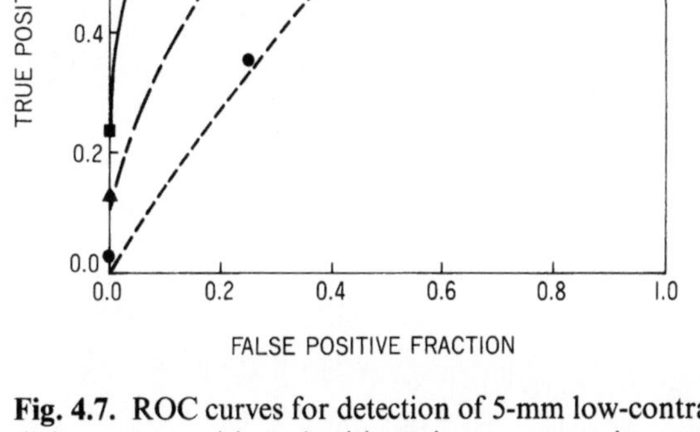

Fig. 4.7. ROC curves for detection of 5-mm low-contrast disk patterns with and without image processing: ●---- without image processing, ▲–-–overall contrast enhancement (× 2), ■——unsharp masking

ditional noise component in the observer's eye-brain system [2–4]. This noise is intrinsic to the observer. The SNR_I when the internal noise N_I is assumed is given by the following equation:

$$SNR_I = \frac{S_R}{(N_R{}^2 + N_I{}^2)^{1/2}} = \frac{SNR_R}{[1 + (N_I/N_R)^2]^{1/2}} \tag{5}$$

Thus, the increased detectability found after contrast enhancement can be explained: When contrast enhancement is applied to a given image, both the signal and the image noise are increased by the same factor, thus causing no change in SNR_R. However, the relative magnitude of internal noise is reduced, thereby providing a greater SNR_I, and thus a greater detectability of a given object contrast, or, equivalently, thereby requiring a lower threshold contrast.

References

1. Ishida M, Doi K, Loo LN, Metz CE, Lehr JL (1984) Digital image processing: effect on detectability of simulated low-contrast patterns Radiology 150:569–575
2. Wickelgren WA (1968) Unidimensional strength theory and component analysis of noise in absolute and comparative judgements. J Math Psych 5:102–122
3. Goodenough DJ, Metz CE (1975) Implications of a "noisy" observer to data processing techniques. In: Raynaud C, Todd-Pokropek AE (eds) Information processing in scintigraphy. CEA, Orsay
4. Swensson RG, Judy PF (1981) Detection of noisy visual targets: models for the effects of spatial uncertainty and signal-to-noise ratio. Percep Psychophys 9:521–534

Chapter 5

Image recording

Hisatoyo Kato[1]

Introduction

The X-ray image signals stored and recorded on the IP and read for digital processing by laser light scanning need to be displayed in the form of a visible image to allow reading and diagnosis. The following displaying methods are available: (1) CRT monitor for direct reading; (2) exposure on a film from a CRT monitor by an optical system like a multiformat camera and display as a film image for reading; (3) direct recording on a film from image signals by a laser printer to produce a film image for reading.

For current diagnosis with radiographs, images of an extremely high quality are required, and hard copies printed on film are needed in actual applications. Therefore, method 1 in the list is not practical at present. With method 2, it is difficult to obtain images of a high quality due to "rough" images caused by the CRT phosphor noise, the flare of the CRT, and optical image distortion. Method 3 is free from the defects of method 2; it can provide images of high quality and the amount of recording light can be directly controlled by digital signals. This method has, therefore, a high potential as a future digital image-recording method.

The principle and structure of a high-quality laser printer as the image recorder (IRC) of the CR system are outlined in the following section.

Principle of laser printer

The laser printer has a similar structure to the laser scanner for reading information from the IP. It is capable of two-dimensional and consecutive scanning of the entire surface of a film with a laser light spot. It scans a focused laser light on a film by means of a vibrating or polyhedral rotating mirror, while transferring the film with high precision perpendicular to the scanning line of laser light. In so doing, an ultrasonic optical modulator is positioned between the laser light source and the vibrating or polyhedral rotating mirror. This modulator is capable of modulating the light intensity proportionally to the external electric signals it receives. In synchronization with the position of the laser light on the film, the CR image signals at the position are sequentially output from the computer; by applying them to the modulator, the CR image signals can be exposed on the film as a two-dimensional CR image. This film is developed to obtain a radiograph. The laser light source used is determined by the characteristics (sensitivity) of the film. With the CR system, an He-Ne laser is used.

Factors of laser printer

To increase the sharpness of a laser printer, i.e., improve its resolution, it is necessary to reduce the laser beam diameter and the scanning pitch. It is not practical, however, to improve the spatial resolution above that of the reading information from the IP. The spatial resolution of the laser printer is 10 pixels/mm and the laser beam diameter in 80 μm. In

Table 5.1. Image size and format

IP size (cm)	Reading spatial resolution (pixels/mm)	Recording spatial resolution (pixels/mm)	Image size reduction	Recording format
43 × 35 (17 × 14 in.)	5	10	1/2	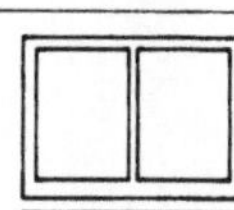
35 × 35 (14 × 14 in.)	5	10	1/2	
30 × 25 (12 × 10 in.)	6.7	10	2/3	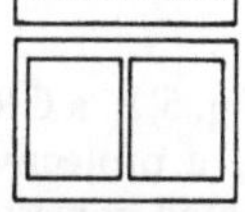
24 × 18 (10 × 8 in.)	10	10	1/1	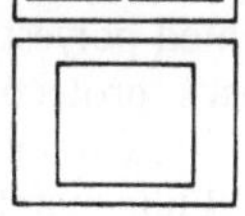

[1] Fuji Photo Film Co., Ltd., Miyanodai Development Center, 798, Miyanodai Kaisei-cho, Ashigarakami-gun, Kanagawa, 258 Japan

Fig. 5.1. Laser printer for CR system

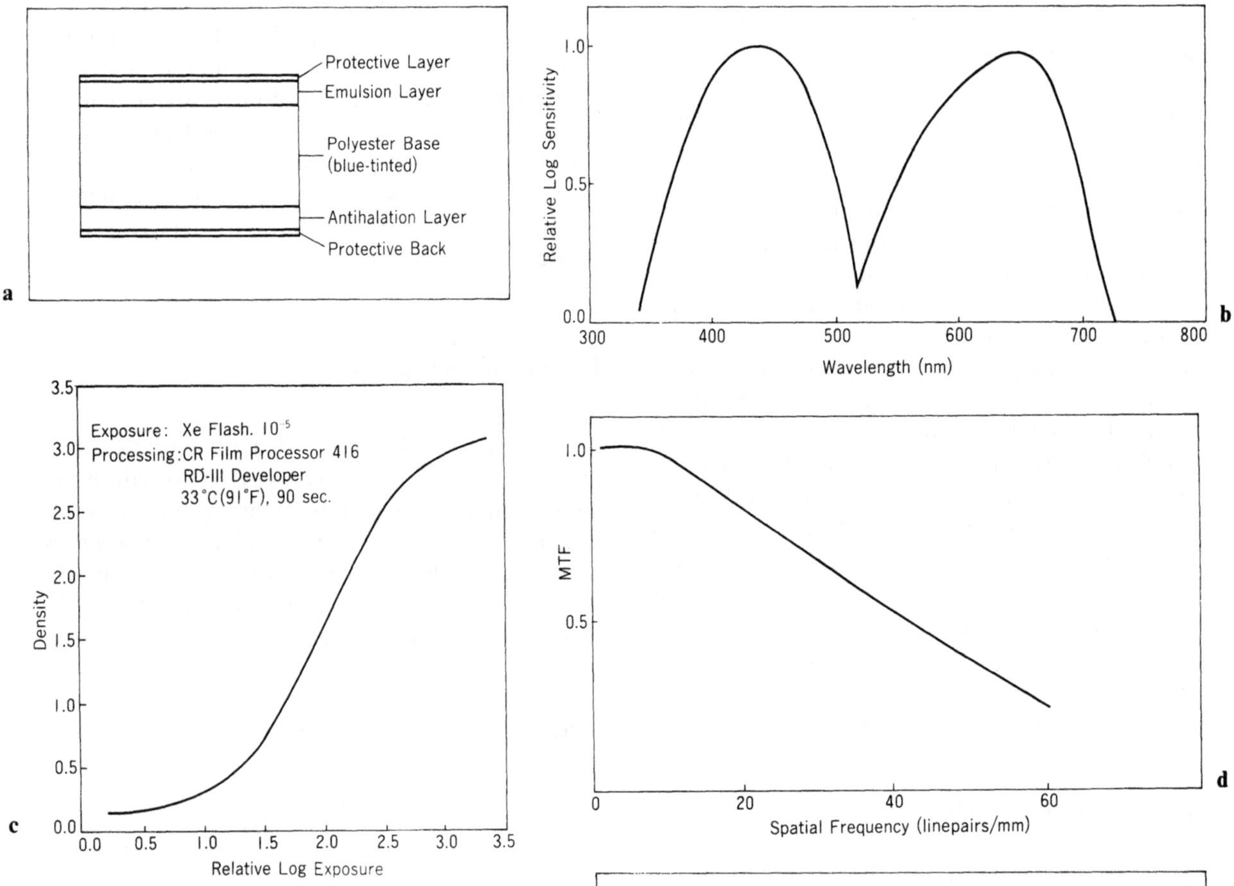

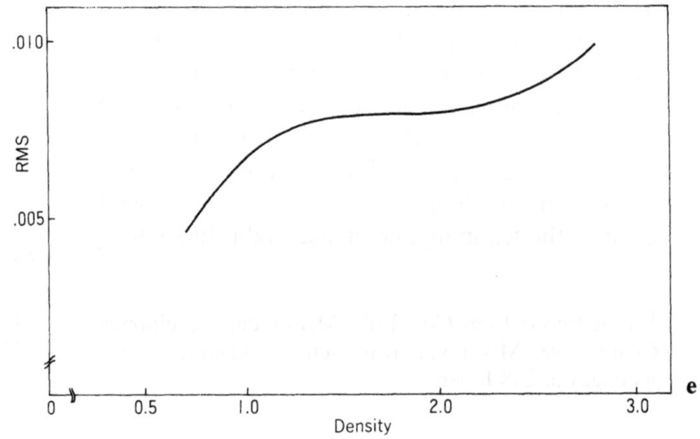

Fig. 5.2. **a** CR film structure. CR film is composed of a protective layer, an emulsion layer, a blue-tinted polyester base, an antihalation layer, and a back protective layer. **b** Relative spectral sensitivity. One of the spectral sensitivity peaks of the CR film matches the wavelength of the He-Ne laser beam (633 nm). **c** CR film characteristic curve. **d** Sharpness. **e** Granularity

conventional screen/film radiography, the X-ray optical pattern on the screen is accurately transferred to the film at the same size as a contact print. With the CR system, on the other hand, there is no such restriction, and an image can be recorded at any enlargement or reduction. Table 5.1 shows the resolution of laser scanning and a typical image size and format. The size of the film itself is 364 mm × 257 mm. A laser printer requires a resolution of 10 bits/pixel. The laser printer prevents false contours caused by lack of density resolution, even if the gradation is changed throguh digital processing prior to recording on the film. If imaging is made on as IP of 43 cm × 35 cm, 35 cm × 35 cm, or 25 cm × 30 cm, the image is reduced on the film. This is not due to loss or absence of information in reading but to the fact that the spatial resolution is higher when recording than when reading. Thus, different images can be recorded on a film as a result of digital processing; two images obtained from the same data are read from the IP. This feature is useful as it allows display-emphasizing characteristics which are clincially different.

Figure 5.1 shows the external appearance of the CR laser printer. A film pack of 100 films can be loaded in daylight. A standby pack can be set up in advance and this allows the film pack to be changed at any time. An automatic developer can be directly connected to the printer.

CR film

The film used with the laser printer is CR film type 633. It has a peak of spectral sensitivity at about 633 nm, which is the wavelength of the He-Ne laser light. It is a newly designed exclusively for this application film, consisting of a single-side emulsion layer of highly sensitive, fine particles. Figure 5.2 shows its layer structure, spectral sensitivity, characteristic curve, sharpness, granularity, etc.

Chapter 6

Technical trends of the CR system

Katsumi Asanuma[1]

The history of the CR system may be briefly summarized as follows: 1981, clinical experiments start, reported at the International Congress of Radiology (ICR; Brussels) and Radiological Society of North America (RSNA) meeting (Chicago); 1983, the first CR system is brought on the market; 1985, the second generation system is brought on the market (a built-in type is also announced). With the full-scale introduction of the second generation system, the advantages of CR technology were more or less established, and the system began to be used not only in research but also in routine clinical applications.

The second generation system is a general-purpose system provided with all the functions required for CR. It is of the centralized type and is capable of processing any anatomical region with any imaging method in a hospital. In view of the present clinical applications of X-ray imaging, however, it is becoming difficult to perform routine processing with a centralized type of system as the processing of individual anatomical regions has to be carried out and a number of rooms are needed for the different imaging techniques, such as angiography, mammography, and ordinary plane radiography.

Looking at the history of the development and spread of various products, it can be seen that in almost every case high-performance general-purpose systems were first brought onto the market, and later more economical user-specific systems were developed to be used in extensive applications. A computer system may be taken as an example. Centralized types of computer system were the first to be developed. Then, there was the move toward minituarization as a result of great advances in semiconductor technology and toward decentralization as a result of the spread of personal computers and progress in communication and transmission technologies, such as local area network (LAN), with the emphasis on optoelectronics, including fiber optics. As a result, today's users can operate computers practically anywhere. CR technology can be seen to have gone through almost the same process; the more popular type of system now seems to be a decentralized processing unit capable of meeting the requirements for individual X-ray rooms. The centralized and decentralized types of processing are presented in Fig. 6.1. As seen in the figure, the decentralized system is far more easy to use since cassettes and/or magazines do not have to be transported a great distance and each processing is performed near the X-ray room.

The future of the CR system may be examined in terms of the following two aspects—characteristics and functions of the CR system and requirements for routine applications.

Characteristics and functions of the CR system

The three most important characteristics of the CR system are considered to be: digital image, wide latitude, and low exposure dose; wide latitude and low exposure dose are the most significant for existing CR systems and will continue to be improved in future. The future trend predicted for the digital image is discussed in the following sections.

Digital image

Digital data processing has undergone rapid progress, as witnessed in the progress in advanced electronics including miniature processing of semiconductors and high-speed technology. As a result, along with the development of the IP and laser scanning technology, commercialization of the CR system centered around digital-processing technology has been brought about in preference to analog recording by the conventional screen/film system. The CR system is expected to advance further, utilizing digital-processing technology which is expected to progress further with the progress in computer technology. There now follows an examination of the prospects for digital technology which may be used in association with the CR system.

[1] Fuji Photo Film Co., Ltd., Miyanodai Development Center, 798, Miyanodai Kaisei-cho, Ashigarakami-gun, Kanagawa, 258 Japan

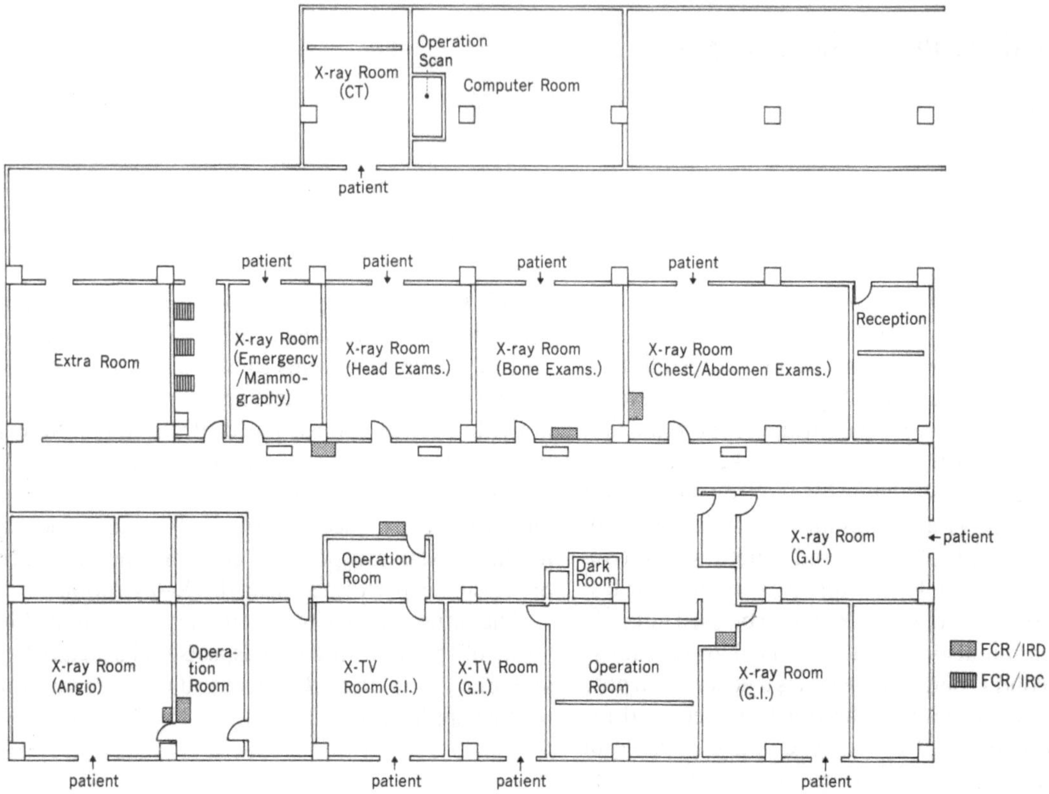

a

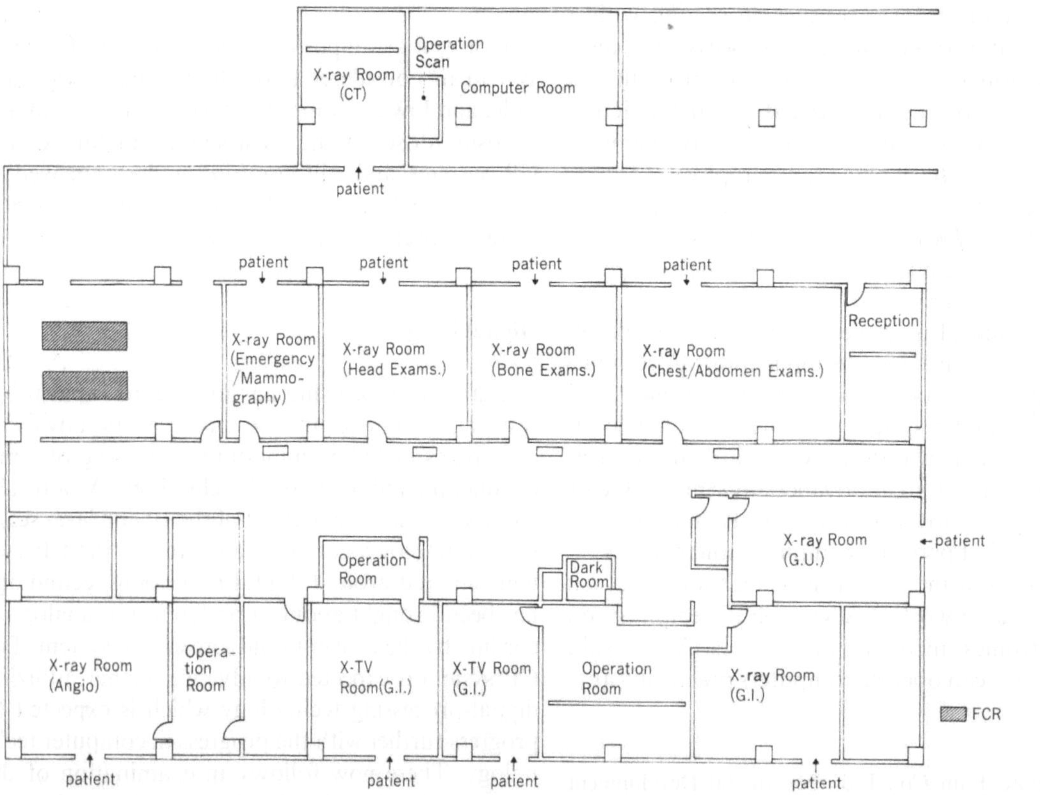

b

Fig. 6.1. a Decentralized and **b** centralized types of processing

Saving of inventory space

There are two key types of technology for saving inventory space. They are reviewed below.

Higher density image data storing unit. The image data of the CR system is a considerable amount, even among medical image data, and each image contains 4×10^7 bits of information. It is, therefore, difficult to use conventional recording media, such as magnetic disc, magnetic tape, or semiconductor memory, as an image file for the CR system. Table 6.1 compares the performances of individual media.

The magnetic disc was made denser and so the distance between head and disc had to be decreased due to the structure. The disc had to be of a closed type; it became larger and difficult to change. With a magnetic tape, data access is consecutive, and the capacity is small compared with other media. Both the floppy disc and the laser card of a vertical magnetic recording system are easy to handle but do not have sufficient capacity, and thus further development efforts are needed. Under such circumstances, the optical disc, which has been put into practical use in recent years, is far better both in handling and capacity than other conventional media and is ideal as a CR file. The non-erasable 30-cm optical disc currently has a capacity of 2–3 gigabytes. In view of past progress, an optical disc of about 10–30 GB will probably have been developed by 1990. Also, just like the magnetic disc, an erasable optical disc is being developed, and this erasable optical disc of about the same capacity as a current nonerasable disc will soon be ready for practical applications. This means that it will be possible within 5 years or so to achieve a five- to tenfold space saving compared with the technology of today.

Image data-compression technology. The simplest way of compressing the data is to make the picture elements larger and to reduce their number. Since this method deteriorates the diagnostic quality, a number of other methods are presently being studied. Two of the most popular methods are Huffman coding and predicative coding. The prediction method indicates a density difference between the preceding point and the present point and takes advantage of the fact that the density of X-ray images does not rapidly change. Huffman coding is a compacting method which depends on the occurrence frequency. The more frequently occurring bit sequence is assigned to the shorter bit sequence, such that the average coding efficiency is improved. This combination is presented in Fig. 6.2. One commercialized system involves a combination of reduction of number of picture elements, prediction, and Huffman coding. This combination achieves data-compression of about one-tenth.

Image compression methods are being developed in a number of countries, and a compression ratio of 1/20–1/30 is likely to be achieved by 1990. Table 6.2 compares the space required for film storage with that for electronic storage. It can be seen that electronic storage today requires 1/20–1/100 the space of film storage; by 1990 or so, this figure will have improved to 1/1000–1/4000 as a result of the increase in optical disc capacity and advances in data compression (Table 6.3). The advantages in converting X-ray images to digital signals will then be fully realized.

Table 6.1. Performance comparison of recording media

Medium	Capacity	Transfer speed (MB/s)	Access time	Remarks
Optical disc	1–4 GB/disc	0.4	250 ms	Large capacity, possible to change disc
Optical disc library	70–80 GB/library	0.4	5 ~ 10 s	Large capacity file
Magnetic disc (large)	1.3 GB/library	3	24 ms	Disc change not possible, high cost
Magnetic disc (small)	0.17 GB/library	0.6	85 ms	Small capacity, not possible to change
Vertical magnetic recording	0.034 GB/face	0.6	85 ms	Small capacity, but handy
Floppy disc	0.001 GB/disc		250 ms	Small capacity, but handy
Magnetic tape	0.18 GB/reel	0.3	over 10 s	Small capacity, low speed
Laser card	0.002 GB/disc			Handy
Semiconductor disc	0.5 GB/library		1 ms	High cost, but high speed

GB gigabyte, *MB* megabyte

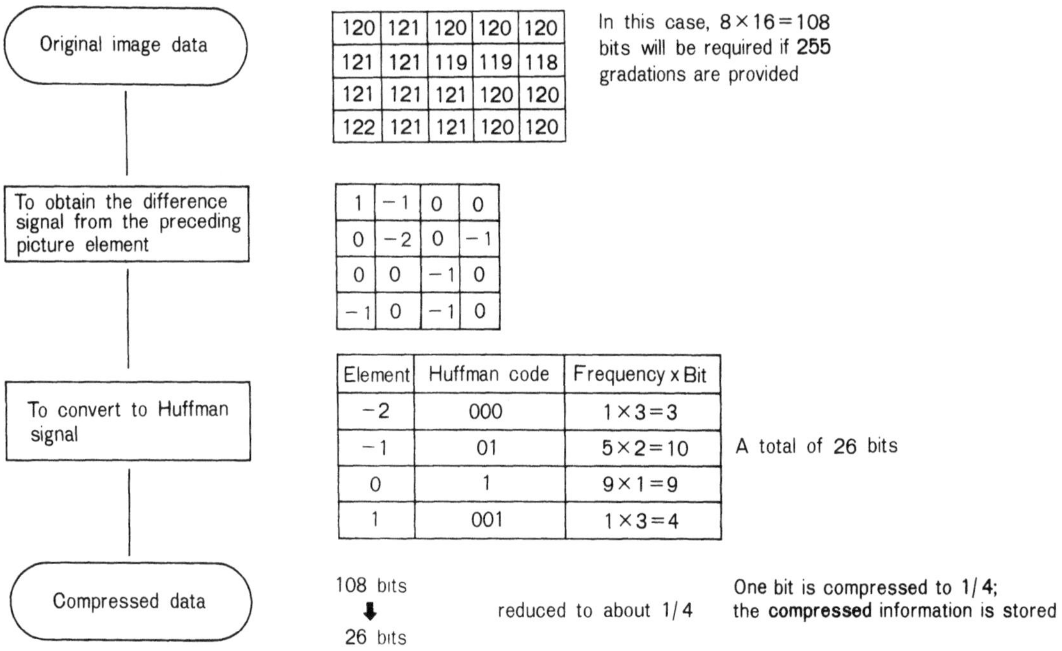

Fig. 6.2. Data compression by prediction and Huffman coding

Long-term stability

Digital and analog data basically differ in that the former is used to determine whether a threshold has been exceeded whereas the latter is used to detect subtle changes. Even with photographic films, some changes in the film quality occur with the passage of time. With digital data, on the other hand, even if some part of the data becomes lost (unreadable), data-correcting methods are available to recover the lost data; with an optical disc, the error rate is about 10^{-12}.

Recording media are believed to be stable for about 10 years. Using glass rather than plastic for the optical discs and/or improving the material used in recording films and sealing, this period could be extended to 20 years or so.

Figure 6.3 [1] shows a comparison of the period of storing X-ray image data and the frequency of their use. Future studies will probably deal with the proper period of storing, permanent storing of key images only, etc.

Improvement of diagnostic ability

With the conversion of image data to digital signals, it is expected that there will be an improvement in the diagnostic ability with automatic diagnosis using the rapid developments in image processing and artificial intelligence. Digital signal processing has developed from advances made in semiconductor technology and algorithms. The past use of computers emphasized the accumulation of knowledge, but the current trend is on how best to utilize the knowledge; a device capable of working along the lines of the human pattern of thinking is required. It is realized that this would be difficult using conventional computers of the Neumann type. The development of the fifth generation computer currently underway is an attempt to produce this new type of computer.

With such new developments, it will be possible to provide images to facilitate diagnosis, measure the

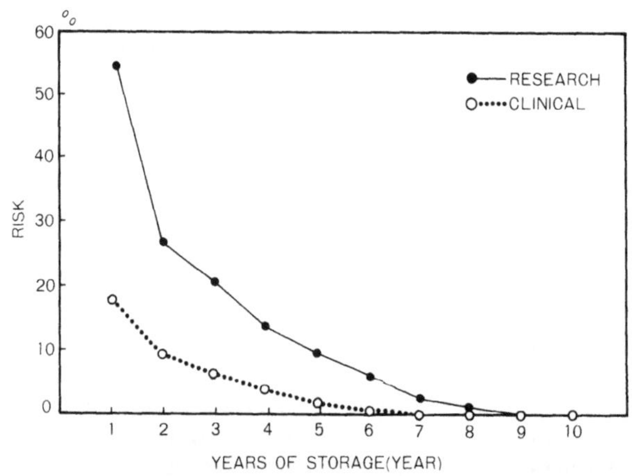

Fig. 6.3. Relationship between the unaccessible risk and the period of storage for X-ray films in clinical (n = 763) and research (n = 83) applications

Table 6.2. Comparison of X-ray image-storing space

	Storage method	
	Optical disc storage	Film storage
Storage form	Optical disc cartridge (20 mm thick)	Film jacket (20 films/jacket = 8 mm thick)
Storage capacity		
Reversible compression (compression to half)	242 000 images/m^3 (20 times file storage)	12 000 images/m^3
Irreversible compression (compression to 1/10)	1 200 000 images/m^3 (100 times file storage)	

Table 6.3. Effects of data compression and OD recording capacity on CR image storage

		Recording capacity		
	Compression ratio	2 GB/disc	10 GB/disc	30 GB/disc
Reversible compression	1/2	20	100	300
Irreversible compression	1/10	100	500	1500
	1/20	200	1000	3000
	1/30	300	1500	4500

extent of calcification, automatically diagnose silicosis, and assist diagnosis through frequency processing, edge emphasis, density windowing, extraction of affected parts by subtraction, etc.

Construction of retrieval and transmission systems

The picture archiving and communication system for medical applications (PACS) is now becoming increasingly common. The electronic components of the system enable efficient utilization of medical images and data. PACS allows a more efficient diagnosis by giving an output of the necessary images when they are required so as to control the rapidly rising medical costs. Such a system has become feasible with the conversion of medical data to digital signals as seen with CR, CT, MRI, and US, the progress in communication and transmission systems such as LAN following advances in computer systems, and the appearance of high-density recording media such as optical discs. The advantages of PACS are enhanced by special hospital information systems, including the medical accounting system. With PACS, it would be possible to set up a comprehensive communication system as shown in Fig. 6.4.

In the USA, the system shown in Fig. 6.5 [2] is being tried at the UCLA. This is a typical example of a mini-PACS which is utilized in the individual departments of a hospital. An example of a big system is that installed at Washington University [3]. It is expected that more systems will be tried and further technological progress made. For the total collection of image data, a standard interfacing of devices is essential. A network architecture has been proposed by the American College of Radiology—The National Electrical Manufacturers Association (ACR-NEMA) [4]. With the growing trend toward PACS, CR, which is capable of handling the maximum amount of X-ray image data, is one of the systems where further progress is most strongly desired, including data-base structure, file stratification, image data-compression technology to reduce the transmission time. A commercial filing system for CR, shown in Fig. 6.6, is an attempt at such an end.

CRT image display should not be neglected in association with PACS. Although further studies are needed, some CRT imge displays are almost up to standard in terms of image quality. Before conversion to CRT diagnosis can be carried out in the clinic, however, certain requirements have to be met: (1) Radiologists must become accustomed to diagnosis with CRT images and gain experience with the new system; (2) a large number of CRT images should be capable of being read at a time; (3) the display must be such that it can be simultaneously observed by a large number of radiologists; (4) it shoulld be transportable; (5) the system cannot be too large. These requirements can only be met through the close cooperation of the medical staff and engineers.

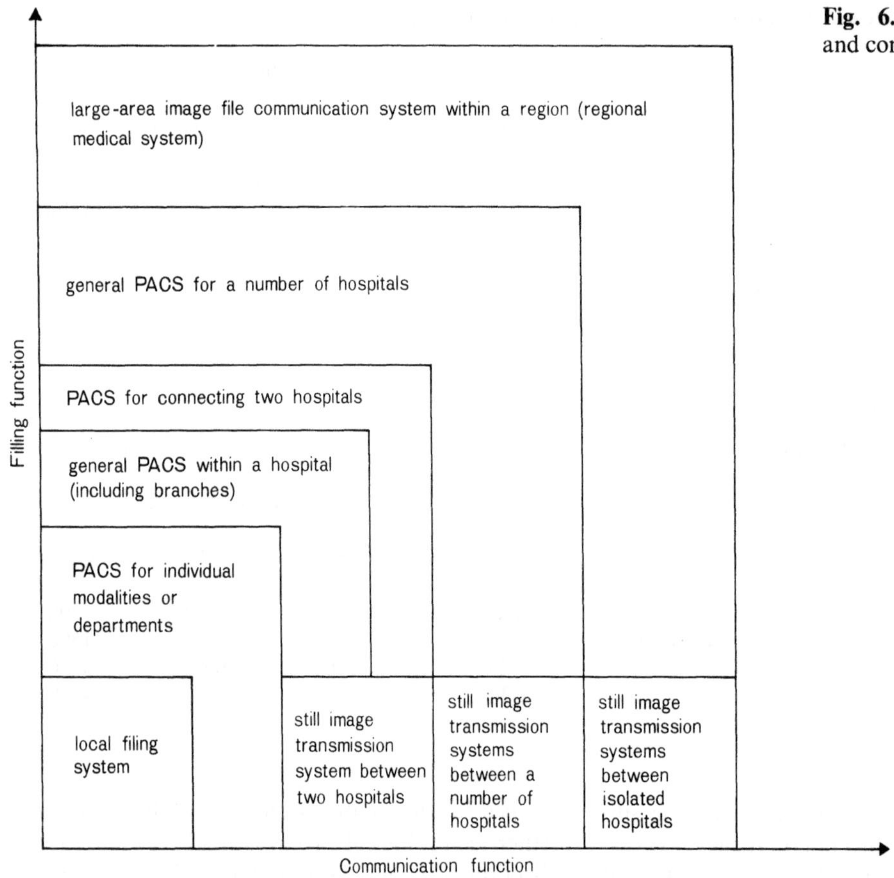

Fig. 6.4. Classification of filing and communication functions

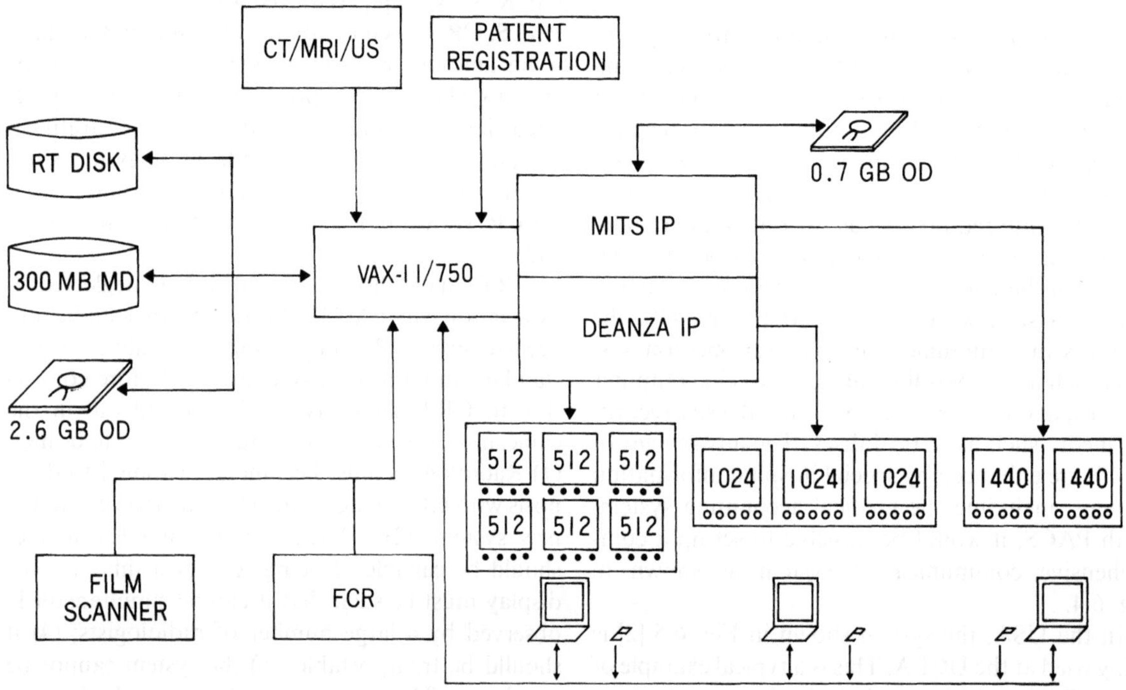

Fig. 6.5. System architecture of digital operation in pediatric radiology

Routine application

Today, the CR system is used in research and in some routine applications. Full utilization of the CR system meeting the needs of future routine medical applications requires correct conditions in the X-ray room. With a decentralized system, the requirements are fairly clear. Some of the more important points are discussed below.

Miniaturization

Hospitals nowadays do not have sufficient space to install large pieces of equipment. Moreover, if a unit is to be provided in each X-ray room, miniaturization of equipment is essential. Thus, just as many devices have been reduced in size so mechanical parts have to be miniaturized, the laser unit needs to be made of smaller, low-power laser, integrated circuits (IC) need to be changed to large-scale integration (LSI)s, and so on.

Fig. 6.6. Filing system for CR images

Simpler functions

A series of systems needs to be developed to meet specific needs, depending on the anatomical region, method, etc. Interfacing with X-ray equipment and small built-in image readers which do not require the handling of the IP are also needed.

Higher performance of automatic sensitivity adjustment mechanism

Wide latitude and image stability are important factors in routine applications and in evaluating the CR system. It is, therefore, necessary to improve the functions and performance of the automatic sensitivity adjustment mechanism which controls them.

Utilization of systems

With the increased use of PACS, there are a number of problems which require consideration, including provision of a standard interface, such as ACR-NEMA, and system construction, as in the CR system.

Greater utilization of the CR system as a file

For the effective utilization of files, it is considered necessary to examine applications and sometimes also to introduce new systems. In routine applications, the designation of key images, the use of data-compressed images, the means of data compression, and reediting from active to permanent files are some of the factors that need to be examined.

The future outlook of the CR system has been reviewed. To meet social demands, such as reduction of medical costs, computerization of the medical system, rationalization of medical services, improvement of diagnostic performance, profitability, better services, the CR system is expected to play an important role, and it is also expected that it will be one of the modalities used in future systems, such as PACS.

References

1. Imamura K et al. (1985) Survey of the access to stored X-ray films in St. Marianna University Hospital. Iryo-johogaku 5(4):358–363 (in Japanese)
2. Huang HK, Mankovich NJ, Taira RK, Kangarloo H (1986) Implementation of a digital multiple viewing station and early clinical experience. SPIE Medicine XIV/PACS IV 626:432–440
3. Jost RG, Hill RL, Blaine GJ, Cox JR (1986) PACS experience as a motivation for a campus-wide picture network. SPIE Medicine XIV/PACS IV 626:549–556
4. ACR-NEMA Digital Imaging and Communications. ACR-NEMA Standards Publication/No. 300-1985. National Electrical Manufacurers Association, Washington

Part II. Clinical aspects

Chapter 7

An introduction to clinical utilization of CR

Yukio Tateno[1]

Through the adoption of a specific sensor i.e., an imaging plate which has a wide dynamic range, computed radiography (CR) has made it possible for digital radiograms with high spatial and high density resolution to be obtained at a relatively low X-ray dose. In order to make this new technique better be utilized, certain problems associated with it will be discussed here.

Introduction

CR is a kind of digital radiography. It is different from other radiographic techniques; it employs a specific sensor called an imaging plate. This chapter, which attempts to evaluate CR from a clinical point of view, will discuss the following three themes: (1) the basic outline of the CR technique based on the characteristics of an imaging plate; (2) image-processing procedures for utilization of the characteristics of an imaging plate in diagnosis; and (3) issues related to digital image data.

Position of CR in digital radiographic techniques

Among the various imaging techniques presently available, the following five fall under the category of digital radiographic techniques.

1) The technique in which two-dimensional images are obtained by means of main scanning linearly across a linear sensor to obtain data in the direction of the *X*-axis and then scanning the object in the direction of the *Y*-axis: This technique can be characterized by its ability to produce an image on a very large screen and by its very high efficiency of X-ray utilization, thanks to the use of an NaI (T1) sensor. However, since the quality of the output image is not high and much time is needed, this technique has not been widely used in the clinical field; rather, it usually has such applications as baggage inspection at airports.

2) The technique using an instrument of X-ray computed tomography (CT): With this technique, a fan beam is applied to a linear sensor and the object is moved in the direction of the body axis to yield a two-dimensional image. This technique has essentially the same advantages and disadvantages as those of technique 1). Although various applications of this technique are now under study, its clinical use is at present found only in radiography to determine position in CT examinations.

3) The technique in which a linear beam is applied to a plane sensor: With this technique, an X-ray beam scans electrically and two-dimensional positioning is done by the X-ray generating unit itself. Since NaI (T1) is used as a plane sensor, range of the field is not large, but the efficiency of X-ray utilization with this technique is as high as that of technique 1). Good operational capacity of the beam is a characteristic of this technique; namely, it allows not only digital radiography but also depiction of the object's contour with the X-ray beam itself and follow-up of moving objects. However, this technique has not yet been commercialized.

4) The technique in which a cone X-ray beam is applied to an X-ray image intensifier—TV (II-TV) camera system: This technique has been applied to digital subtraction angiography (DSA). When the advantages and disadvantages of CR are discussed, this technique is often used as a reference.

5) CR: If X-ray data are evaluated by the factors of spatial resolution, contrast detectability, and time, CR is generally superior to the II-TV technique in terms of spatial resolution and contrast detectability but is inferior in terms of the time factor.

[1] Division of Clinical Research, National Institute of Radiological Sciences, 9–1, Anagawa 4-chome, Chiba, 260 Japan

Dynamic range of imaging plate

An imaging plate as an X-ray sensor has a fairly wide dynamic range. According to published data, incident X-rays (log X) correlate linearly with the luminescence (log E) of the imaging plate for a wide range (10^4 or more), from a very low to a very high dose level. Due to its wide dynamic range, the CR technique is characterized by its low X-ray does and wide latitude.

Radiography at low X-ray doses

Extent of X-ray dose reduction

The lowest X-ray dose level with this system varies considerably with different film readers. Several users of CR have reported their estimates of the lowest X-ray dose possible with this system based on their experience with conventional radiographic systems. For example, in the field of chest radiography, Yamada (National Cancer Center Tokyo., Japan) determined that radiography with CR was possible at an X-ray dose equivalent to 1/20 of the conventional dose level [1]. Further, Matsubara (Nagoya University; Part II Chap. 17) noted that the X-ray dose could be reduced to approximately one-seventh of the dose required for the conventional screen-film system (Sakura A type film + Fuji STD intensifying screen), without causing deterioration in image quality which might hamper diagnosis.

For use in stomach radiography with contrast medium, Yamada found that CR required only 1/20 of the conventional X-ray dose. [2].

In connection with excretory urography, Nishitani et al. (Kyushu University; Part II Chap. 13) claimed that compared with the conventional rare-earth intensifying film-screen system, CR provided imaging of the same quality at half the X-ray dose. When the X-ray dose was reduced to one-quarter or less of the conventional dose, granulation became more evident on the output image; however, they found that the image was clear enough to allow diagnosis even when the dose was reduced to one-eighth of the conventional one.

Further, Watanabe et al. (1983) (Kyorin Univ. Japan) [3] experimentally took radiograms of a pelvic phantom using the Martius technique at different exposure times and evaluated the quality of the images obtained based on the degree of erroneous information and noise. This study disclosed that the X-ray dose may be reduced to one-eighth of the conventional level in the single-plate CR technique and to 1/18 when three imaging plates are employed.

Diagnostic advantages

In the routine radiological procedures, the benefits of such X-ray dose reduction may be found only in a limited number of cases. Good psychological effects and other favorable influences would be expected only in critical examinations such as those performed by Aoki (pelvic measurement; Part II Chap. 18) or Doi (i.e., patients with an implanted pacemaker or valvular disease who require chest radiography every 1 or 2 months, patients after pulmonary or cardiac surgery being treated in ICU who require radiography two to three times a day). However, if we consider the influences of such reduced X-ray doses in the wider sense of society as a whole, the impact appears to be fairly large.

Mass examination

In radiographic mass examinations, the risk from X-rays sometimes makes the mass examination itself impossible or limits the number of subjects for examination; breast cancer examinations using mammography fall under the former case and mass examinations for stomach caner under the latter. In these cases, large reduction in X-ray dose can favorably affect the risk-benefit balance in mass examinations, thus making it easier to perform such examinations or to increase the number of subjects examined. Therefore, through the wide application of CR conventionally impossible examinations may become feasible and the number of subjects examined may be increased.

Before CR can be widely used in mass examinations, it will be necessary to make a detailed assessment of the X-ray dose, rate of detection, cost, and other factors, assuming that this technique is to be applied to a wide variety of a group examinations. The results of such an assessment may indicate the necessity to formulate new rules for mass examinations.

From the viewpoint of effective utilization of medical resources, steps toward the wide application of CR should definitely be taken.

Population X-ray dose

It is also important to assess the X-ray dose in terms of the entire nation. This issue refers to the concept of population X-ray dose. Problems involved here are the genetically significant dose and the leukemia significant dose. In 1978, the former was considered to be 10.3 mrad person^{-1} year^{-1} from radiography and 5.0 from fluoroscopy; the leukemia significant dose was 29.9 from radiography and 52.3 from fluoroscopy.

If the adoption of CR reduces the radiography X-ray dose by one-fifth on average, the genetically significant dose becomes 2.1 mrad person^{-1} year^{-1}, and the leukemia significant dose 6.0. Thus, the genetically significant X-ray dose would become only 2%–3% of the natural radiation dose, and the leukemia significant dose would be lowered to 10% of the natural radiation dose. Therefore, radiography could be applied without fear of the radiation practically.

Automatic setting of sensitivity and latitude and its effects

Automatic setting of sensitivity and latitude

Before converting the latent image on the wide dynamic-range imaging plate into a final image on the film, the following two issues need to be settled: (1) which region of the dynamic range is to be selected?; (2) in which relationship are the data on the imaging plate to be converted into density levels on film? In connection with these two issues, CR has a function (auto-mode) for setting the sensitivity and latitude automatically by determining the maximum and minimum intensities of the signals necessary for imaging (i.e., the maximum and minimum X-ray levels which pass through the object to reach the imaging plate).

Clinical effects of automatic setting

The automatic sensitivity and latitude setting mechanism of CR has taken a step forward from the "automatic camera" of conventional radiographic instruments. This CR mechanism has already received good reviews based on its use for radiography with a portable machine in hospital wards or ICU where appropriate setting of exposure conditions is often difficult. Where radiography is carried out at well-equipped facilities by well-trained technologists, this feature of CR is not particularly important. However, now that radiography is commonly practiced in various situations, this mechanism has taken on greater importance.

Facilities for developing imaging plates

To enable this advantage of CR to be utilized on a nation wide basis, a large number of imaging plate developers (IPD) will be needed. Of all radiograms taken in Japan, one-third each is accounted for by large hospitals (with 200 or more beds), medium-sized or small hospitals (less than 200 beds), and small clinics (less than 19 beds). Since only large hospitals can afford to buy CR systems, a service system for medium-sized or small medical institutions is indispensable to maximize the benefits of CR. In this sense, IPD would be expected to provide not only imaging plate developing but also, when necessary, imaging plate leasing and delivery/collection for medical institutions where only a few radiograms are taken in a day.

Development of automatic radiographic instruments

If manufacturers improve radiographic instruments further by unifying the various units (X-ray generator, photographic stand, etc.), it may, for example, become possible to take an appropriate chest film just by pushing the button "frontal chest film." The technician would not have to measure body thickness or determine whether the patient had fluid in a chest or whether the patient's body was rich in fat or muscle; the technician could take a good radiogram as long as he properly positioned the patient and the X-ray beam. In this way, the system more closely approaches the stage of complete automatization.

Image processing

Before a final X-ray image is obtained with CR, image-processing procedures such as tonal conversion are needed, although the number of procedures needed may vary with each case. The clinical influence of these processing procedures is closely connected with the above-mentioned issue of latitude.

Tonal conversion and frequency processing

With CR, the so-called γ characteristic (which refers to the relationship of the X-ray dose at each minute part of the imaging plate with the level of density on the final image) can be freely changed. For tonal conversion with CR, several kinds of gradation curves are prepared in advance, and a standard density level is set for each. The γ characteristic is then changed around this standard level or the standard level is moved in parallel. Although there has been much discussion, no clear answer has been given to the question of which one of these various combinations of the parameters of tonal conversion is best suited to a given type of radiography. One reason for the absence of a clear-cut answer is the different preferences of film readers; another is the recently disclosed fact that the image whose quality suits the preference of a given film reader is not always the image which allows the film reader to make the best diagnosis. However, in chest radiography, for example, where both the relatively dark lung field and the relatively bright mediastinum need to be evaluated, an appropriate gradation curve is perhaps a two-phase curve smoothly connecting two curves (one a curve of high γ corresponding to the dark parts and the other a curve of low γ corresponding to the bright parts).

In spatial frequency processing, a freely selected range of spatial frequency of an X-ray image can be enhanced. With this selection, there is freedom of choice for two factors—frequency and degree of enhancement. If enhancement is made at a relatively high frequency range, an image resembling a xeroradiograph is obtained. Although this kind of xeroradiographic image is not useless, enhancement is generally made at a relatively low frequency range (0.1–0.5 cycles/mm) in the case of chest radiography.

Although there seems to be room for further discussion on the question of which mode of image processing is most appropriate for a given type of radiography, it is clear that easily readable films can be obtained with CR thanks to the image-processing procedures and the wide dynamic range of the imaging plate.

Image subtraction technique

Digital subtraction angiography

In the field of digital subtraction angiography (DSA), CR is competing with the older I.I.-TV technique. CR is superior to I.I.-TV in terms of the following three properties:

Wider field of image. In the case of the I.I.-TV system, the field of image is limited by the diameter of the I.I. tube. In the case of CR, an Imaging plate (IP) rapid changer has been developed to allow the use of an imaging plate of a size equal to the film usually used in angiography.

Better spatial resolution. CR has the resolution of 3.3 lp/mm when an imaging plate of 12 × 10 in. is used, while the resolution of the I.I.-TV system is approximately 1.8 cycle/mm.

Wider dynamic range. With CR, the dynamic range of the imaging plate is wide, which works favorably when regions with a large difference in contrast (e.g., lung field and mediastinum, or lung field and abdomen) need to be examined.

On the other hand, in the field of DSA, CR is inferior to the I.I.-TV system in terms of real-time imaging ability. At present, processing of DSA images with imaging plates takes 30–60 min for one patient (about ten images). This means that CR has not truly arrived at the stage where it can be used in daily diagnosis with DSA. However, if this problem is solved, enabling the utilization of the superior properties of CR, DSA of the abdomen, extremities, head, neck, and other organs (excluding the heart) can be conducted with this system.

Energy subtraction

With the development of CR, the energy subtraction technique now under study has taken a step forward toward the use in daily diagnosis. The K absorption edge subtraction technique, which performs subtraction between two images taken at two energy levels above and below the K absorption edge of iodine (33 keV as X-ray energy), makes it possible to leave only the image of an iodine contrast medium. Further, it is also possible with this technique to produce the images of bones, calcification, and contrast medium in the output image or to produce only the image of soft tissues, eliminating the other images from the output image.

Other image-processing techniques

Easy digitalization of X-ray image data with CR has resulted in substantial progress in studies related to X-ray image processing. (In the past, the data once recorded on an X-ray film were digitalized, reguiring much time and labor.) Among the studies where progress has been accelerated by the use of CR are a study to detect automatically the time-course changes in a series of health check radiograms taken each year, a study to detect coin lesions automatically, and a study to diagnose pneumonoconiosis. A detailed account of these studies is omitted in the present paper, though many of them are promising.

Impact on picture archiving and communication system

CR involves the procedure of temporarily converting X-ray image information into time-series electric signals. This procedure has been instrumental in the incorporation of radiograms (which have a central role in medical imaging) into a digital imaging system. CR is now an important cornerstone in the realization of the picture archiving and communication system (PACS), which is seen as a potential medical imaging system.

References

1. Yukio Tateno (ed.) (1984): Clinical application of FCR Journal of Medical Imagining, Vol. 4, Suppl. 1, 7–15
2. Yukio Tateno (ed.) (1984): Clinical application of FCR Journal of Medical Imagining, Vol. 4, Suppl. 1, 34–40
3. Yoshiji Watanabe, Eriko Ito, Masayoshi Kitamura (1983) Dose reduction in X-ray pelvimetry (Martius method) using Fuji Computed Radiography (FCR) system Fuji Medical Forum 143:15–18

Chapter 8

Chest

Tsutomu Takashima[1]

Introduction

The recent advances in medical imaging systems owe much to the digitalization technique made possible by the application of electronics and computer technologies to the medical field. X-ray computed tomography (CT) represented the first application of the digitalization technique and is often referred to as the greatest technological development. Since then, a variety of digital radiographic systems have been introduced and utilized in medicine. In this wave of technological innovation, computed radiography (CR) was developed as a new system of standard radiography. Unlike the conventional film/screen system (CFS), CR uses an imaging plate (IP) capable of storing X-ray energy [1]. In this chapter, an account is given of the application of this system to chest radiography, reporting the author's clinical experience with this new system.

Exposure settings

Chest radiography with CR was performed with an X-ray generating unit (Shimadzu, Japan) installed at the Department of Radiology of Kanazawa University Hospital. Posteroanterior views of the chest were taken with exposure settings of 90 ± 5 kV_p, 100 mA, 0.05 s, and 150 cm FFD and with a Bucky diaphragm. Lateral views of the chest were taken at 100 kV_p, 200 mA, and 0.1 s. For each radiograph, an imaging plate of 354×354 mm was used.

CFS radiography was performed with a condenser-type X-ray generating unit (Hitachi, Japan) at 120 kV_p and 250 cm FFD with the use of Kodak XS-1 film, a Toshiba EM intensifying screen, and a 10:1 grid.

[1] Department of Radiology, Kanazawa University School of Medicine, 13-1, Takamachi, Kanazawa, 920 Japan

Recording

In the preceding chapter, a detailed description of technical aspects of CR, such as detection, reading, processing, and recording of an image from the imaging plate, was given. These aspects will thus be omitted from this chapter. The size of each CR film presented to the physician as a final output image after automated film processing is fixed at 364×257 mm, irrespective of the varying sizes of the IP. Currently, a dual-image display system is being used; namely, one film simultaneously displays two images of the chest, which are processed under different image-processing conditions. Of these two images, the left image is processed in much the same way as in conventional radiography, and the right is processed according to the purpose of the examination. The dual-image display format aims at improving the diagnostic capability of a radiograph.

The chest X-ray films consist of quite a wide range of shadows from low to high density, e.g., an air density shadow of the lung field (filled with ordinary air, which acts as a negative contrast material), a central water density shadow, including the heart and large vessels, and a metallic density shadow of the surrounding skeleton. Therefore, abnormal shadows emerging in these fields have a characteristic appearance.

The purpose of chest X-ray examination varies; in some cases examination is performed as a routine study (e.g., routine chest radiography), in other cases it is done as a detection study (e.g., mass screening for lung cancer). In addition, the purpose of radiography for closed examination is not confined to detection of lesions but also aims at qualitative diagnosis of the lesions. Thus, chest radiography is characterized by quite a wide range of applications. For this reason, it seems to be too early to write an article titled "Systematic review of chest radiography using the CR technique," answering various questions related to the above aspects. Therefore, the scope of this paper is confined to a comparison between variously processed CR and CFS radiographs of types of pulmonary lesion taken at the same time, namely, segmental pneumonia (parenchymal shadows), lung

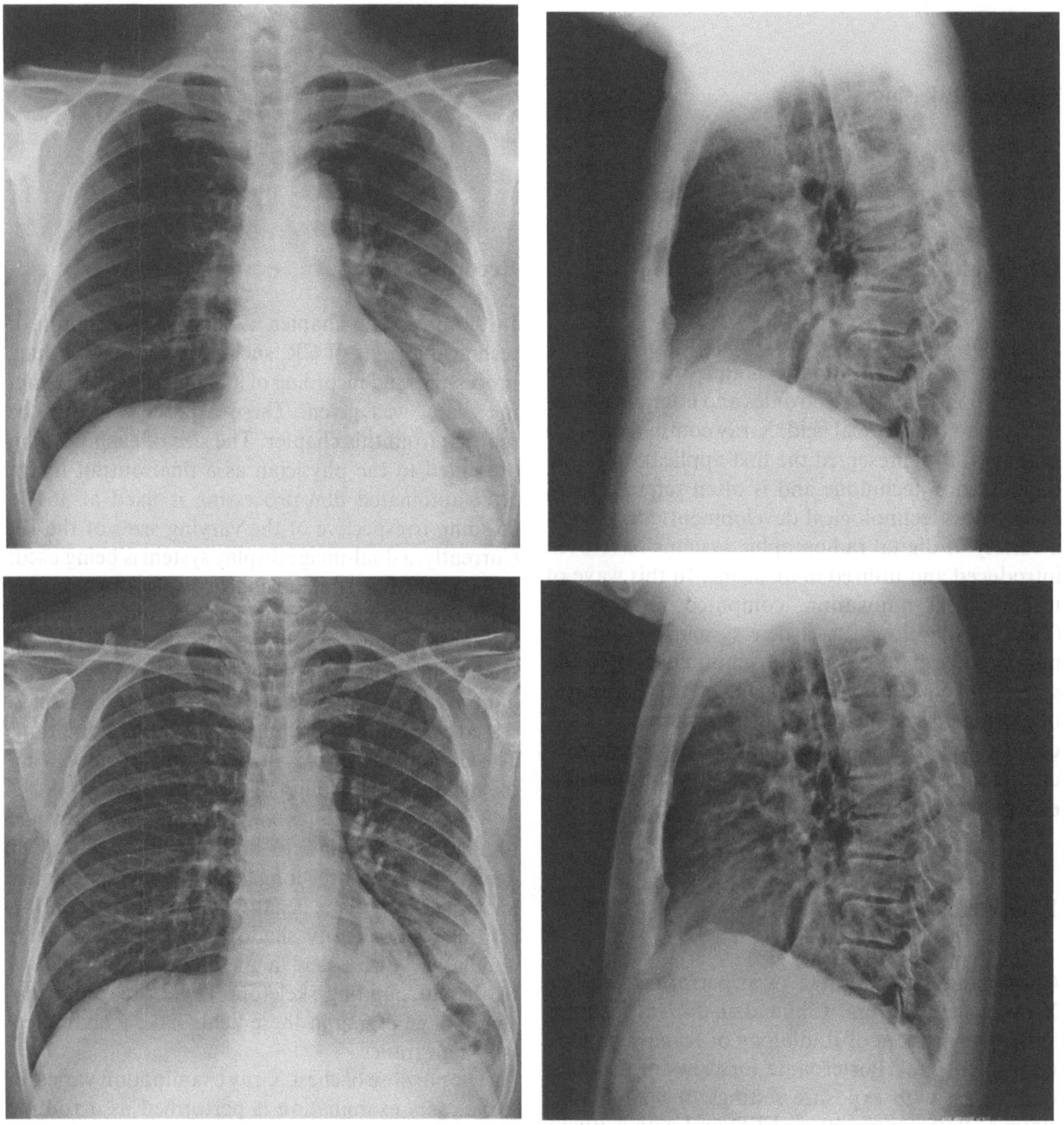
a b c d

Fig. 8.1a–d. Case 1. Pneumonia. **a** CFS (frontal view). **b** CFS (lateral view). The frontal view reveals consolidation in the left middle and lower lung fields, and the lateral view shows the presence of it behind the heart shadow. **c** CR (frontal view). **d** CR (lateral view). The left diaphragm shadow is obliterated and the descending aorta is clear on the frontal view of CR

fibrosis (interstitial shadows), heart disease showing Kerley's B lines, hematogenous metastatic lung cancer (rounded shadows), tumor in the mediastinum, and metastatic bone cancer.

Case 1: A 60-year-old male with pneumonia in the left lower lobe

A frontal radiographic view of the chest (Fig. 8.1a) reveals consolidation in the left middle and lower lung fields. Consolidation is found also at the site superimposed on the heart shadow. A lateral view (Fig. 8.1b) clearly shows the presence of consolidation behind the heart shadow. CR films which were processed to make the radiographic latitude wider (Fig. 8.1c, d) reveal that the left diaphragm was obliterated and the left heart border and descending aorta were not evident on frontal view. Therefore, we surmised that the shadow lay in a lateral basal segment, which was confirmed by the lateral view (Fig. 8.1d).

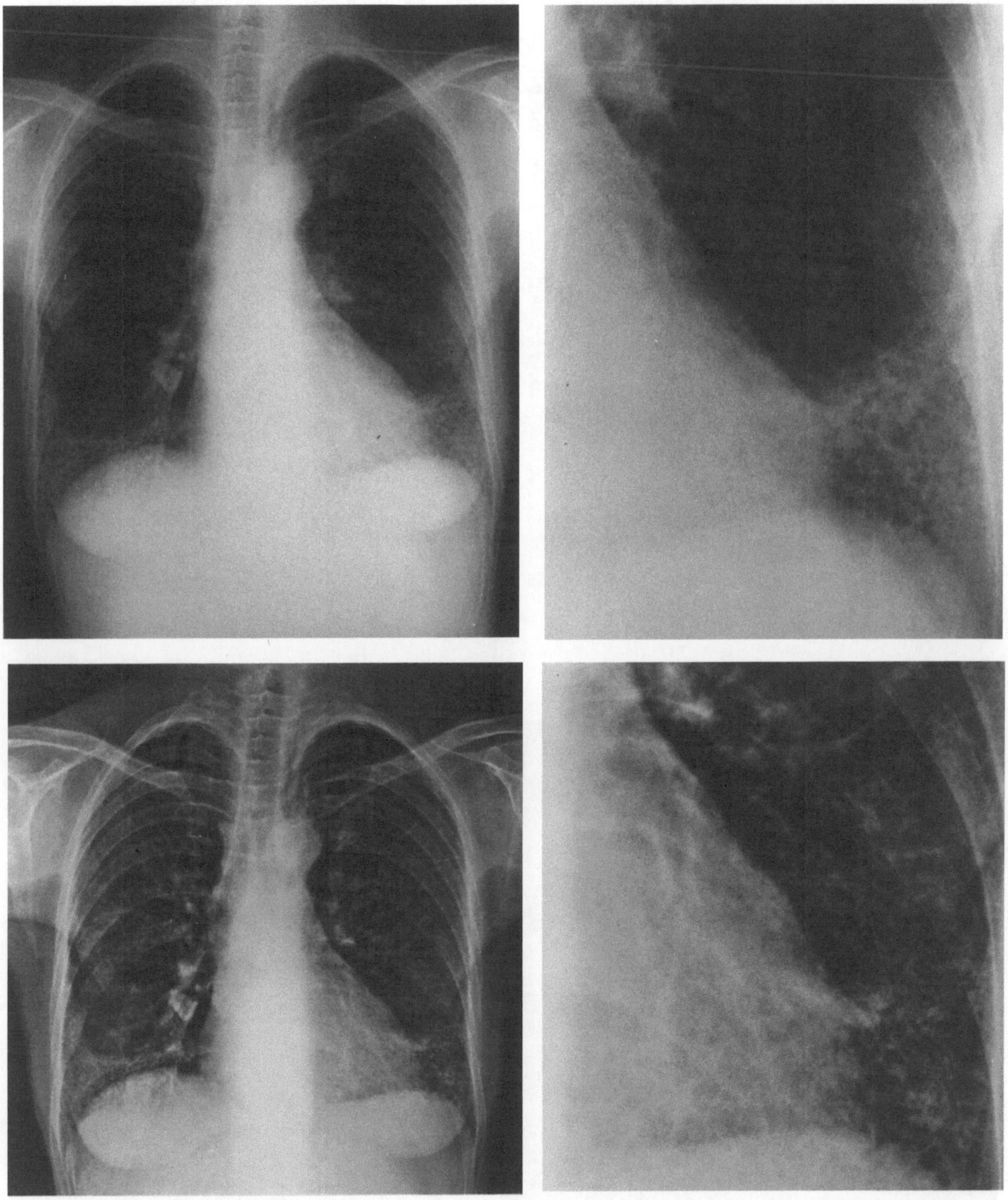

a b c d

Fig. 8.2a–f. Case 2. Lung fibrosis. **a** CFS image. **b** The left lower lung field of **a**. (CFS). **c** CR image (a lower frequency range enhanced). **d** The left lower lung field of **c** (CR). **e** (*see over*) CR image (a high-frequency range enhanced). **f** The left lower lung field of **e** (CR)

Case 2: A 63-year-old female with lung fibrosis

A frontal CFS image of the chest (Fig. 8.2a, b) reveals reticular, interstitial shadows in the bilateral lower lung fields. Comparison between a low-frequency range-enhanced CR image (Fig. 8.2c, d) and a high-frequency range-enhanced CR image (Fig. 8.2e, f) reveals that on the latter image the abnormal shadows consist of linear opacities and that the presence of such a shadow can be observed

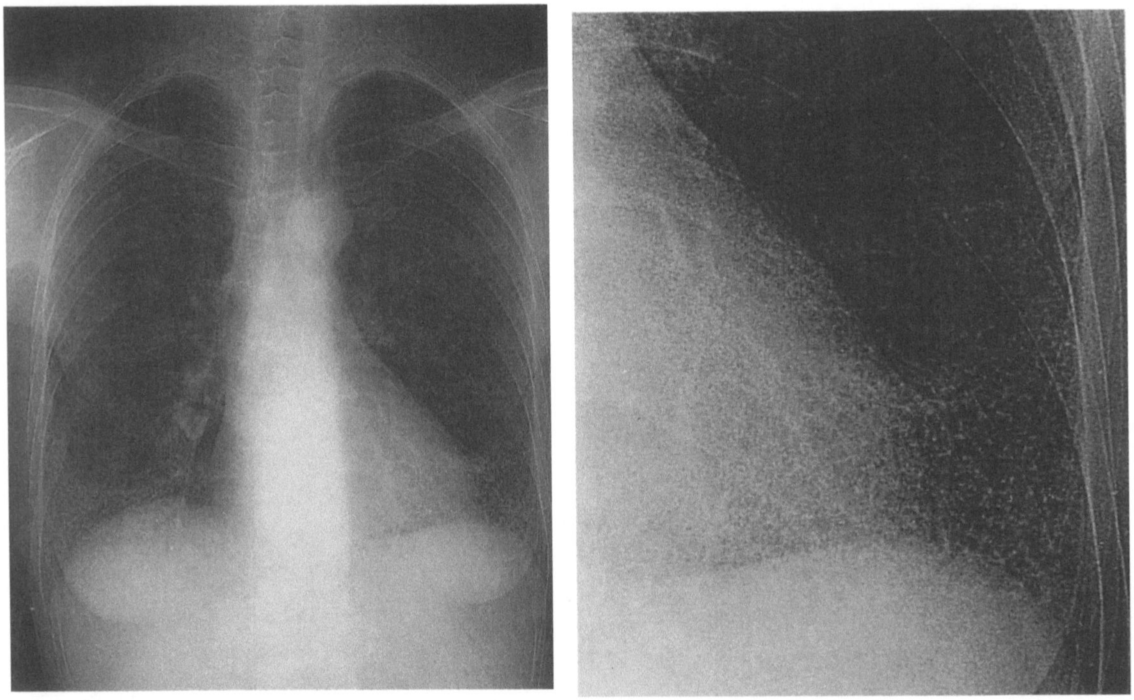

Fig. 8.2e, f.

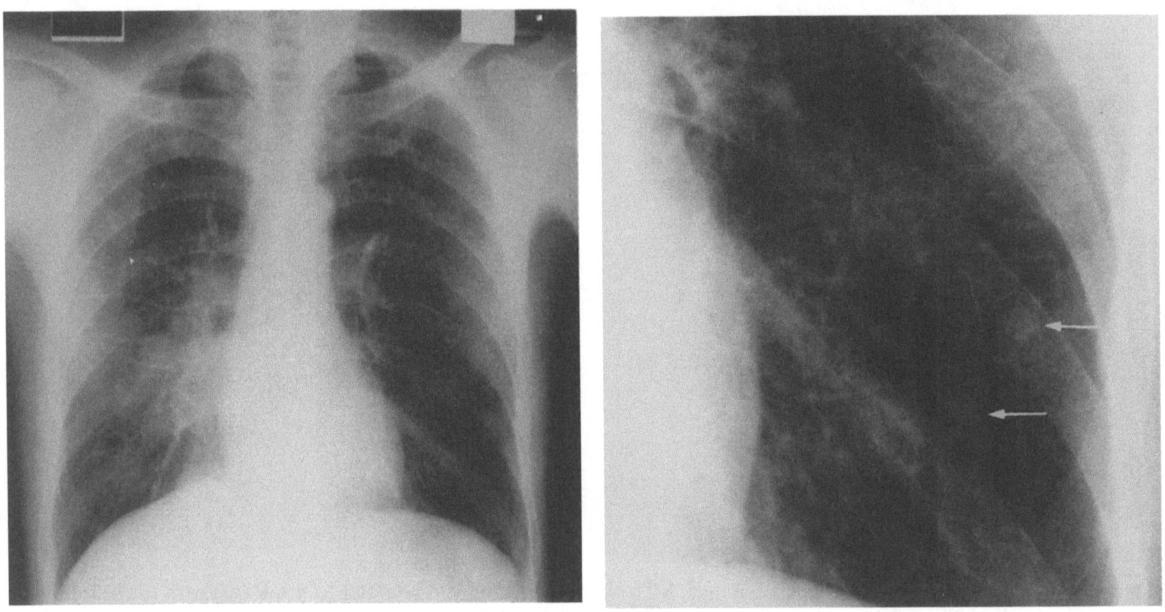

Fig. 8.3a–f. Case 3. Primary lung cancer. **a** CFS image. **b** The left lower lung field of **a**. (CFS). Two rounded shadows (*arrows*) are revealed. **c** CR image (a low-frequency range enhanced). **d** The left lower lung field of **c** (CR). The new (*black arrows*) rounded shadows are seen in **c**. **e** CR image (a high-frequency range enhanced). **f** The left lower lung field of **e** (CR)

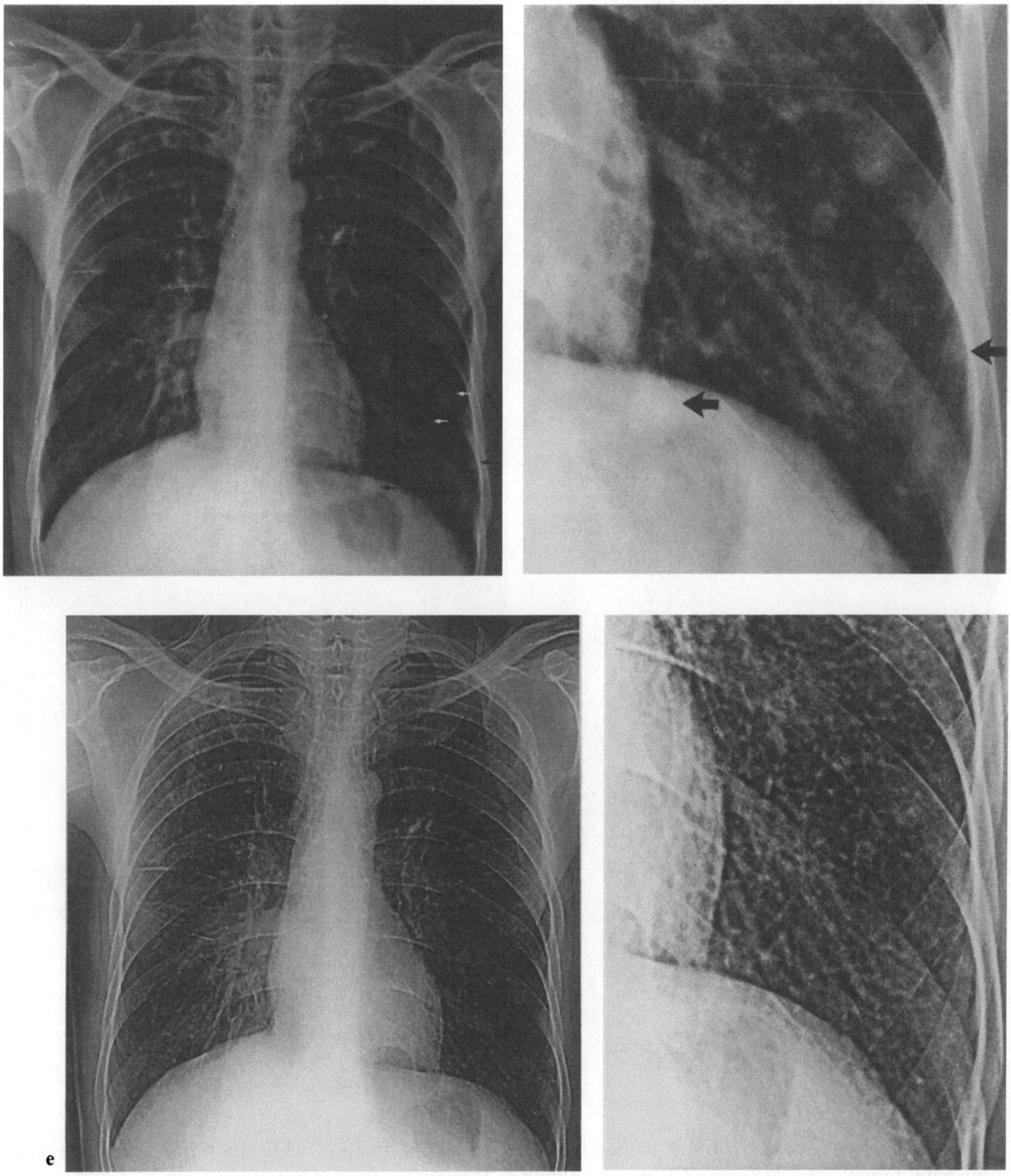

c d e f

Fig. 8.3c–f.

in a wider area on the latter (Fig. 8.2e) than on the former image (Fig. 8.2c). Furthermore, on both images (Fig. 8.2c, e), the presence of the abnormal shadows behind the heart or in the subphrenic area can be detected more easily than on the CFS image (Fig. 8.2a).

Case 3: A 54-year-old male with primary lung cancer (small cell carcinoma)

As is clear from the frontal CFS image of the chest (Fig. 8.3a, b), this patient has a primary focus in the lower part of the right hilus, and two rounded shadows (about 5 mm in diameter), suggestive of

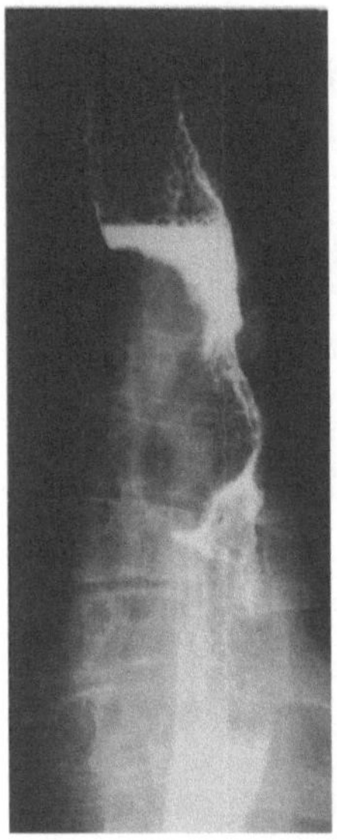

a

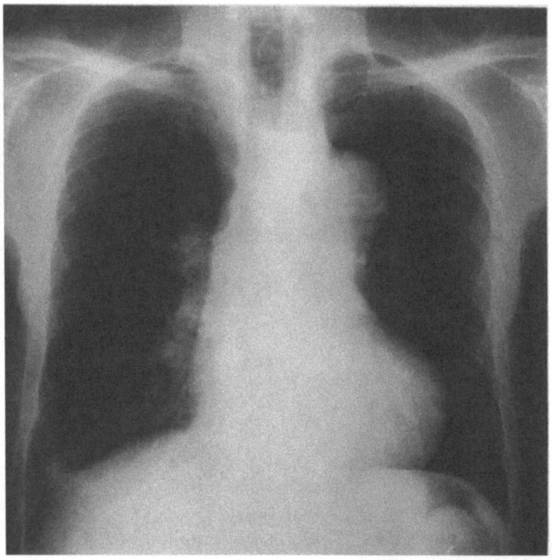

b

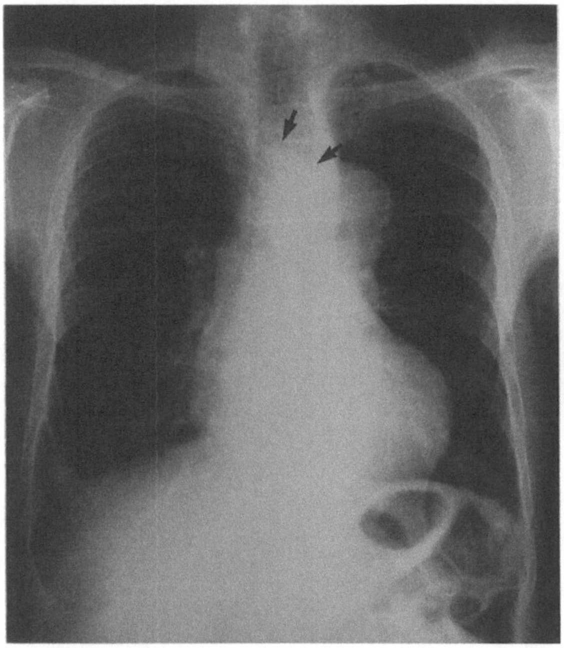

c

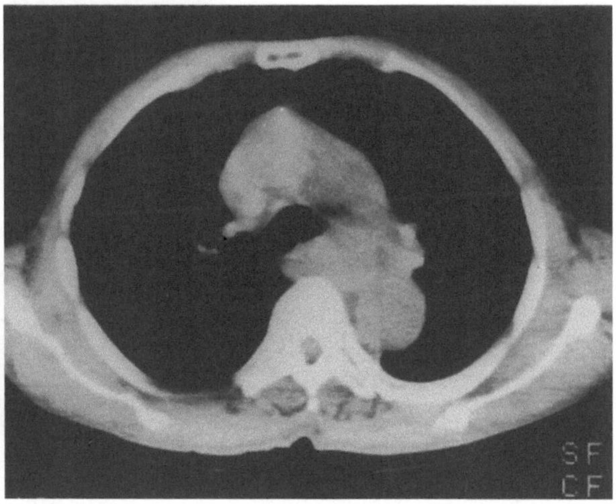

d

Fig. 8.4a–d.

hematogenous metastases, are found in the left lower lung field. Figure 8.3c, d shows a CR image enhancing the low-frequency range, and Fig. 8.3e, f is a CR image enhancing the high-frequency range. In Fig. 8.3e, the two rounded shadows are obscure, while they are evident in Fig. 8.3c. Furthermore, Fig. 8.3c shows two small rounded shadows in the lower outer side and under the diaphragm. Only careful inspection of Fig. 8.3a reveals the presence of such small rounded shadows in the same area.

Putting aside the question as to whether these rounded small opacities are a metastatic focus, they are easily detectable on low-frequency range-enhanced images, while they are absent on high-frequency range-enhanced images.

Case 4: A 68-year-old male with esophageal cancer (squamous cell carcinoma)

This patient has esophageal carcinoma, showing a large filling defect in the superior esophagus on the esophagram (Fig. 8.4a). A frontal view of the chest radiograph (Fig. 8.4b) shows an air density area in both sides of the trachea suggestive of dilatation of

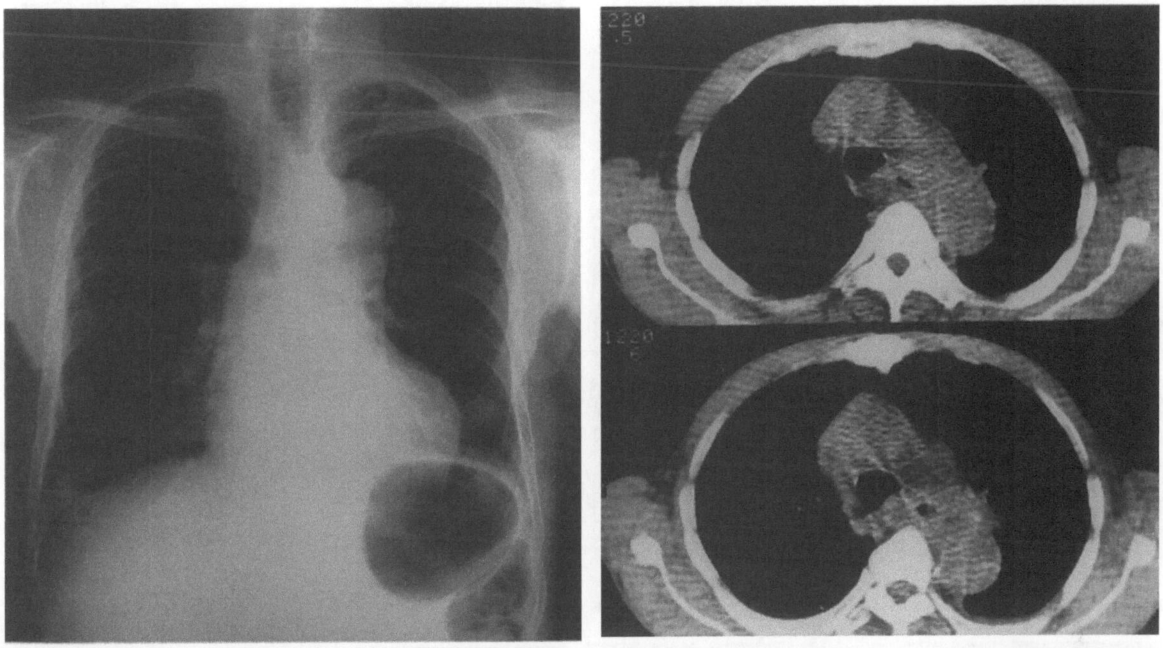

e f

Fig. 8.4a–f. Case 4. Esophageal cancer. **a** A frontal view of the esophagram. (CFS). **b** CFS image of chest. **c** CR image. The half-spherical shadow (*arrows*) suggestive of tumor mass is revealed at upper mediastinum. **d** CT image. **e** CR image taken after radiation therapy. **f** CT image taken after radiation therapy. Reduction of the tumor mass and presence of air in esophageal lumen are evident

the superior esophagus. A CR image (Fig. 8.4c) after graduation processing demonstrates not only the above characteristics but also the presence of a tumor mass at the same site. A CT image obtained at the same time (Fig. 8.4d) shows the presence of a giant tumor mass at this site. A CR image taken after radiation therapy (Fig. 8.4e) does not show any tumor mass in the superior mediastinum as seen in Fig. 8.4c. Reduction of the tumor mass was evident on the CT image obtained after radiation therapy (Fig. 8.4f) compared with the CT image before therapy (Fig. 8.4d).

Case 5: A 63-year-old female with primary lung cancer (squamous cell carcinoma) and hematogenous metastases to both lungs and left scapula

A frontal view of the chest radiograph (Fig. 8.5a) clearly reveals a tumor mass in the right hilus and multiple nodular shadows in both lungs. However, it does not show any clear defect in the left scapula. A CFS image taken about 1 month later (Fig. 8.5c) reveals abnormal opacity in the outer side of the right middle lung field (suggestive of pneumonia) in addition to the shadows indicative of progression of the primary focus. Close inspection of the left scapula on this film reveals the presence of a small defect in the bone's outer side. An edge effect-enhanced CR film taken at the same time (Fig. 8.5d) shows the defect in the left scapula more clearly. With enhancement of the edge effect of the CR film taken 1 month earlier (Fig. 8.5b), a small defect is detectable in the same site of the left scapula.

Figure 8.5b, d shows a linear opacity of high density, several centimeters long, at the dome of the diaphragm, suggestive of calcification, while none of the CFS images demonstrate such a shadow. Therefore, this shadow was considered to be an artifact arising from image processing.

Case 6: A 49-year-old female with mitral stenosis

The cardiac silhouette seen on the standard chest film allows easy diagnosis of mitral stenosis. Kerley's B lines, which are not clear on a CFS image (Fig. 8.6a), can be detected on a high-frequency range-enhanced CR image (Fig. 8.6b). On a low frequency-enhanced CR image (Fig. 8.6c), it is possible to detect these lines, but it is not easy to conclude that such linear shadows are not the peripheral shadows of pulmonary vessels. Redistribution of pulmonary flow, which is indispensable for diagnosis of the left heart overload, is difficult to see on CR images; CFS images allow it to be seen more easily.

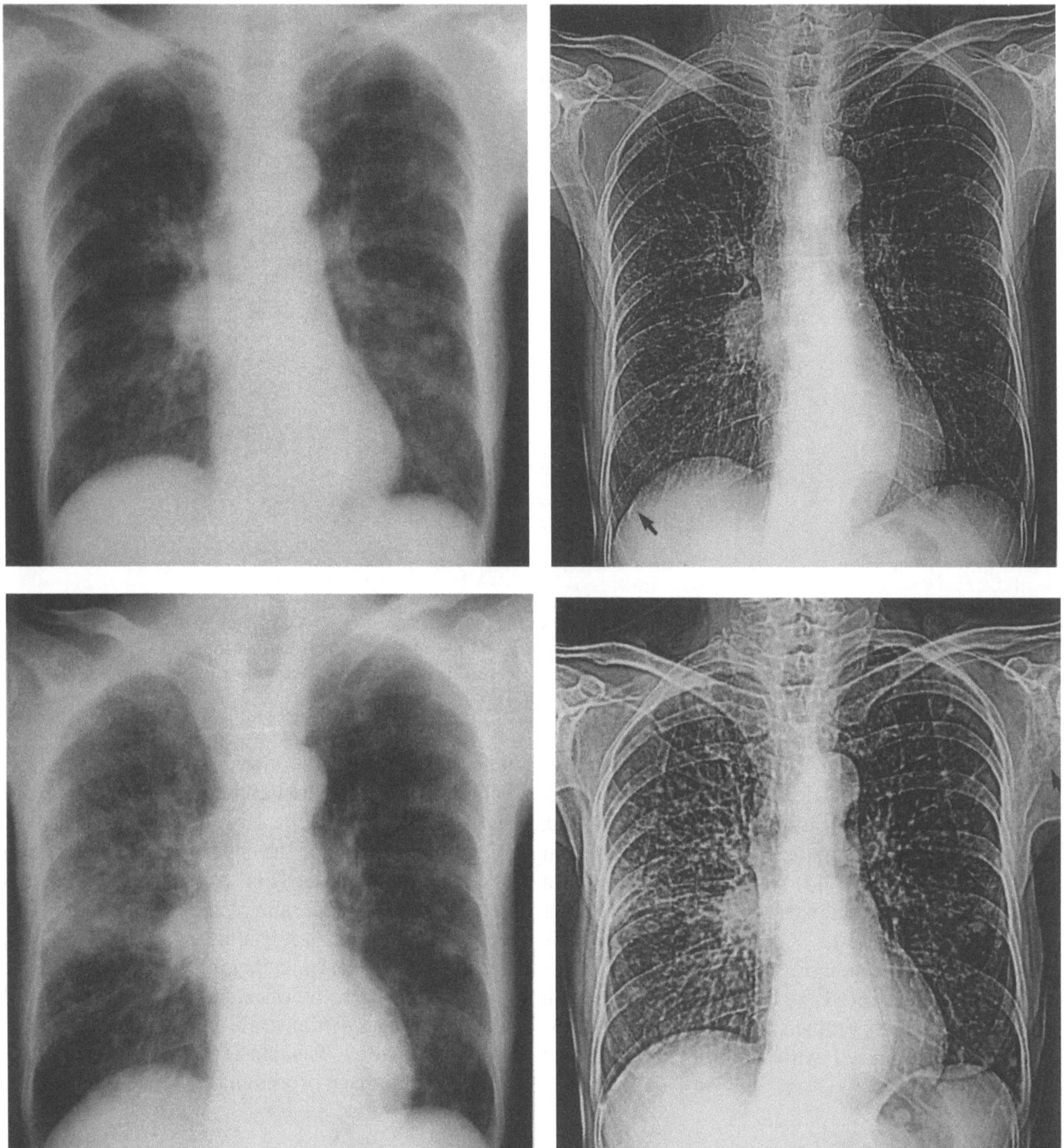

a b c d

Fig. 8.5a–d. Case 5. Primary lung cancer. **a** CFS image. **b** CR image. A small defect is detectable in the left scapula. A long linear shadow at the right upper diaphragm (*arrow*) is an artifact arising from image processing. **c** CFS image taken about 1 month later. A defect in the outer side of the left scapula was revealed in the original film (not clear in this image). **d** CR image taken about 1 month later. The defect in the outer side of the left scapula is clearer

Discussion

The greatest advantage of CR for radiologists is that the information obtained with one-time X-ray exposure can be processed in various ways to yield X-ray images of various quality, from which new roentgenological findings can be derived and utilized for diagnosis. The methods of image processing for the CR system are gradational enhancement and spatial frequency enhancement. In this connection, consideration should be given to which method of image processing is appropriate for observation of a given

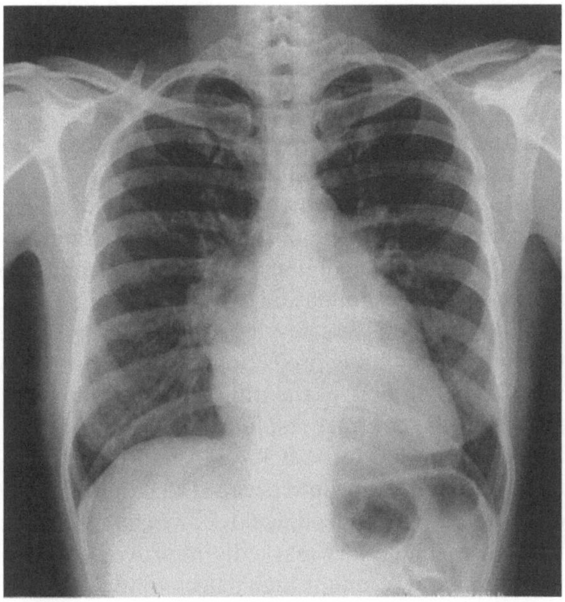
a

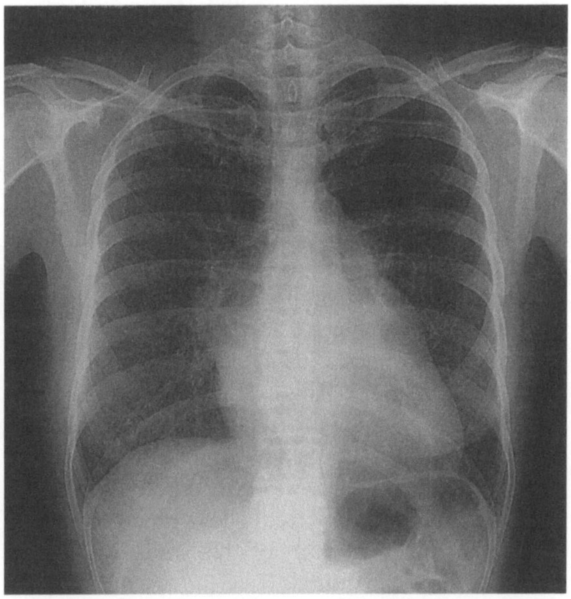
b

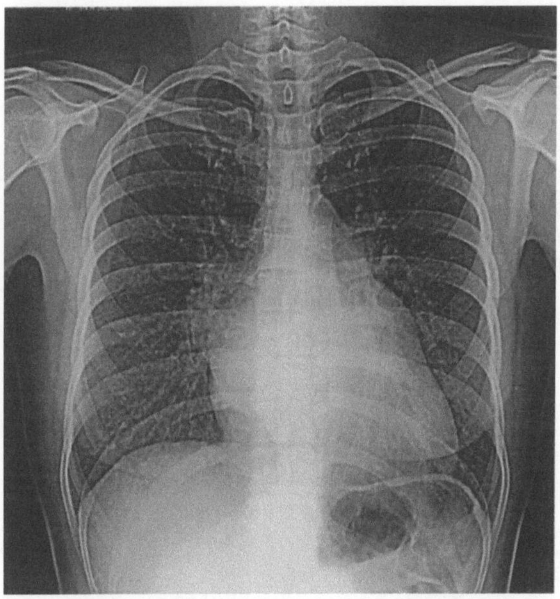
c

Fig. 8.6a–c. Case 6. Mitral stenosis and Kerley's B lines. **a** CFS image. The diagnosis of mitral stenosis is easy, but Kerley's B lines are not clear. **b** CR image (a high-frequency range enhanced). The presence of Kerley's B lines is clear, but the shadows of the blood vessel are indistinct. **c** CR image (a low-frequency range enhanced). It is difficult to distinguish between Kerley's B lines and the peripheral shadows of pulmonary vessels

abnormal shadow.

An image with widened exposure latitude utilizing the wide dynamic range of IP can usually be obtained by gradational enhancement. On CR images allowing observation of the lung field (a chief purpose of chest radiography), we could easily trace an abnormal shadow which was overlapped by the central shadow (cases 1, 2, and 4), an abnormal shadow below the diaphragm (case 3), and abnormalities in the skeleton (case 5), when the images were processed as for gradation. On standard CFS images, only one characteristic curve, which determines the relationship between exposure levels and density levels, is available for each CFS system, while CR enables free modification of the characteristic curve by means of computerized nonlinear gradation enhancement. This capability was clearly indicated in the report of Sakurai et al. [2], who compared the detectability of various structures in normal chest films taken by CFS and by CR.

Another method of image processing is spatial frequency enhancement, which amplifies the responses to a certain range of frequency. With this method, linear shadows can be enhanced as shown in the images for cases 2 and 6. However, the ability to evaluate pulmonary vessels was reduced by this type of processing. Furthermore, this method of processing may adversely affect the detectability of low-contrast shadows, such as small rounded opacities and parenchymal shadows. Consequently, we realize the necessity to clarify the relationship between the nature of abnormal shadows and the spatial frequency enhancement or the degree of enhancement.

In patients with primary lung cancer and intrapulmonary hematogenous metastasis, we compared the detectability of small rounded shadows in the lung field by standard CFS images and variously processed CR images, using a Receiver Operating Characteristics curve. This examination revealed that high frequency-enhanced images had poor detectability, while low frequency-enhanced images had good detectability; there was little difference between the detectability of low frequency-enhanced images and CFS images. Details of this study will be reported in the near future.

As mentioned above, the wide exposure latitude of CR images is a great advantage in their clinical use. However, we may say that, in general, in examination of abnormal shadows in the lung field parenchymal shadows and small rounded shadows are more appropriate for low-frequency enhancement and gradation enhancement for the purpose of contrast improvement and that interstitial shadows are suitable for high-frequency enhancement alone for the same purpose. This aspect, however, requires further study.

It is questionable whether processed CR images can show abnormal shadows in the lung field more clearly than CFS images taken under optimum conditions. The chest is a characteristic field where even very small abnormalities can be seen on X-ray films with the help of air, which works as a negative contrast material. The spatial resolution of the CR system is 2.5 cycles/mm, which is undoubtedly excellent compared with the resolution of 1 cycle/mm reported for another digital radiographic technique for use in chest examination called the Digital Chest Unit (*Picker*); this was developed on the basis of the technique of scanning projection radiography (SPR) used for positioning of the transverse section in CT. However, the contrast resolution with the CR system was reported to be 0.5%/3–5 mm in a experimental study. Taking into account also the influence of scattered X-rays in clinical use, we cannot expect from this technique a markedly improved contrast resolution and resultant high detectability of tissues comparable with that of X-ray CT.

At present, our CR system is not connected with a high-quality cathode ray tube (CRT) display device (does it require 2000 or more scanning lines?), so image processing with this system is not totally free or flexible.

Depending on the nature of the shadows, image processing can adversely affect their diagnostic value. Such a tendency was also reported for SPR images by Fraser et al. [3] and can be considered a disadvantage inherent in the image-processing technique. The image in case 5 is an example of

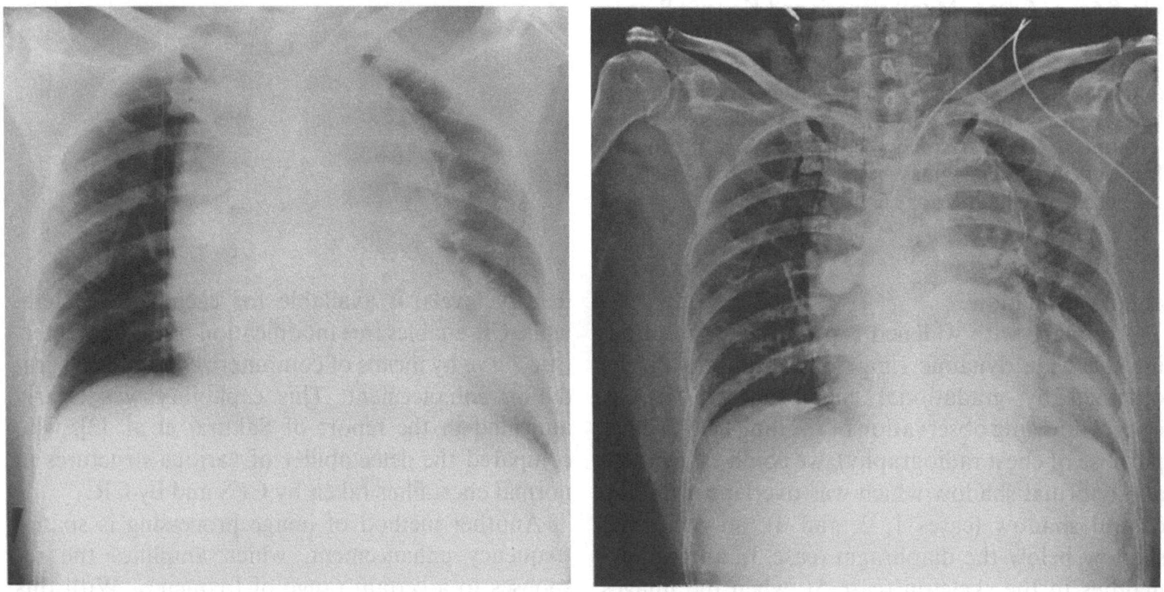

a b

Fig. 8.7a, b. Dissecting aneurysm. **a** CFS image with portable X-ray equipment. **b** CR image with the same exposure time as **a**

an artifact (appearance of false abnormal shadows on an image), which sometimes occurs with the CR system. To make up for such disadvantages, a dual-image display format has been adopted, which reveals two output images of different quality on a film. At present, the left image on the film represents a weakly processed image (close to a CFS image) and the right represents a strongly processed image. However, there is no established view as to the appropriate image-processing conditions for clinical routine use of this dual-image format. This aspect requires further study.

At any rate, compared with the CFS technique, CR enables easier detection of abnormalities on chest film. This is undoubtedly a great advantage of CR for the radiologist, who has to observe a large number of radiographs every day [4]. This property of the CR system is more valuable than any other for the radiological department of a large hospital like ours where many patients have to be examined.

Since the quality of CFS X-ray images varies according to the exposure conditions, it is not unusual that the radiologist is not satisfied with the quality of the images provided. Such unsatisfactory images are particularly likely when chest radiography is performed with a portable machine in the ward. In this respect, the CR technique is superior because it provides output images of acceptable quality even under slightly unfavorable exposure conditions, and further it can provide output images of constant quality (Fig. 8.7).

Dual-energy subtraction. Physicians are expecting much of the application of the energy subtraction technique to the CR system [5]. A possible method of energy subtraction CR imaging currently under consideration is to take two images of the same site at a different kilovoltage, followed by subtraction, making use of the difference in the X-ray attenuation between the two energy modes, for the purpose of reconstructing a skeleton-cancelled image of the lung field or a soft tissue-cancelled image of the thoracic skeleton. However, there are unresolved questions with this technique, e.g., how to cope with the three-dimensional shift in the position of organs between the two modes and how to take two images with different energies in a rather short space of time.

In an effort to clarify such points, we conducted an in vivo experiment in which we tried to take dual-energy radiographs with **one time exposure** using a highly sensitive IP of wide exposure latitude and a metallic filter. In this study, we obtained subtracted images which were substantially free of artifacts arising from movement of the organs (Fig. 8.8). A key factor for this successful dual-energy imaging is the use of two highly sensitive IPs behind the metallic filter so as to make up for attenuation of the X-rays due to absorption by the filter. Dual images obtained with this technique are presented in Fig. 8.9. The clinical significance of these dual-energy subtraction images is now under evaluation and will be reported at a later date.

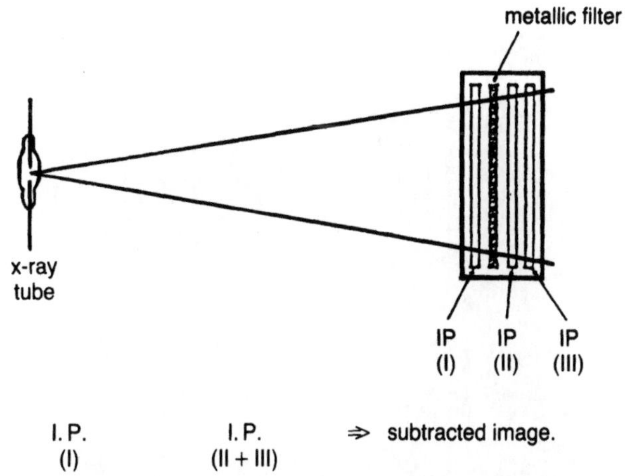

Fig. 8.8. Dual-energy subtraction system with one-shot exposure

As another advantage of the CR system, the skin X-ray dose can possibly be reduced by the use of the highly sensitive IPs and the computed image-processing technique. Furthermore, it is highly probable that a system for storage, screening, and delivery of standard X-ray images will be developed by the application of a high-density large-capacity recording system. Thus, the CR system will represent a substantial innovation to the systems available in the radiological department of large hospitals.

Conclusion

In recent years, a variety of new digitalized devices for diagnostic imaging have been introduced in the clinical field. However, standard chest radiography is still contributing much to clinical practice. In this paper, an account has been given of the CR system for use in chest radiography, mainly based upon the author's clinical experience with this technique.

Wide exposure latitude, which is made possible by image processing of the CR system, is undoubtedly an advantage of the CR system.

Another great advantage of the CR system is that image processing by this technique can constantly

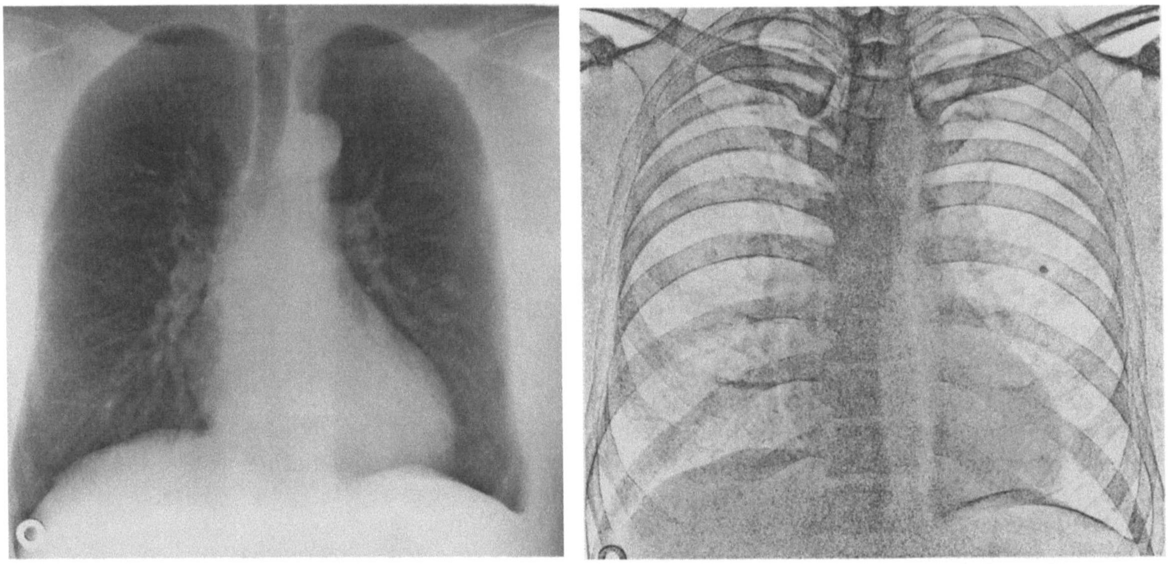

a b

Fig. 8.9a, b. Dual-energy subtraction images. **a** A skeleton-cancelled image. **b** A soft tissue-cancelled image

supply homogeneous, excellent radiographs of the chest without being much affected by different exposure conditions.

Concerning abnormal shadows in the lung field, further study is needed, chiefly with regard to the relationship between the quality of the abnormal shadows and the image-processing method.

Some new techniques are available whose combined use with CR seems to be of value (e.g., energy subtraction). However, to compete with the conventional chest examination, several technical points need to be improved, including the processing speed.

References

1. Sonoda M, Takano M, Miyahara J, Katoh H (1983) Computed radiography utilizing scanning laser stimulated luminescence. Radiology 148:833–838
2. Sakurai K, Hachiya J, Korenaga T, Miyasaka Y, Huruya Y, Nitatovi T (1984) Digital radiography of the chest: Evaluation of normal anatomical structure and low dose radiography. Nippon Acta Radiol 44: 11–22
3. Fraser RG, Breatnach E, Barnes G (1983) Digital radiography of the chest: Clinical experience with a prototype unit. Radiology 148:1–5
4. Kundel HL, Revesz G (1976) Lesion conspicuity, structured noise, and film reader error. AJR 126:1233–1238
5. Brody WR, Cassel DM, Sommer FG, et al. (1981) Dual-energy projection radiography: Initial clinical experience. AJR 137:201–205

Chapter 9

Thoracic tomography by CR: Investigation of bronchial shadows

Ryoichi Kamimura[1], Tsutomu Takashima, and Hiroshi Itoh

Introduction

In cases of suspected lesions on plain radiographs such as tuberclelike shadows in the lung field, thoracic tomography has many applications for use in daily clinical practice as a follow-up step to X-ray diagnosis, with the aim of confirming the presence of such lesions or analyzing their nature in detail. Even following the appearance of CT, this modality is still significant in view of its simplicity for verification of bronchial lesions, analysis of the nature of shadows, and so on. On the other hand, the characteristics of computed radiography (CR) include: (1) images with a wide latitude by tonal conversion, (2) edge enhancement by spatial frequency enhancement, and (3) reduction of exposure dose. Thoracic tomography with CR has been used in our hospital since April 1984, leading to the conclusion that it offers many advantages, such as evaluation of both the mediastinum and lung field with one image, which is especially useful in the diagnosis of bronchial lesions. In this chapter, a comparison is made between the ability to describe normal bronchial shadows obtained in thoracic CR tomography (CRT) and those obtained in conventional film/screen tomography (CFST); the clinical usefulness was investigated using actual cases.

Subjects and methods

Translucent images and peripheral streaky images of the bronchus formed by air within the lumen can be identified by thoracic tomography and considered as shadows of the bronchial wall. CRT was compared with CFST regarding the describability of such shadows. Among 155 cases in which CRT was performed from April to May 1985, 37 underwent CFST simultaneously because of suspected peripheral coin lesions in the lung field and formed the subjects of this study. The clinical course and endoscopy, however, revealed that there were no bronchial lesions. CFST, obtained with the X-ray system for linear tomography under exposure conditions for the mediastinum or by using a compensating filter, was used only in cases evaluable on the mediastinal side. CRT was used under almost the same conditions as those for CFST, and the type of image processing was specified by Fuji Photofilm Co. Ltd. A combination of the left image using mainly tonal conversion to resemble a CFS image with the right image using mainly spatial frequency enhancement was evaluated as a routine image.

The determinative criterion (Fig. 9.1) was assigned as three points in cases presenting clear translucent images of the bronchial lumen and shadows of its wall on both sides, as zero points in cases presenting translucent images but not shadows, and as one or two points according to the various degrees between zero and three points. According to this criterion, the left and right main bronchi, intermediate bronchus, and lobar bronchus were investigated by planigraphy.

Results

Table 9.1 shows the mean points value estimated for each bronchus. CRT was evaluated as 2.2–2.8 points, revealing relatively good shadows of the bronchial wall, while with CFST there was much diffusion due to exposure conditions, individual differences at the area proximal to the mediastinal side, and because the area overlapped shadows of the blood vessel (pulmonary artery). Figure 9.2 presents a comparison of CRT with CFST in the left main bronchus of the same case. With CRT, shadows of the wall at the mediastinal side were easy to evaluate. With respect to the intermediate bronchus overlapping the right lower pulmonary trunk at the external side, the wall shadows were difficult to identify with CFST; streaky shadows were clear in detail with CRT, also allowing easy evaluation at the mediastinal side and yielding the best results for each bronchus (Fig. 9.3; Table 9.1). In the right and left main

[1] Department of Radiology, School of Medicine, Kanazawa University, 13-1, Takamachi, Kanazawa, 920 Japan

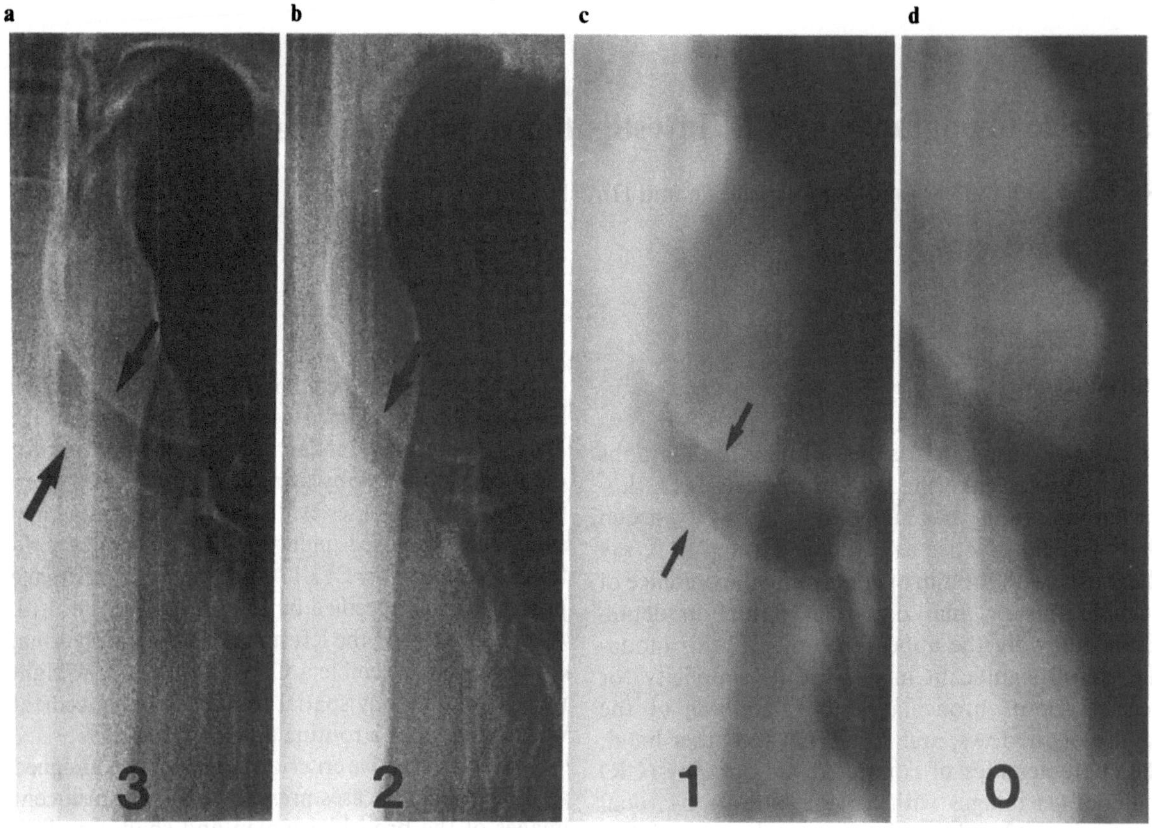

Fig. 9.1. **a** Three points: cases presenting clear translucent images of the bronchial lumen and shadows of its wall on both sides (*arrows*). **b** Two points: cases presenting clear shadows of the bronchial wall on one side (*arrows*) or revealing relatively good shadows of its wall on both sides. **c** One point: unclear cases of wall shadows. (*arrows*). **d** No points: cases presenting translucent images but wall shadows not difficult to identify

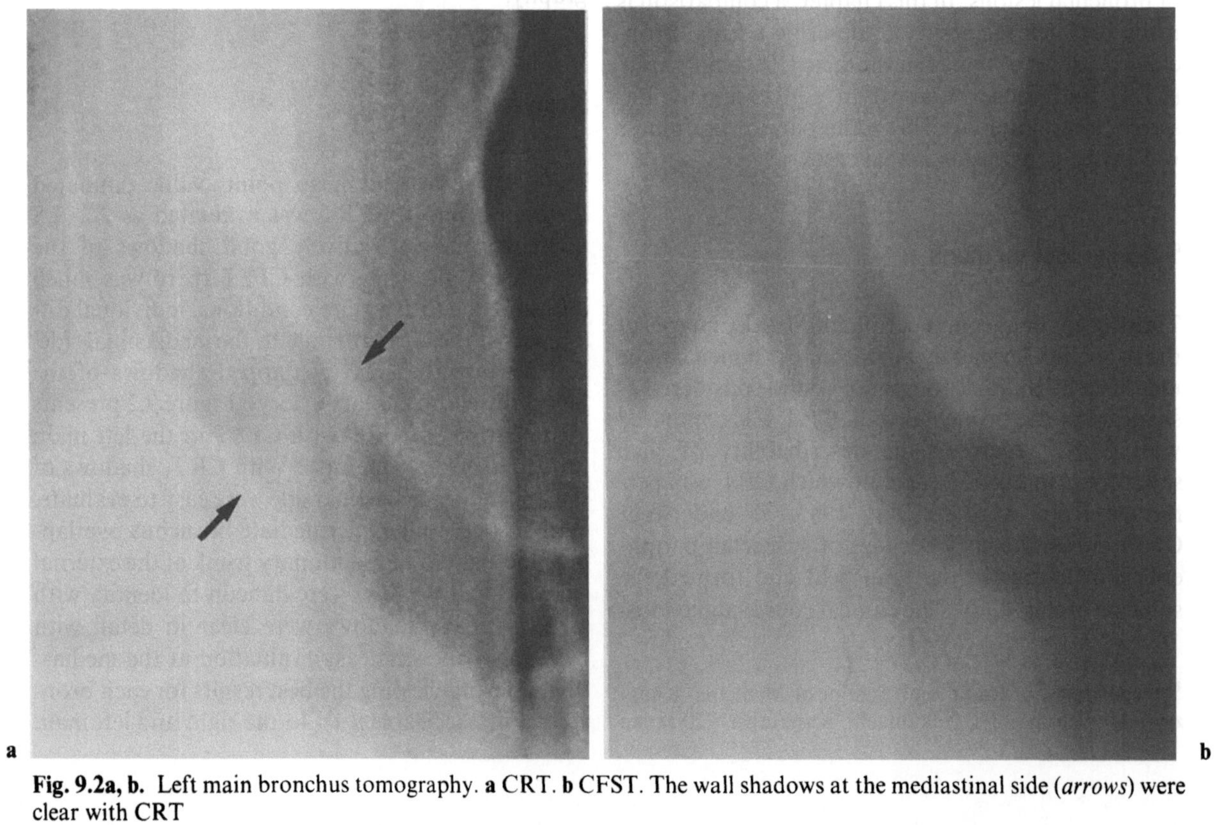

Fig. 9.2a, b. Left main bronchus tomography. **a** CRT. **b** CFST. The wall shadows at the mediastinal side (*arrows*) were clear with CRT

Table 9.1. The mean points value estimated for each bronchus

	Main bronchus	Intermediate bronchus	Lobar bronchus		
			Upper	Middle	Lower
CRT	2.6	2.8	2.3	2.5	2.2
CFST	1.5	1.5	1.1	1.5	1.1
Number of cases	32	20	29	15	30

bronchi, there was a reduction of describability; for CRT this was due to spinal shadows but was improved by changing the manner of image processing. Figure 9.4a shows unclear shadows of the main bronchus due to spinal shadows in the CR right image of the routine images, while Fig. 9.4b shows improvement of the wall shadow due to changing the density shift and the degree of spatial frequency enhancement. The lobar bronchial diameter was poorest in the lower lobar bronchus, probably as a result of overlapping of the lower pulmonary trunk, lower pulmonary vein, and so on.

Clinical case reports

Case 1

M.Y., a 53-year-old man, had primary lung cancer (adenocarcinoma, which had developed in the bronchial lumen). The posteroanterior (P-A) view of the chest (Fig. 9.5a) at the first medical examination presented a tumor shadow 5 cm in diameter in the upper left lung field. CFST (Fig. 9.5b) revealed that there was a tumor in S^3 (anterior segment) with the formation of a notch in its periphery. The hilar side

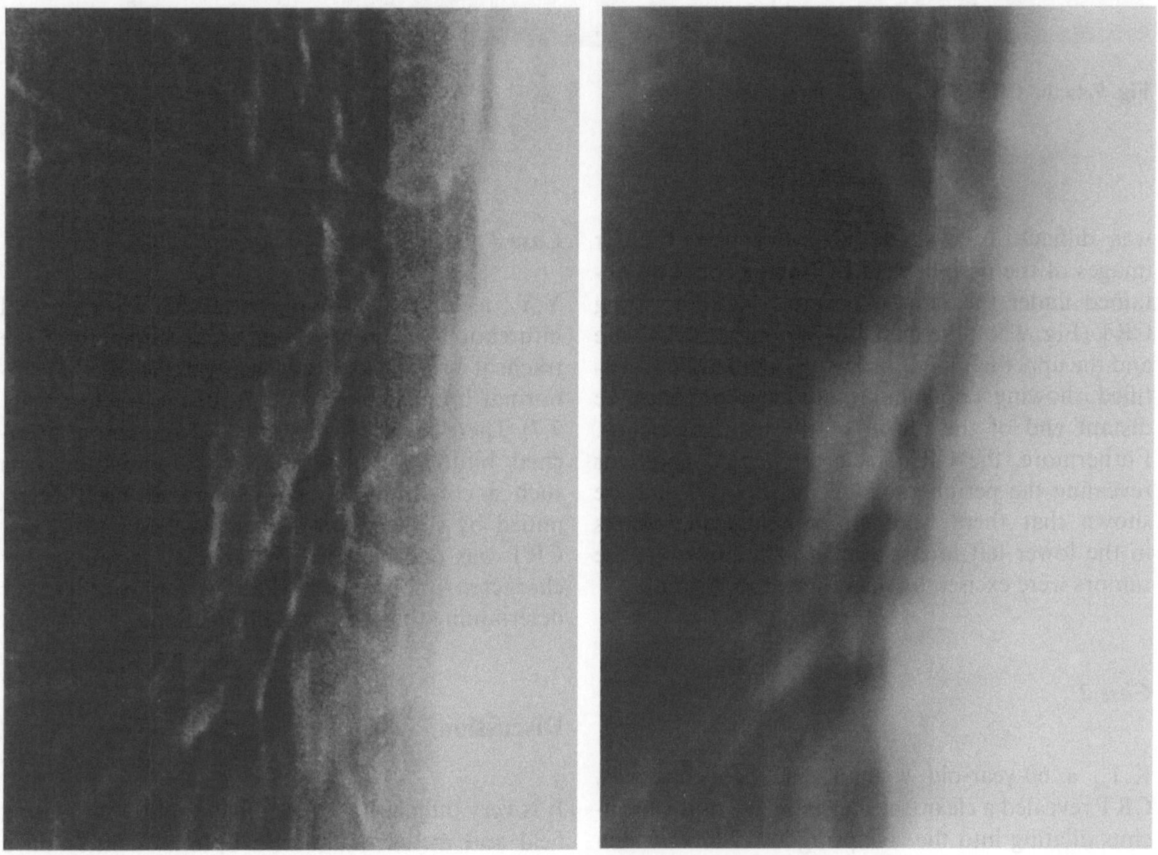

a b

Fig. 9.3a, b. Right hilar pulmonic tomography. **a** CRT. **b** CFST. In part of the area overlappe by shadows of the blood vessel (posteroanterior), the wall shadows were clearer with CRT

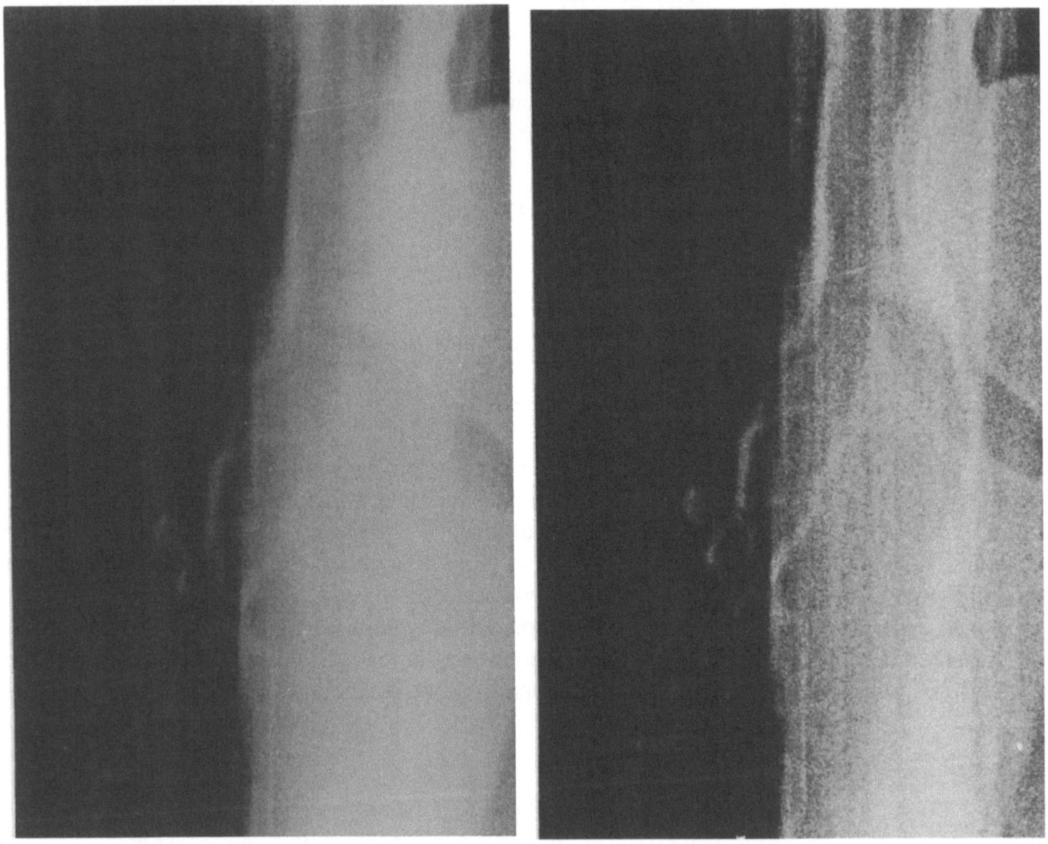

a b

Fig. 9.4a, b. CRT. These images show the same case

was difficult to evaluate with this film. Clearer images of the periphery of the tumor could be obtained under the same exposure conditions using CRT (Fig. 9.5c); the hilar side was easy to evaluate and the upper left lobar bronchus could not be identified, showing a polypoid tumor projecting into the distant end of the lumen of the main bronchus. Futhermore, the CRT image proved so good for revealing the peripheral bronchus that it could be shown that there were no pathological findings in the lower left lobar bronchus. In this case, the tumors were excised by sleeve upper lobectomy.

Case 2

K.T., a 60-year-old woman, had bronchiectasis. CRT revealed a clear translucent image of the bronchus dilating into the collapsing lower lobe, which facilitated the diagnosis of bronchiectasis (Fig. 9.6a, b.)

Case 3

Y.Y., a 75-year-old man, had either an abnormal bifurcation of the bronchus or an anomaly of the tracheal bronchus. CRT clearly revealed that abnormal bronchi bifurcated from the trachea (Fig. 9.7). There have been no reports of anomalous bronchial bifurcation determined by tomography, as such a condition has conventionally been determined by surgery, bronchography, or endoscopy. CRT was considered to be an easy technique for characterizing bronchial shadows and excellent for determining the abnormal bifurcation.

Discussion

It is very difficult to obtain images of both the lung field and mediastinum with optimal density on a sheet of film using CFST, as there are marked differences in density between the lung field and the media-

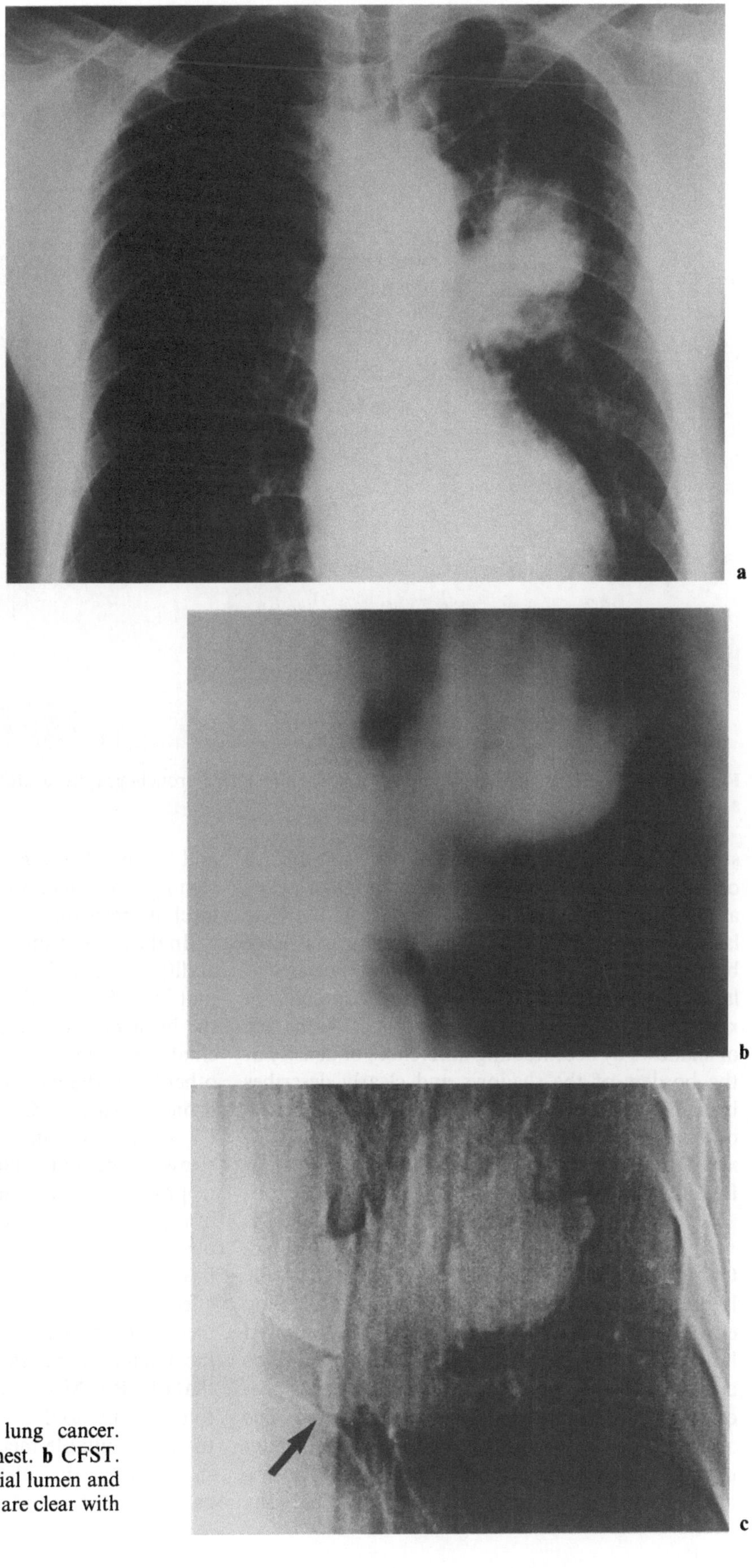

Fig. 9.5a–c. Case 1. Primary lung cancer. **a** Posteroanterior view of the chest. **b** CFST. **c** CRT. The tumor in the bronchial lumen and the peripheral bronchus (*arrows*) are clear with CRT

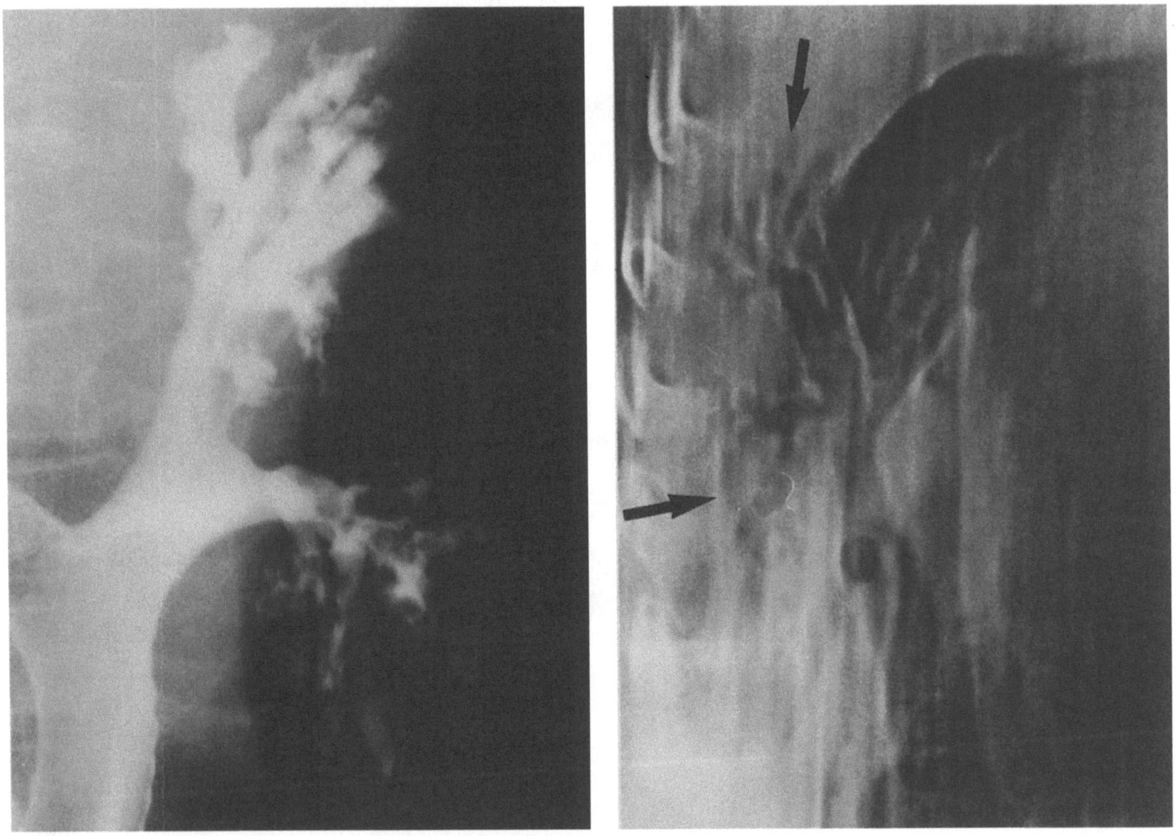

a b

Fig. 9.6a, b. Case 2. Bronchiectasis. **a** Conventional film/screen bronchography. **b** CRT. The dilating bronchus (*arrows*) are clear with CRT

stinum. As means of resolving this problem, the use of a compensating filter made of such materials as aluminum and the usefulness of images with a wide latitude obtained from xerotomography, etc. have been reported [1–3]. It has been said that the particular edge enhancement effect obtained, especially in xerotomography, facilitates diagnosis of the nature of shadows by accentuating differences of density at the borders of the shadows and clearly describes bronchial shadows, resulting in excellent detection of bronchial lesions. However, the sensitivity of xerotomography is low, necessitating about a tenfold increase in dose and a very high degree of exposure, as compared with that used for conventional X-ray films, and so the technique is not commonly used. On the other hand, CRT enables images with a wide latitude to be obtained by tonal conversion, allowing evaluation of images of both the mediastinum and lung field with one image and yielding images as clear as those of xerotomography by frequency processing, with no problems regarding exposure. The ability to describe the shadows of an object under optimal conditions by controlling image processing is one of the characteristics of digital radiography and is an advantage not shared by conventional radiography, facilitating X-ray diagnosis with a high level of precision.

In the present study, a comparison of the describability of normal bronchial shadows in CRT with that in CFST revealed that clearer shadow strips of the bronchial wall could be described in the bronchial lumen and both its sides with CRT. There are other X-ray diagnostic techniques for evaluating the bronchus such as CT scan [4] and bronchography. CT scan, an examination method that does not cause stress to patients, facilitates diagnosis of hypertrophy of the wall or stenotic deviation of the lumen in a particular region of the bronchus, such as the main bronchus or upper lobe bronchus. However, it has limits of spatial resolving power, and it is easily affected by artifacts derived from respiratory movement or heart beat, leading to the consideration that the diagnostic precision of the CT scan is lower than that of CRT. Moreover, CT scan is limited to cross-tomography, so that it is difficult to say whether it is the most suitable method for visualizing bronchial shadows. On the other hand, bronchography is an examination method which subjects the patients to

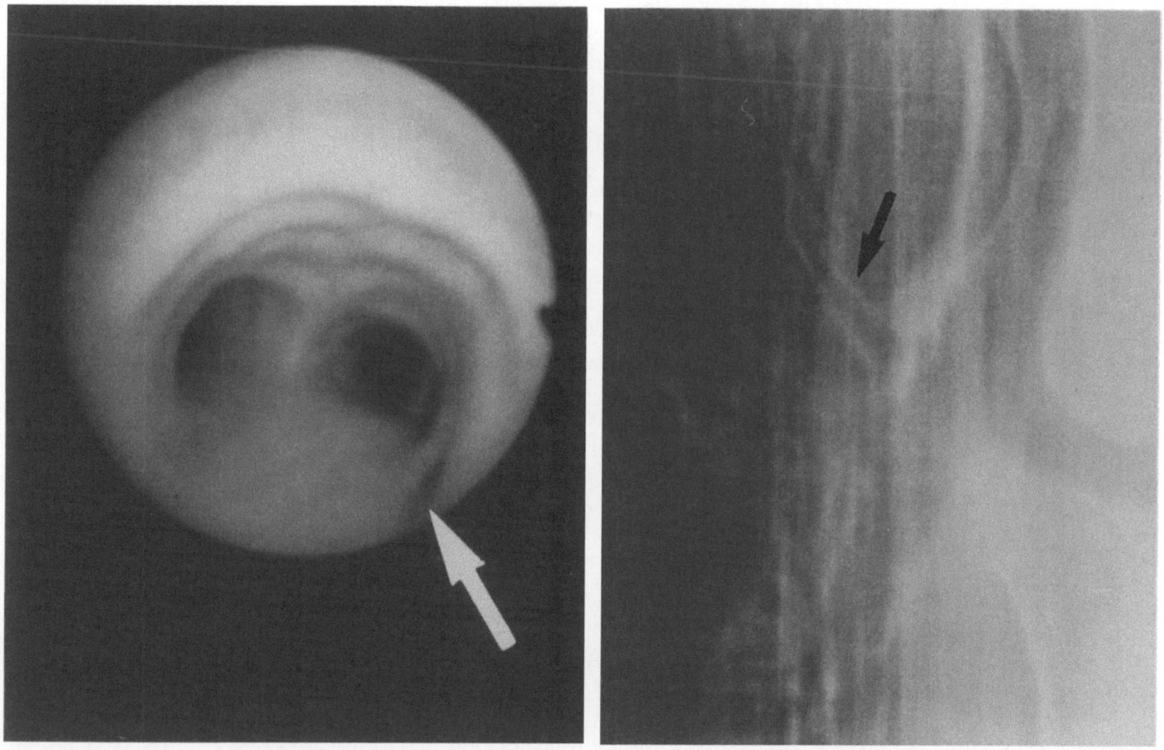

a b

Fig. 9.7a, b. Case 3. Abnormal bifurcatation of the bronchus. **a** Endoscopy. **b** CRT clearly reveals the abnormal bifurcation of the bronchus (*arrows*)

stress, and its application range has been somewhat narrow to date except for diagnosis of such cases as bronchiectasis, which commonly also requires the use of an endoscope. The application range of bronchography is expected to become increasingly narrow as CRT can clearly evaluate dilated bronchial shadows as described in case 2 (bronchiectasis).

Bronchoscopy is the most precise examination method for fine lesions in the tunica mucosa bronchiorum at present but provides information only from the lumen on the central side. In cases of high-degree airway stenosis, it is important to assess the degree of lesion expansion in a peripheral direction in relation to the surgical technique for bronchoplasty. CRT presents excellent information as in case 1, in which clear bronchial wall shadows were shown, resulting in the possibility of detecting lesions on the bronchial wall or its outer side. The combination of CRT with bronchoscopy is therefore considered to be the most effective examination method for evaluation of bronchial lesions.

Conclusion

Thoracic tomography using CR is excellent for describing the normal bronchus and for detecting bronchial lesions in comparison with conventional film/screen tomography.

References

1. Edward FH, Pfeil CE: Use of a compensating filter in lung hilus tomography. Radiology 1976; 121:745–747
2. Ting YM, Doust BD, Chuang VP: Xerotomographic diagnosis of central bronchogenic carcinoma. Chest 1975; 67:172–175
3. Chuang VP, Doust BD, Ting YM: Xerotomography of the mediastinum and tracheo-bronchial tree. Radiology 1974; 111:475–477
4. Webb WR, Gamsu G, Speckman JM: Computed tomography of the pulmonary hilum in patients with bronchogenic carcinoma. J. Comput. Assist. Tomogr. 1983; 7:219–225

Chapter 10

Upper airway in children: Medical application of CR

Mutsuhisa Fujioka[1]

Introduction

There are several methods to evaluate the upper airway in children. Direct observations can be made by laryngoscopy or endoscopy. Bronchography with the injection of contrast material is also a method for making a definitive diagnosis. However, these methods are rather invasive and some risks and complications cannot be avoided. The indication of these methods is, therefore, rather limited. The conventional radiographic examination taking the lateral nasopharyngeal view is noninvasive, but the diagnostic accuracy does not seem to be sufficient because of the poor resolution in observing the entire airway. Special techniques, including the high-voltage technique [1] and xeroradiography [2], have thus been introduced to provide better evaluation. However, especially in the case of small children with an abnormal upper airway, the choice of imaging method is limited and only radiographs utilizing a portable X-ray unit can be obtained in the ward or ICU because of the requirement of respiratory management. Therefore, it is rather difficult to obtain sufficiently good images to make the correct diagnosis.

Recently, computed radiography (CR) has been introduced with high resolution and wide latitude [3], moreover the images can be obtained using a portable X-ray unit. At our medical center, this method is considered suitable for the lateral nasopharyngeal view in children, whereby a very thin portion of the neck and a very thick portion of the thorax are included in one exposed field. We have used CR to make many evaluations of the upper airway in children [4].

Demonstrability of tracheal stenosis utilizing CR

When utilizing a regular film/screen system exposed by a portable X-ray unit, the range for adequate exposure is very narrow. Because a poorly exposed radiograph is inadequate for such an examination, radiologists often request technicians to use somewhat overexposed radiographic conditions. However, even utilizing a slightly overexposed technique, it is almost impossible to observe the entire upper airway including the mouth through to the lower trachea in one field. Generally speaking, an overexposed image is fairly easily corrected utilizing CR and the edge effect can be obtained using frequency enhancement. These two factors provide sharp uniform images of the entire upper airway (Fig. 10.1). We feel at our medical center that if we are able to obtain these images consistently at this quality, CR can replace tracheography, which has a potentially high risk (Fig. 10.2).

Anatomy of the upper airway

The narrowest portion of the upper airway is usually held to be at the level of the cricoid cartilage [5]. This has been documented with CR when the neck is positioned by pulling down the chin to reduce the lordotic curvature of the cervical spine (Fig. 10.3). However, when the chin is pulled up in the supine position, as endotracheal intubation is applied, the narrowest portion becomes at the level of the thyroidal isthmus. This is because of anterior bowing of the upper trachea with increased lordosis of the cervical spine associated with external indentation by the fixed thyroidal isthmus (Fig. 10.4). This is obviously the effect of physiological narrowing and the narrowest portion remains anatomically at the level of the cricoid cartilage.

Relationship between the endotracheal tube and upper airway

Endotracheal intubation is a standard procedure for general anesthesia and respiratory management of critically ill children and is widely applied. However, it is not known with certainty how the tube is situated in the airway, its attitude upon introduction,

[1] Department of Radiology, Saitama Children's Medical Center, 2100 Magome, Iwatsuki, 339 Japan

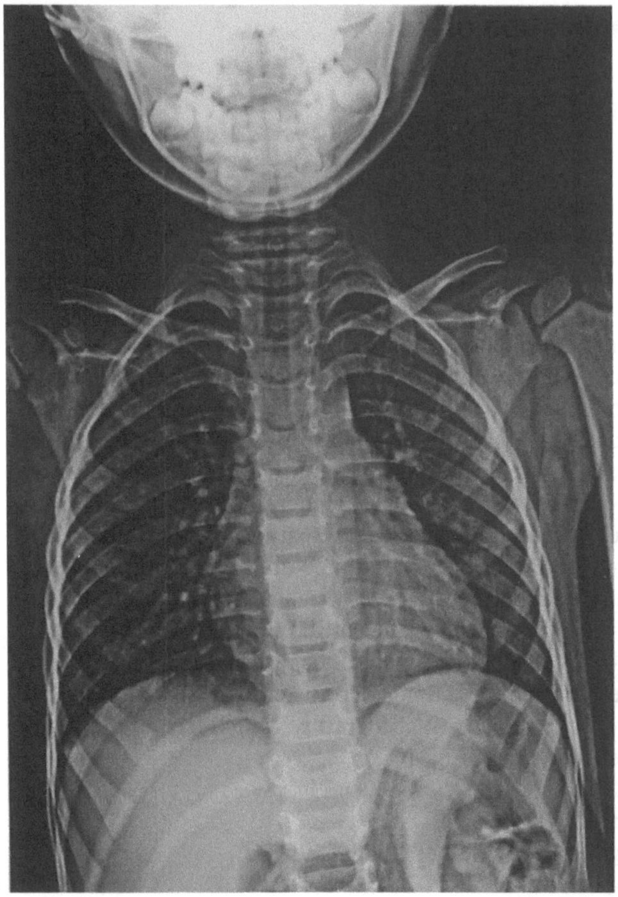
a

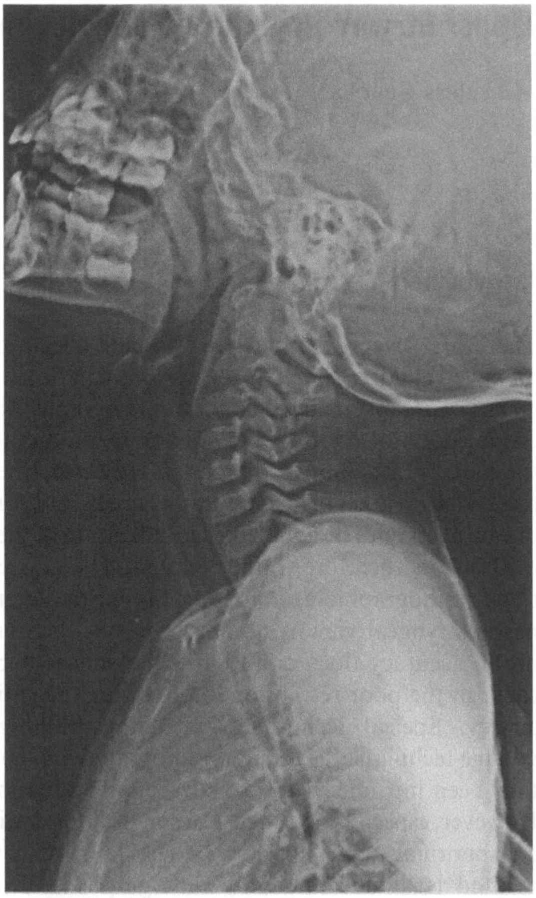
b

Fig. 10.1a, b. CR images of the upper airway in children. The images were taken with frequency enhancement. The upper airway including the subglottic region to the tracheal bifurcation is well demonstrated on the frontal image **a**. On the lateral image **b**, the soft palate is well outlined; the structure of the cervical upper airway and the intrathoracic portion of the trachea are well demonstrated. The gas within the esophagus can be easily recognized. Both images were obtained on the operating table utilizing a portable X-ray unit

Fig. 10.2. a, b CR images of tracheal stenosis. A 2-month-old girl with congenital stenosis of the trachea, which is well demonstrated on the CR images, with irregularity along the right side wall on the frontal image and narrowing at the level of the thoracic inlet on the lateral image. An area of atelectasis in the left lower lung is also well recognized. These CR images were obtained by a portable X-ray unit in the ward. **c, d** Tracheography. Irregular narrowing demonstrated on the frontal image and smooth narrowing at the level of the thoracic inlet are exactly the same as those seen on the CR images ▷

how it bends, or if it scratches or compresses portions of the wall of the airway.

The upper airway is bent at 90° from the oral cavity through the oropharynx to the larynx. Therefore, the endotracheal tube is also bent at this level at 90°. This produces a recoiling force toward the posterior wall of the upper trachea. This pressure was shown to be as high as 300–400 g with a PVC tube and 1000 g with a rubber tube [6]. The relationship between the endotracheal tube and the upper airway is elegantly demonstrated with CR (Fig. 10.5). The same endotracheal tube was introduced to the same patient through alternately the nasal and oral cavities, and the relationship between the upper airway and the endotracheal tube was evaluated with CR. With the tube introduced through the oral cavity, the anteroposterior (AP) diameter of the trachea increases compared with that with nasal intubation (Fig. 10.6). The more the endotracheal tube bends, the greater is the recoiling force toward the posterior wall of the anteriorly convex cervical trachea. The recoiling force is conducted to the tip of the tube against the anterior wall of the trachea, widening the AP diameter at this level.

When the endotracheal tube is introduced, the tube descends vertically to the glottic portion and the

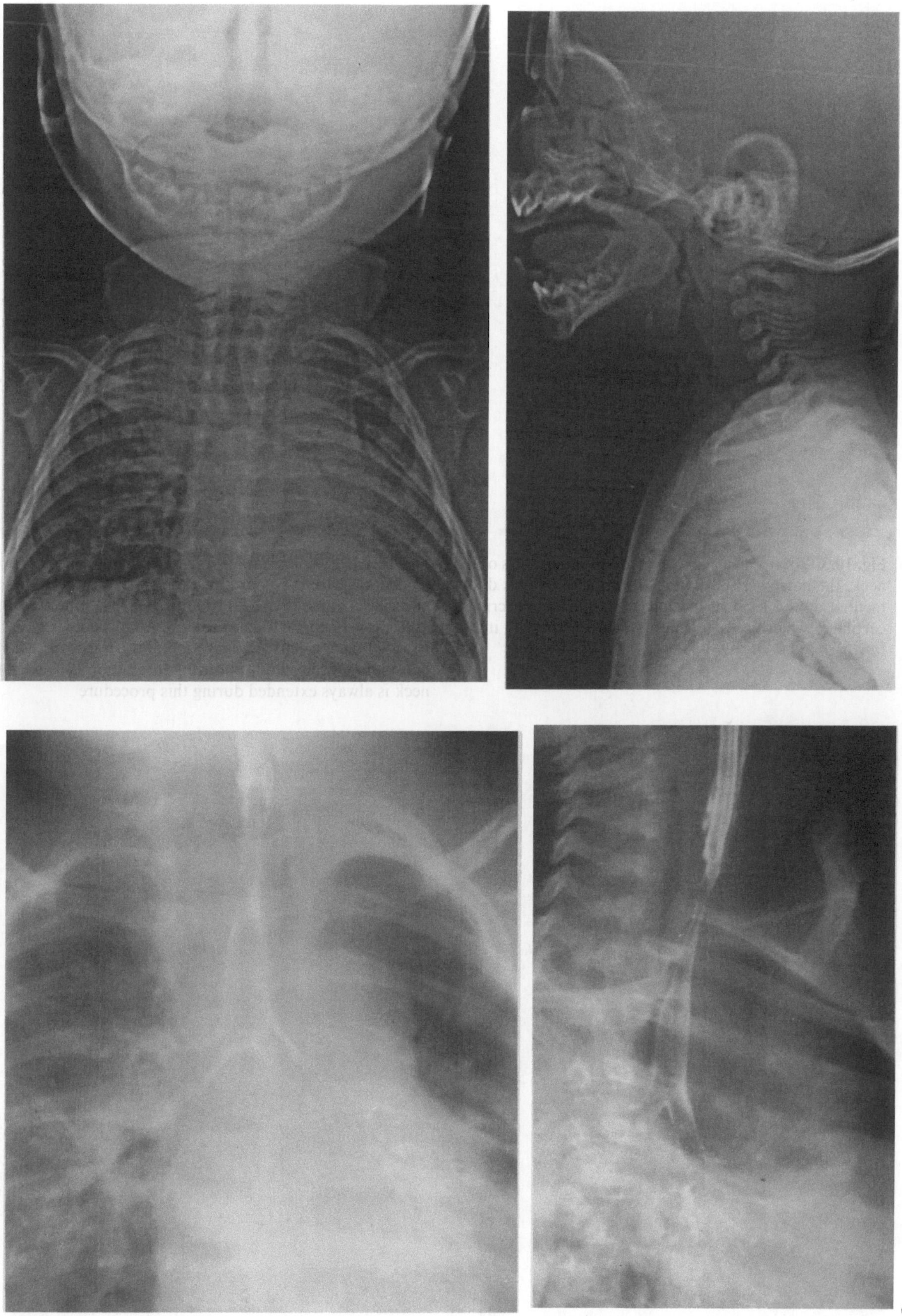

a b c d

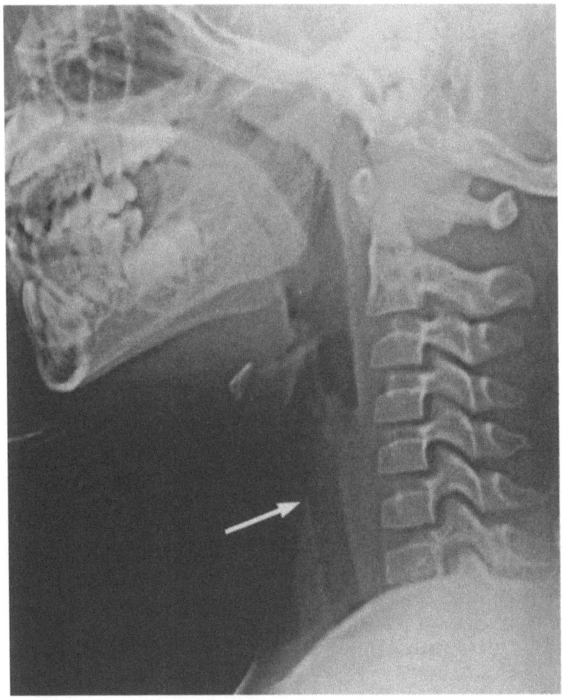

Fig. 10.3. Anatomical narrowing. Without lordosis of the neck, the narrowest portion of the upper airway is documented at the level of the lower margin of the cricoid cartilage, which is well demonstrated on the CR image (*arrow*)

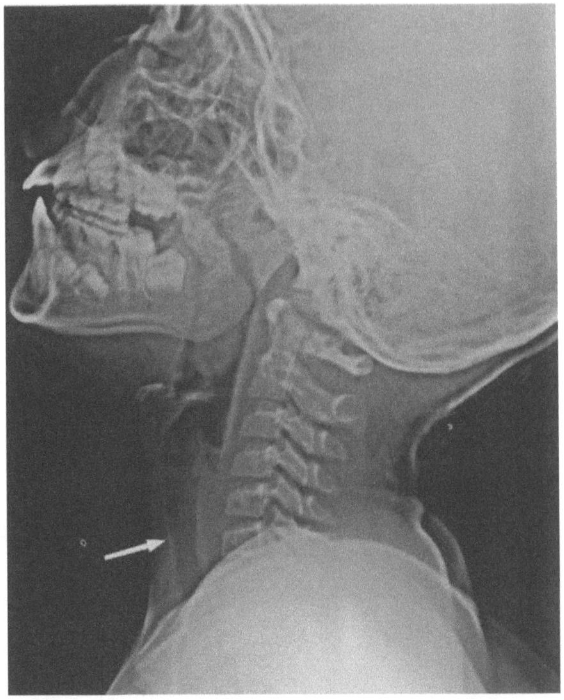

Fig. 10.4. Physiological narrowing. When the neck is extended, the narrowest portion of the upper airway becomes at the level of the thyroidal isthmus, which compresses the anterior wall of the trachea to produce physiological narrowing (*arrow*). This configuration of the upper trachea is considered to be standard during intubation because the neck is always extended during this procedure

a

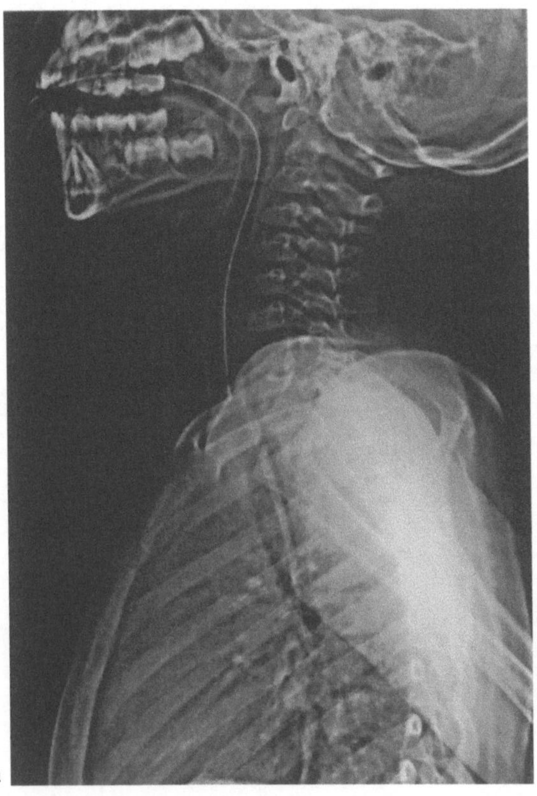

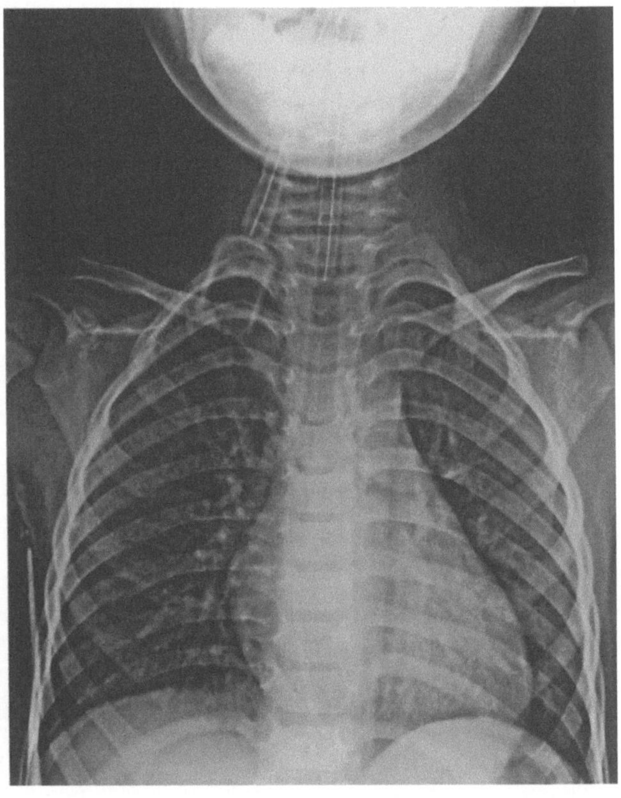

b

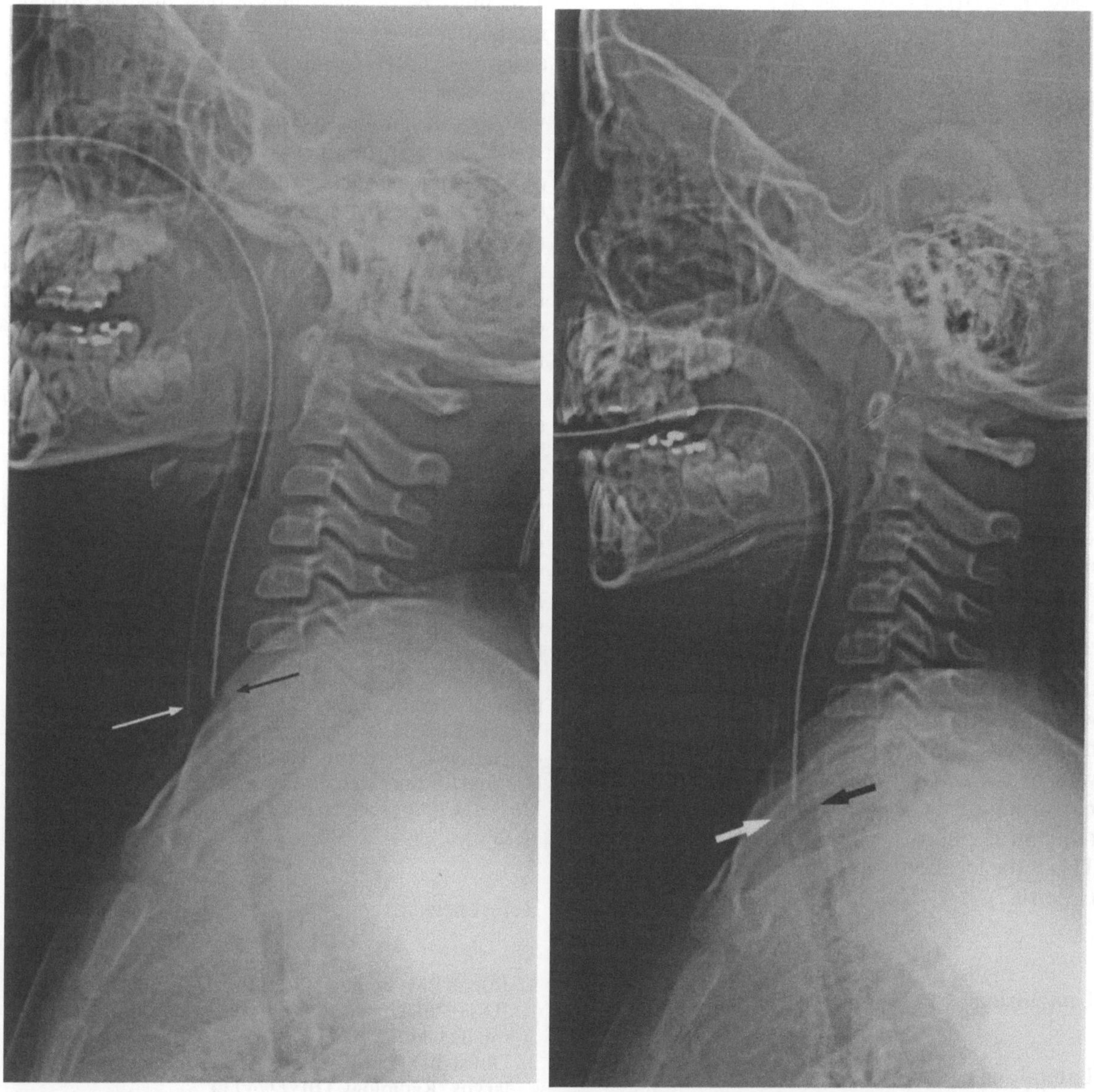

a b

Fig. 10.6a, b. Flexion and recoiling of the endotracheal tube in the upper airway. **a** Intubation through the nasal cavity; **b** intubation through the oral cavity. The tube is more flexed in **b** and the tip of the tube compresses more the anterior wall of the trachea, resulting in an increase in the AP diameter of the trachea at this level in comparison with that seen in **a**. The recoiling force produced within the oral cavity is considered to be conducted indirectly toward the tip of the tube

◁ **Fig. 10.5a, b.** Relationship between the endotracheal tube and upper airway. Not only the inclusive metallic marker but also the vinyl portion of the endotracheal tube are well demonstrated on the CR images. The tube is flexed at 90° in the oropharynx and bends anteriorly just below the glottic portion; the tip of the tube touches the anterior wall of the trachea, which is well demonstrated on the lateral image **a**. The configuration of the tip of the tube is well recognized on the frontal image **b**

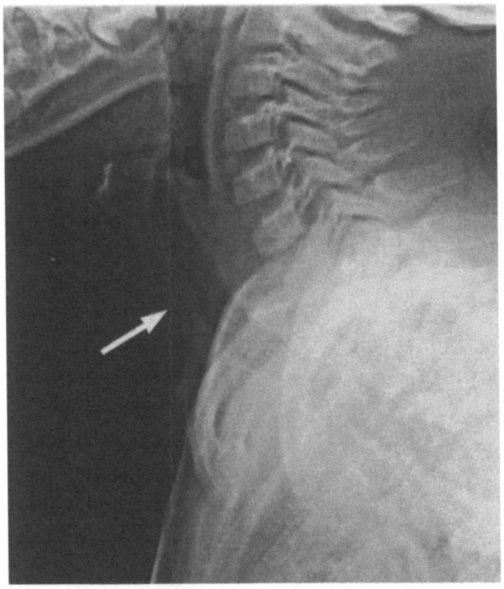

Fig. 10.7. Subglottic granuloma. The granuloma is formed along the anterior wall of the upper trachea at the level of the physiological narrowing in this patient with a history of a single endotracheal intubation

tip of the tube touches and scratches the anterior wall of the trachea at the level of the physiological narrowing. The anterior bulging of the posterior wall of the cervical trachea and the physiological narrowing of the trachea increase, especially when the chin is pulled up to extend the upper airway. This is the location where granuloma tends to occur (Fig. 10.7).

Conclusion

Lateral nasopharyngeal radiography is used for evaluating the upper airway in children and is very important in pediatric radiology [7]. However, by means of conventional radiography, direct observation of the vinyl portion of the tube is almost impossible. The radiographic density varies considerably in the oropharyngeal, cervical, shoulder, and intrathoracic portions. It is, therefore, almost impossible to visualize the entire airway in one image without using special filters or equipment. To the best of the author's knowledge, no paper regarding an evaluation between the endotracheal tube and the upper airway with radiographic images has been reported.

During this investigation, with respect to the anatomy of the upper airway, although its narrowest portion is certainly at the level of the cricoid cartilage anatomically, we found the existence of a physiological narrowing related to the thyroidal isthmus. This must be of great importance because during intubation the cervical trachea is forced to be convex anteriorly. This is certainly related to the etiology of granuloma formation in this location, which was visually documented by our study.

Diagnostic procedures with tracheal stenosis are difficult and the method utilizing contrast material obviously involves a high risk. If the high-quality images can be consistently obtained, especially by utilizing a conventional portable X-ray unit, the benefit for such patients will be immeasurable. Our study obviously suggests such a possibility with this new method.

Acknowledgments. This study was supported by cancer research fund (57–20) from MHW and was performed at the Departments of Radiology and Anesthesiology, National Children's Hospital. This paper was presented at the scientific exhibits of the 69th Meeting of Radiological Society of North America, Chicago in 1983 and awarded "Magna Cum Laude." Co-workers of this study included: Eiichi Kohda M.D., Department of Radiology, Keio University Hospital, Hidekazu Masaki M.D., Department of Radiology, National Children's Hospital, Katsuyuki Miyasaka M.D., and Hiroshi Sankawa M.D., Department of Anesthesiology, National Children's Hospital. We thank technicians from the Departments of Radiology of National Children's Hospital and Keio University Hospital for their technical assistance and also thank Fuji Medical Co. for their technical support.

References

1. Joseph PM, Berdon WE, Baker DH, Slovis TL, Haller JO (1976) Upper air way obstruction in infants and small children. Radiology 121:143–148
2. Doust BD, Ting YM (1974) Xeroradiography of the larynx. Radiology 110:727–730
3. Sonoda M, Takano M, Miyahara J, Katoh H (1983) Computed radiography utilizing scanning laser stimulated luminescence. Radiology 148:833–838
4. Fujioka M, Kohda E, Masaki H, Miyasaka K, Sankawa H (1983) Upper air way evaluation in children by Fuji Computed Radiographic System: Preintubation and postintubation, presented at 69th Scientific Assembly and Annual Meeting of the Radiological Society of North America, Chicago, Nov. 13–18. Radiology 149(P):165, 291
5. Eckenhof J (1951) Some anatomic considerations of the infant larynx influencing and endotracheal anesthesia. Anesthesiology 12:400–410
6. Lindholm CE, Carroll RG (1975) Evaluation of tube deformation pressure in vitro. Critical Care Medicine 2:96
7. Capitanio MA, Kirkpatrick JA (1968) Upper respiratory tract obstruction in infants and children. Rad Clin NA VI(2):265–277

Chapter 11

Breast and thyroid

Akiko Narimatsu[1], Chisato Kawai[1], and Junichi Hachiya[2]

Since the images provided by computed radiography (CR) are generally easier to evaluate than the images obtained with the conventional X-ray technique, both mammography and neck roentgenography would appear to be good applications for CR. However, in a strict sense, CR is poor in its capacity to depict the fine detail of microcalcifications in the breast compared with the nonscreen technique. It seems that this problem comes from the limitation of the spatial resolution with the present equipment.

Introduction

Detailed reports on the characteristics of the CR digital system have been published [1, 2] and articles evaluating the diagnostic value of this system in some clinical fields have also been published [3, 4].

This chapter deals with the breast and the thyroid. In diagnosis of diseases of these organs, manual palpation is of great importance, and in qualitative diagnosis of lesions in these organs by radiological techniques, detection of microcalcifications as well as their detailed analysis plays an important role. We compare the images of the breast and thyroid obtained by CR and those obtained by the conventional X-ray technique.

Subjects and methods

Breast

The subjects of the study were five patients with breast cancer and eleven with benign breast diseases (total 16 patients, 17 examinations). In three patients with benign diseases, mammary duct roentgenography with a contrast medium was performed.

The equipment used was Mammodiagnost (Philips); conventional radiography was conducted by the nonscreen technique using X-Omat MA films (Kodak). Radiographs were taken at two projections (vertical and horizontal) for each patient with a compression technique, under the conditions of 28 kV_p, 170 mA, and 40 cm FFD, using a photo-timer. Then, of the two projections, the one presenting abnormal shadows more clearly was selected for the subsequent radiography, which was conducted using an imaging plate of 8 × 10 in. and under the conditons of 28 kV_p, 110 mA, and 45 cm FFD, whereby the X-ray dose was possibly half or less that of the conventional one, although no strict comparison was made. Therefore, each examination consisted of three radiographs—two standard ones and one CR. The maximum spatial resolution was 5 cycles/mm (c/mm) with the 8 × 10-in. imaging plate.

Processing of images was conducted with a slightly flatter gradation and wider latitude than the characteristic curve of standard X-ray films. Spatial frequency enhancement, a process characteristic of this system, was conducted mainly at 0.5 c/mm, thus emphasizing the edges, which consist of the components adjacent to this spatial frequency.

Compression spot radiography was excluded from this study.

Thyroid

Of the three patients with thyroid carcinoma and eight patients with benign thyroid diseases (total 11 patients), eight lateral soft X-ray examinations and six posteroanterior examinations were subjected to the study. Standard lateral soft X-ray radiographs were taken with the use of Kodak MRI films (Kodak Min-R system) under the conditions of 120 cm FFD, 50 kV_p, 100 mA, and 0.4 s. Thereafter, radiography with the use of an imaging plate of 8 × 10 in. was conducted under the same conditions. Standard posteroanterior radiographs were taken using Fuji RX films and an intensifying screen (FS) with Bucky Blende 8 : 1 under the conditions of 135–140 kV_p, 100 mA, 0.032 s, and 120 cm FFD. Thereafter,

[1] Department of Radiology, Tokyo Women's Medical College, 8-1, Kawada-cho, Sinjuku-ku, Tokyo, 162 Japan
[2] Department of Radiology, Kyorin University School of Medicine, 20-2, Shinkawa 6-chome, Mitaka, Tokyo, 181 Japan

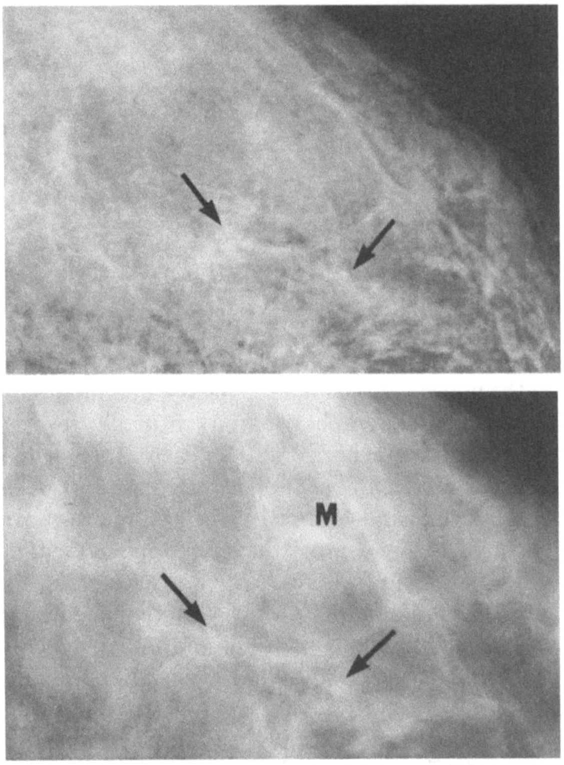

Fig. 11.1a, b. Microcalcification in a case of impalpable cancer. Many shadows showing microcalcifications of uneven size are seen (*arrows*). The CR image **a** is superior in detectability of microcalcification but is less sharp at the contour of such shadows than the nonscreen image **b**. *M* marking with minute lead pieces

radiography with the use of an imaging plate of 10 × 12 in. was conducted under the same conditions. Therefore, each examination consisted of one standard radiograph and one CR radiograph. The limitations of spatial resolution for the 8 × 10-in. imaging plate and the 12 × 10-in. plate were 5 and 3.3 c/mm, respectively.

Image processing of lateral radiographs was conducted with the tonal conversion using a slightly flatter gradation than the characteristic curve of films; the spatial frequency enhancement was around 0.7 c/mm. Posteroanterior radiographs were processed in two different ways and these two processed images were displayed on a CR film (hereinafter referred to as "left image" and "right image"). The tonal conversion for the left image was performed with the gradation closer to the characteristic curve of standard X-ray films and for the right image it was performed with a flat gradation; frequency enhancement was done around 0.35 c/mm. In addition, the degree of enhancement is weak on the left image and strong on the right.

Results

Breast

Image of calcification

Of the 17 examinations, nine showed an image of calcification. Of these nine examinations, two presented the image of microcalcification characteristic of breast cancer. In all nine examinations (nine patients), CR images were superior to conventional radiographs in terms of detecting calcifications. On the other hand, since the spatial resolution with this system was limited, the contour of the microcalcifications was blurred on the CR images, which were thus poor in the capacity to depict microcalcifications of irregular contour or uneven size characteristic of carcinoma (Fig. 11.1a, b). Furthermore, on the CR images enhanced strongly, it was more difficult to discriminate between dust and calcification than on conventional radiographs, owing to the strong effect of such processing on the boundaries. Thus, the discrimination was difficult in 6 of 17 CR images.

Shadow of the tumor

Of the 17 examinations, seven showed a shadow of a tumor. Of these seven examinations (seven patients), four had breast cancer. In terms of the capacity to demonstrate the contour of the tumor, CR images were better in all seven patients than the conventional radiographs (Fig. 11.2a, b). Not only the image of an irregular contour (so-called spicula) or blurred margin, which is characteristic of breast cancer, but also indirect findings like nipple retraction and skin thickening were obtained in two of the four CR images, indicating the marked superiority of CR over conventional X-ray images.

Mammary duct roentgenography with the use of a contrast medium

Mammary duct roentgenography using a contrast medium was conducted in three patients with benign breast diseases and revealed that CR images were on the whole easier for diagnosis than conventional radiographs.

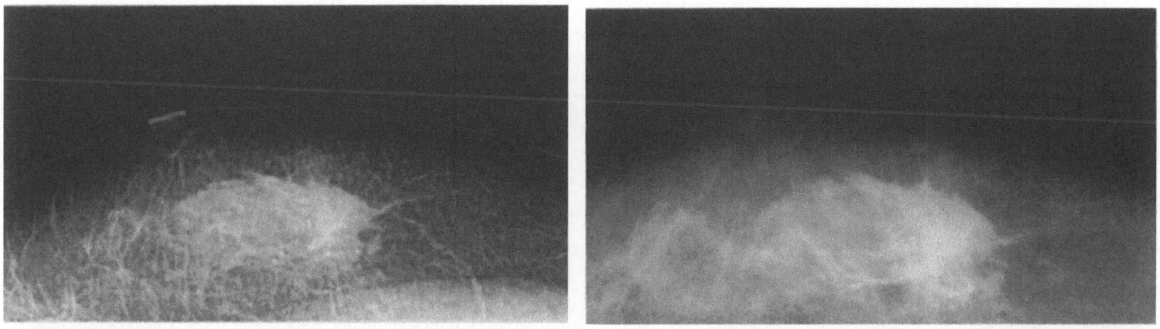

Fig. 11.2a, b. Tumor and calcification shadows in medullary tubular carcinoma case. The capacity to depict tumor shadows and their spiculae with irregular contours is markedly improved in the CR image **a** compared with the nonscreen image **b**

Thyroid

Lateral images

Image of calcification. Of the eight examinations, six showed calcification. Of these six examinations (six patients), one had papillary adenocarcinoma and another had follicular adenocarcinoma. In terms of detectability of calcifications, CR images proved to be superior to conventional X-ray images in all six patients. In the present cases, a shadow of coarse calcification was the main finding; therefore, CR images were not particularly inferior to standard X-ray images in terms of the capacity to represent the shape of the calcification (Fig. 11.3a, b).

Compression of the trachea. In terms of the capacity to represent the contour of the trachea, CR images were superior to conventional X-ray images in all eight patients.

Frontal images

Compression of the trachea. In terms of the capacity to depict the contour of the trachea from the glottis to the intrathoracic portion of the trachea, CR images were superior to conventional radiographs in four of the six left CR images and three of the six right CR images (the right CR images were processed with the spatial frequency enhancement at a higher level than the left and CR images). No marked difference was noted between left and right CR images in the degree of improvement over conventional X-ray images. If anything, left CR images seemed to be easier for diagnosis than right CR images, because the contrast in density between the target organ and the surrounding organs was emphasized in left CR images (Fig. 11.4a–c).

Image of calcification. Of the six examinations, four showed an image of calcification. One of these four examinations was a case of papillary adenocarcinoma having characteristic psammoma bodies. These psammoma bodies, however, were not clearly shown on any of the conventional or CR images, thus indicating no improvement by CR over conventional X-ray images. However, for coarse calcification of adenomatoid goiter or calcification of the lymph nodes (not overlapping with the cervical vertebrae), most of the right CR images were superior to conventional X-ray images.

Discussion

The CR system is capable of effectively visualizing small differences in density between soft tissues even when applied to mammography and lateral soft X-ray examination of the neck. On the whole, the images obtained with this system are much easier for diagnosis than the ones obtained with the conventional system.

A limitation to this system is that images can be evaluated only in a restricted range of spatial resolution (5 c/mm is the maximum with the present equipment). Diagnosis for microcalcifications is not optimal because the contour of the shadow is blurred by the limited spatial resolution, although this kind of blur does not prevent diagnosis in other fields.

Figure 11.5a is a model of microcalcification made of very small glass beads (about 100–500 μm in diameter). In this model, the thickness of the acrylic plate inlaid with glass beads is 3 cm. Radiography was conducted with either an imaging plate of 8 × 10 in. or a Kodak MA film for the nonscreen technique under the conditions of 40 cm FFD, 30

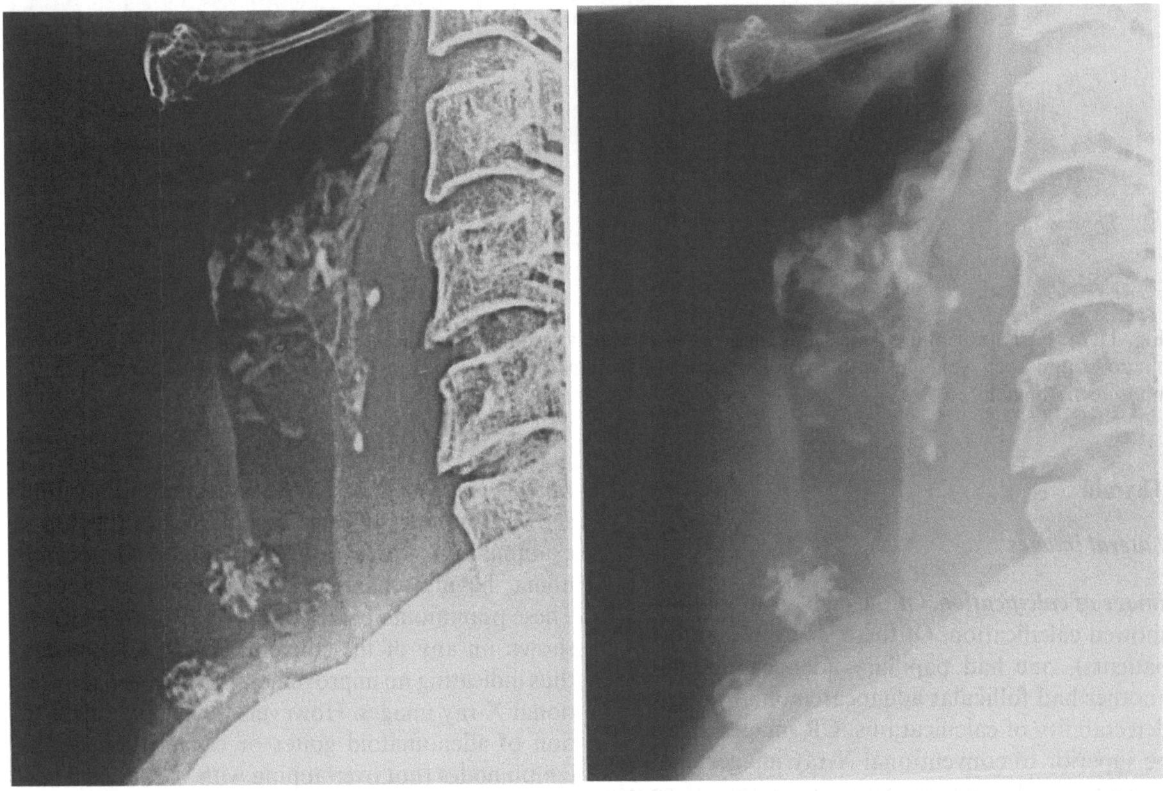

a b

Fig. 11.3a, b. Shadows of coarse calcification in papillary adenocarcinoma case. The CR image **a** is superior in terms of the capacity to demonstrate the irregularly outlined and acutely angled shadow of coarse calcification. In addition, the CR image **a** presents swelling of the trachea or soft tissues more clearly than does the conventional X-ray image **b**

kV_p, and 170 mA. Figure 11.5b is the CR image after digital image processing and Fig. 11.5b is the image obtained with the nonscreen technique. In terms of the detectability of beads with a size between 300 and 500 μm, the CR image was superior to the nonscreen X-ray image: however, the contour of the shadow was less sharp on the CR image, thus confirming that CR images are poor in representing the edge sharpness. In terms of the detectability and edge sharpness of the beads with a size of 177–250 μm, the image obtained with the nonscreen method was superior to the CR image. When it is recalled that about 20% of the mammographic diagnosis of breast cancer is based on microcalcification [5] and that the shadow of microcalcification is the only sign of the disease in about one-third of the impalpable cancer cases [5], it is desirable that this system should be further improved to allow more minute image analysis.

Generally, the shadow of a psammoma body, which is characteristic of papillary adenocarcinoma of the thyroid, is found in about 30% of the cases; the majority of the cases showing such a shadow tend to have a shadow of coarse calcification mixed with it [6]. In terms of the capacity to detect such calcifications, CR was on the whole superior to the conventional technique. Therefore, this type of cancer seems to be a good application for the CR technique.

In addition, one of the advantages of the CR system is the fact that it enables low-dose radiography. In the case of mammography, excellent images could be obtained at about one-half or less of the X-ray dose. In the case of neck radiography (not evaluated in this study), we estimate that the X-ray dose may be reduced to about one-fifth, and the same reduction would apply to other fields.

On CR images, dust or a scratch on the imaging plate is sometimes indistinguishable from the shadow of microcalcification. Therefore, it is necessary to take care when handling the imaging plates.

Conclusion

In this study, 17 mammographic examinations (16 patients) and 14 neck examinations (eight lateral soft X-ray examinations and six posteroanterior examinations) were made. Each examination consisted of

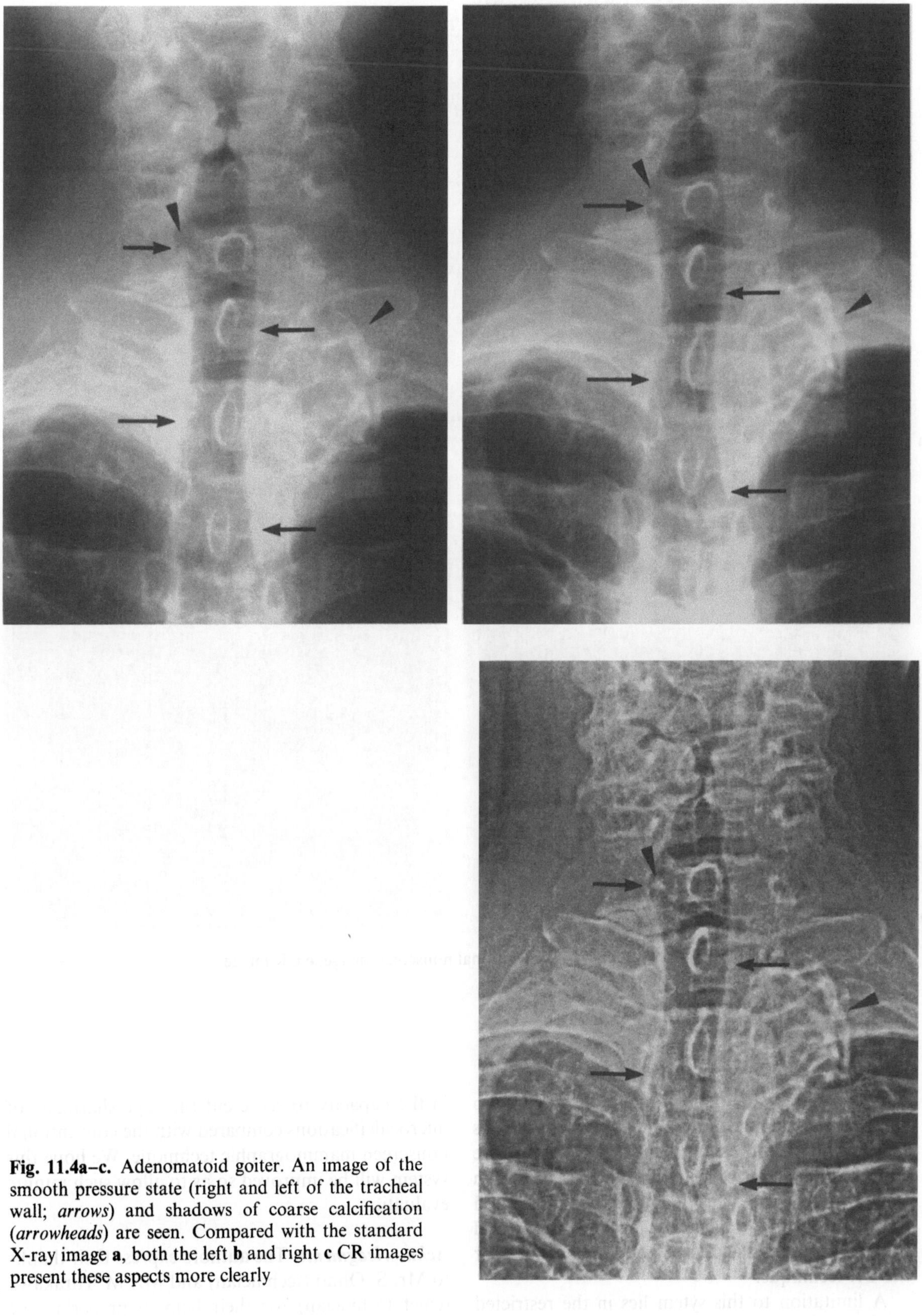

a b c

Fig. 11.4a–c. Adenomatoid goiter. An image of the smooth pressure state (right and left of the tracheal wall; *arrows*) and shadows of coarse calcification (*arrowheads*) are seen. Compared with the standard X-ray image **a**, both the left **b** and right **c** CR images present these aspects more clearly

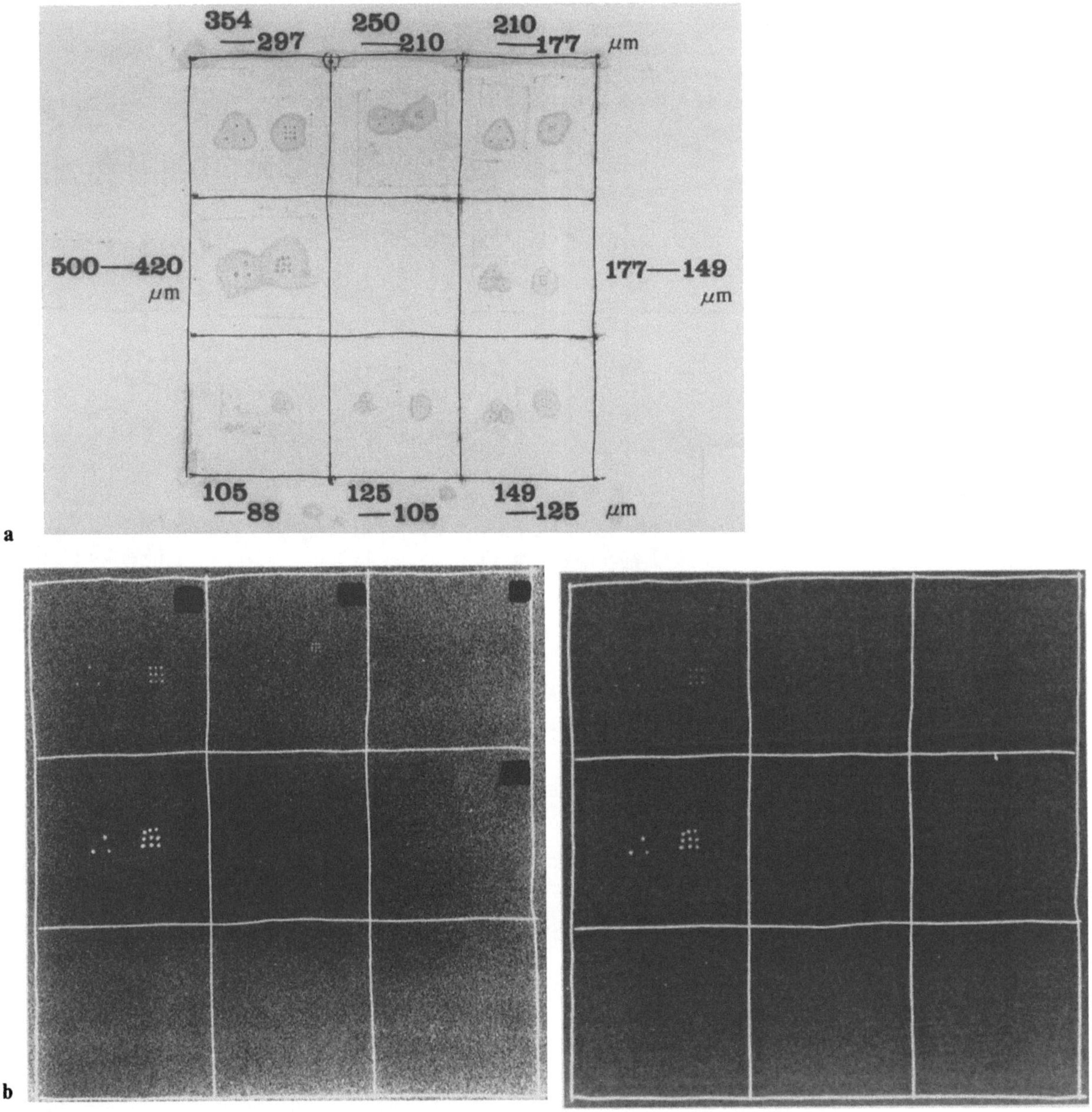

Fig. 11.5. **a** A model of minute glass beads; **b** conventional nonscreen image; **c** CR image

both conventional and CR radiographs in order to compare their diagnostic qualities. Through various types of image processing with the CR system, we could obtain images which effectively presented even slight differences in density among soft tissues and were quite clear for diagnosis. Mammography and neck radiography seem to be good applications for the CR technique.

A limitation to this sytem lies in the restricted spatial resolution; in a strict sense, this sytem is poor in the capacity to represent the edge sharpness of microcalcifications compared with the conventional nonscreen mammographic technique. We hope this system will be improved so as to allow such minute evaluation.

Acknowledgments. The authors express their thanks to Mr. S. Ohno (technician) and Mr. M. Hirakawa (chief technician) for their help in preparing the model of the glass beads and photography.

References

1. Takano M (1982) New computed radiography—Fuji Intelligent Diagnostic X-ray system. Journal of Medical Imagings 2:202–207
2. Hachiya J, Kato H (1982) Digital radiography using imaging plate.: Digital radiography—Perspective for clinical applications. Akisada M (ed.) Life science center Co. Tokyo
3. Hachiya J, Koranaga T, Sakurai K, Sakai F, Kato K, Takano M (1982) New method of digital angiography. Japanese Jaurnal of Clinical Radiology 27:705–710
4. Sakurai K, Hachiya J, Korenaga T, Nitatori T, Miyasaka Y, Furuya Y (1984) Digital radiography of the chest—Evaluation of normal anatomical structure and low dose radiography Nippon Acta Radiologica 44: 11–22
5. Okazaki M (1982) Mammography. Medical imaging lecture series [II] Nagai T (ed.) Asakura shoten 3–40
6. Ohara T, Fujimoto Y (1983) Imaging diagnosis of discase of thyroid Journal of Medical Imagings 3: 128–135

Chapter 12

Upper gastrointestinal tract

Fumio Unayama[1]

X-ray examination of the upper gastrointestinal (GI) tract by computed radiography (CR) is effective with the use of the double-contrast method, by which fine mucosal patterns can be clearly visualized. In particular, the effects of the wide latitude and image processing of CR facilitate simple recognition of lesions upon double-contrast radiography of the anterior wall or with the oblique projection. CR may also allow a reduction in the exposure dose to about one-tenth of the usual amount. This would be further reduced if the examiner could obtain better adhesion of contrast medium to the gastric wall.

Introduction

Since the CR system was introduced to our hospital in October 1984, it has been widely used for diagnosis of lesions in various fields. In our hospital, CR examination was applied to about 27000 subjects during the 18-month period between October 1984 and March 1986.

This chapter will deal with indications for double-contrast radiograhy of the stomach by CR with several CR examinations of the GI tract. Double-contrast radiography is good for visualization and qualitative diagnosis of lesions in the stomach and plays an important role in the diagnosis of early gastric cancer. Double contrast examination by CR was, therefore, compared with that by the conventional film/screen system in the present study. The results are reported with some descriptions of clinical experience with CR.

The results of a study on reduction in the exposure dose for double-contrast radiography are also reported.

[1] Department of Gastroenterology, Kanagawa Cancer Center, 54-2, Nakao-cho, Asahi-ku, Yokohama, 241 Japan

Double-contrast images by CR

Double-contrast radiography by CR facilitates a clear visualization of microlesions and allows wide diagnostic applications compared with double-contrast radiography with the conventional system. Therefore, in general, diagnosis by CR is easier and more accurate. The results of diagnosis by CR are compared below with those by the conventional system with reference to clinical cases.

In image processing of CR, negatives or positives were chosen according to the recognition of the respective lesions.

Case 1: A 56-year-old woman with early gastric cancer (type IIc)

Figure 12.1 is a double-contrast radiograph obtained by CR with the patient in the prone position. A shallowly depressed lesion can be noted along the anterior wall of the lower part of the gastric body. Areae gastricae in the depressed part are distorted, showing a frosted glasslike shadow. Tapering of the converging folds is seen in the proximal part of the lesion. The distal portion is depressed. This is an early gastric cancer (type IIc).

Figure 12.2 is the resected specimen of the early gastric cancer (type IIc) shown in Fig. 12.1. The irregular excavated lesion, 32 × 25 mm in diameter, is located on the anterior wall near the great curvature of the body. Rugal effacement of the surrounding mucosa and discoloration are evident.

Figure 12.3 is a double-contrast radiograph obtained by the conventional system with the patient in the prone position. A comparison of Figs. 12.1 and 12.3 reveals that the radiograph obtained by CR provides clearer and more detailed information, including the converging fold pattern and distorted areae gastricae.

Figure 12.4 is a double-contrast CR image with the patient in the supine position. Figure 12.4 allows recognition of the lesion in the same manner as Fig. 12.1. The details of the lesion are rather easier to

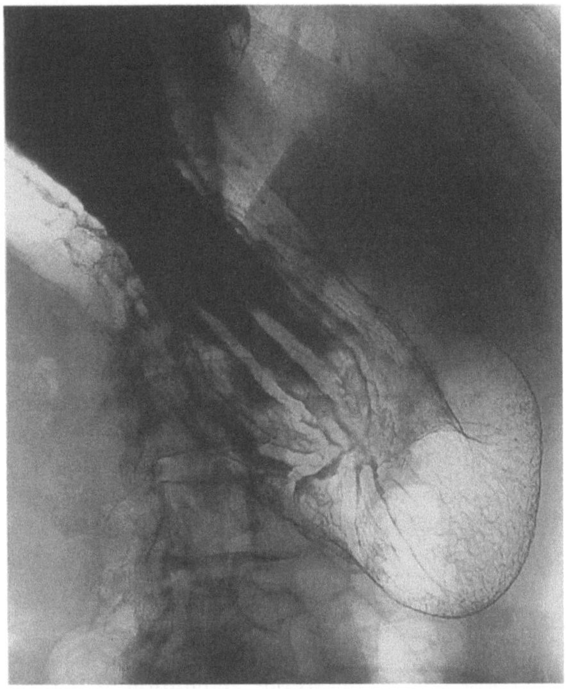

Fig. 12.1. Case 1. CR; double-contrast radiograph with the patient in the prone position

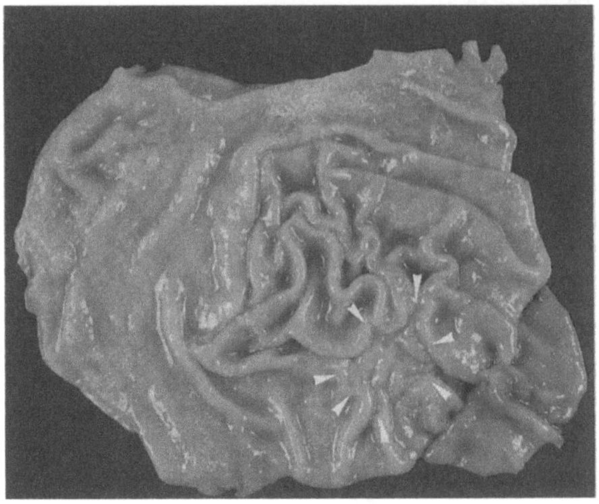

Fig. 12.2. Case 1. Resected specimen of an early gastric cancer (type IIc)

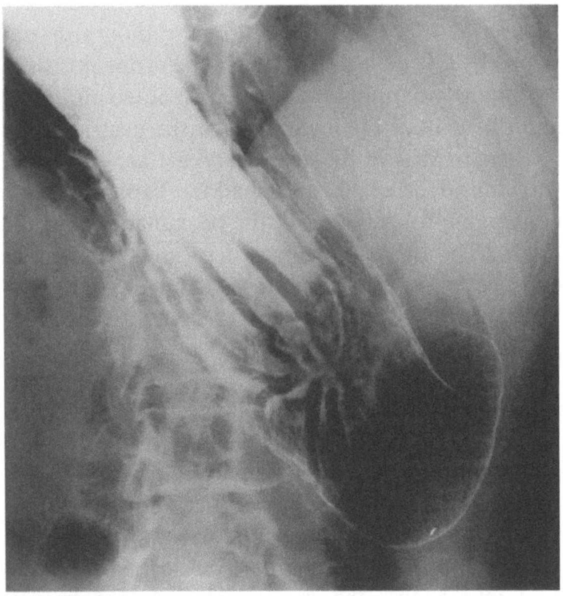

Fig. 12.3. Case 1. Conventional system; double-contrast radiograph with the patient in the prone position

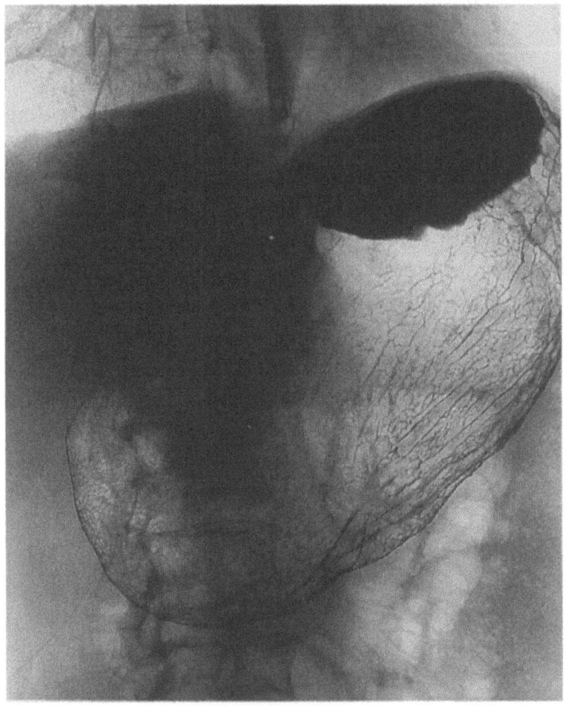

Fig. 12.4. Case 1. CR; double-contrast radiograph with the patient in the supine position

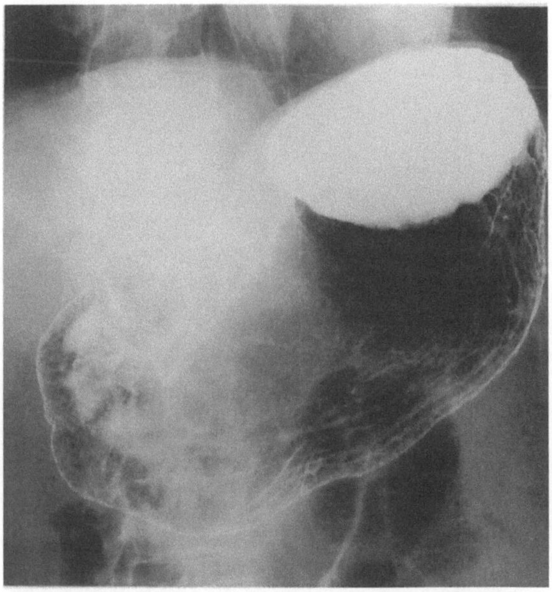

Fig. 12.5. Case 1. Conventional system; double-contrast radiograph with the patient in the prone position

distinguish in Fig. 12.4 than in Fig. 12.1.

Figure 12.5 is a double-contrast image obtained by the conventional system with the patient in the supine position. This picture barely allows recognition of the lesion. Qualitative diagnosis is difficult.

Case 2: A 76-year-old man with early gastric cancer (type IIa + IIc)

Figure 12.6 is a double-contrast radiograph of the anterior wall obtained by CR. An irregular star-shaped crater can be distinguished along the anterior wall of the gastric antrum. In addition, it is obvious that elevated granular shadows surround the crater.

Figure 12.7 is a double-contrast radiograph of the anterior wall obtained by the conventional system. The crater is recognizable, but the elevated lesion surrounding the crater cannot be seen.

A resected specimen of this case is shown in Fig. 12.8.

Figure 12.9 is a double-contrast CR image with the patient in the supine position. The crater is clearly detectable. The elevated lesion surrounding the crater is not clear. However, double-contrast radiography with the patient in the supine position is a simpler manipulation than double-contrast radiography of the anterior wall. Such a simple manipulation providing an image sufficient for diagnosis facilitates diagnosis and is useful in that it minimizes failure to recognize the abnormal lesion.

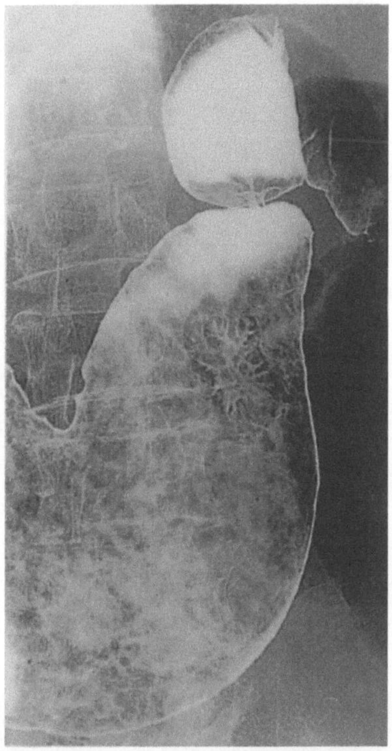

Fig. 12.6. Case 2. CR; double-contrast radiograph of the anterior wall

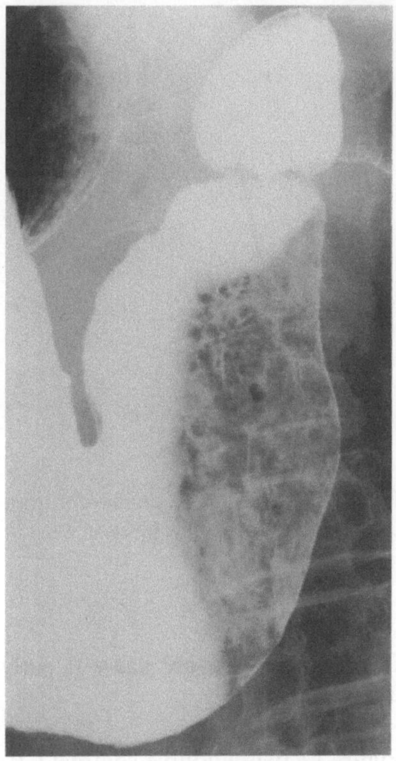

Fig. 12.7. Case 2. Conventional system; double-contrast radiograph of the anterior wall

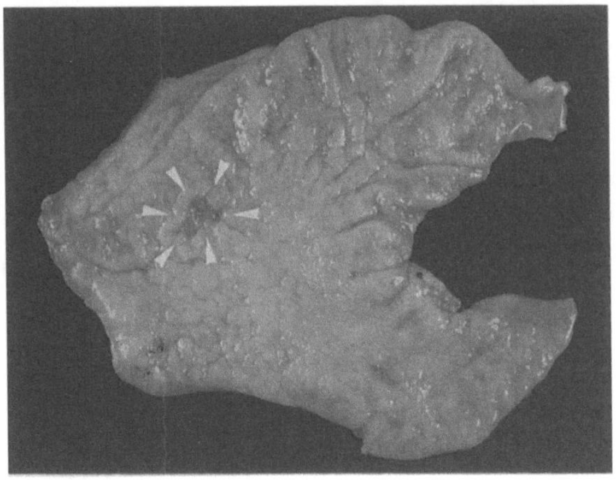

Fig. 12.8. Case 2. Resected specimen of an early gastric cancer (type IIa + IIc)

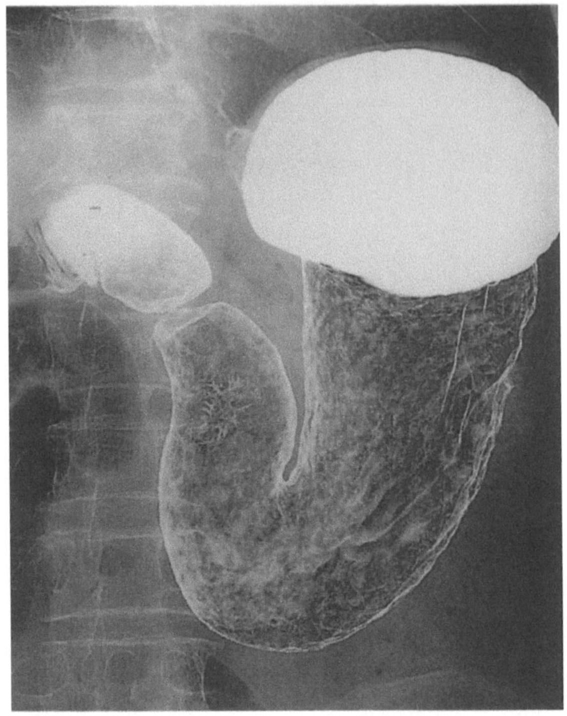

Fig. 12.9. Case 2. CR; double-contrast radiograph with the patient in the supine position

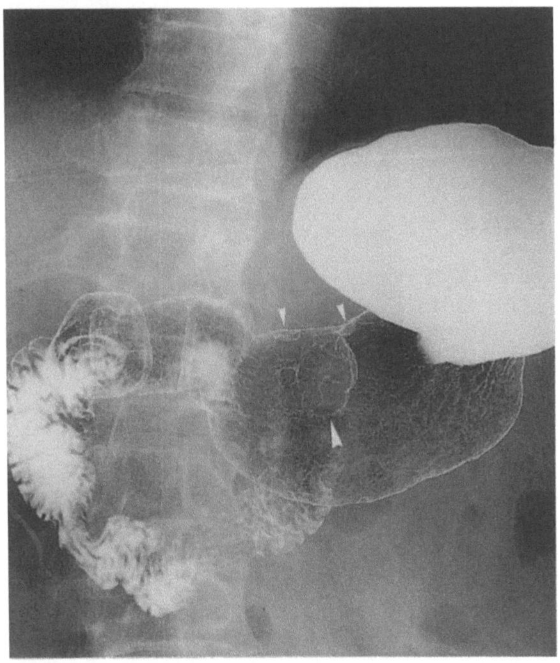

Fig. 12.10. Case 3. CR; double-contrast radiograph with the patient in the supine position

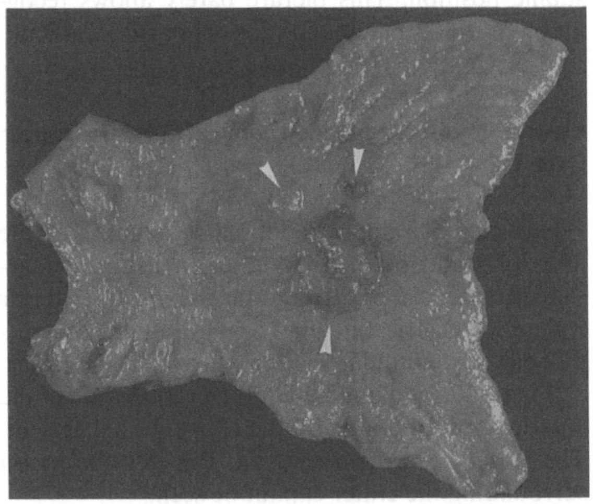

Fig. 12.11. Case 3. Resected specimen of an early gastric cancer (type I & IIa)

Case 3: A 69-year-old man with early gastric cancer (type I & IIa)

Figure 12.10 is a double-contrast CR image with the patient in the supine position. A large, type I lesion (big arrow) and two type IIa lesions (small arrow) adjacent to the type I lesion are seen near the lesser curvature of the ventricular angle. The lobulated surface and changes in the fine areae gastricae are visualized in the type I lesion. The adjacent type II lesions are clearly visualized and easy to distinguish.

Figure 12.11 is the resected specimen of the lesion

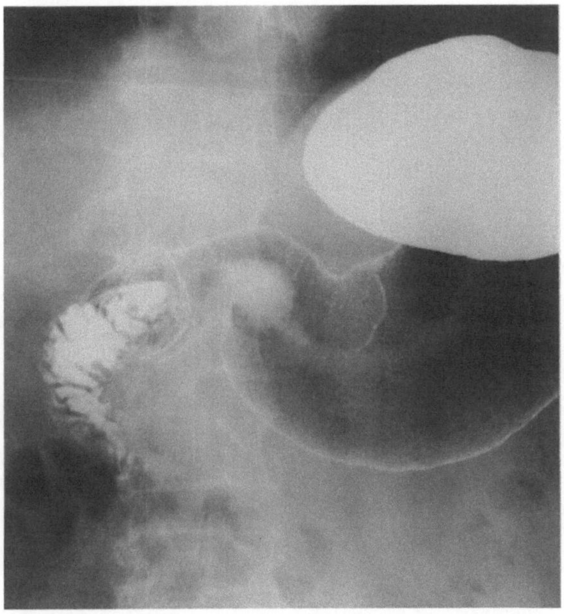

Fig. 12.12. Case 3. Conventional system; double-contrast radiograph with the patient in the supine position

shown in Fig. 12.10.

Figure 12.12 is a double-contrast radiograph obtained by the conventional system with the patient in the supine position. Figure 12.12 presents less clear outlines and a more uneven surface of the type I lesion than does Fig. 12.10. The outlines on the anal side, in particular, are unclear. The type IIa lesions are insufficiently visualized.

Case 4: A 57-year-old man with gastric cancer (Borrmann type II)

Figure 12.13 shows a resected specimen of a small Borrmann type II lesion along the lesser curvature just under the cardia. This is a case of a gastric cancer located in a region where it is difficult to obtain good visualization of lesions in X-ray films.

Figure 12.14 is a radiograph obtained by the conventional system with the patient head up in the right prone oblique (RPO) position and the table tilted at 40°–60°. In this image, the presence of a lesion can be confirmed, but qualitative diagnosis is difficult.

Figure 12.15 is a CR image with the patient in the same position as in Fig. 12.14. This obviously reveals the distorted and slightly elevated mucosal pattern (big arrow) and an irregularly shaped ulcer that has formed in the center (small arrow). Converging folds going toward the lesion are also clearly seen.

Figure 12.16 is an image of Fig. 12.15 after reversal processing. Thus, in some cases, reversally processed films are more advantageous in that the adverse effects of granularity are suppressed.

Case 5: A 56-year-old man with leiomyoma

Figure 12.17 is a double-contrast CR image with the patient in the supine position. A shadow of an irregular protruded lesion is observed on the side of greater curvature of the lower part of the body. In order to obtain a frontal view of the lesion, radiography by a lateral projection is required.

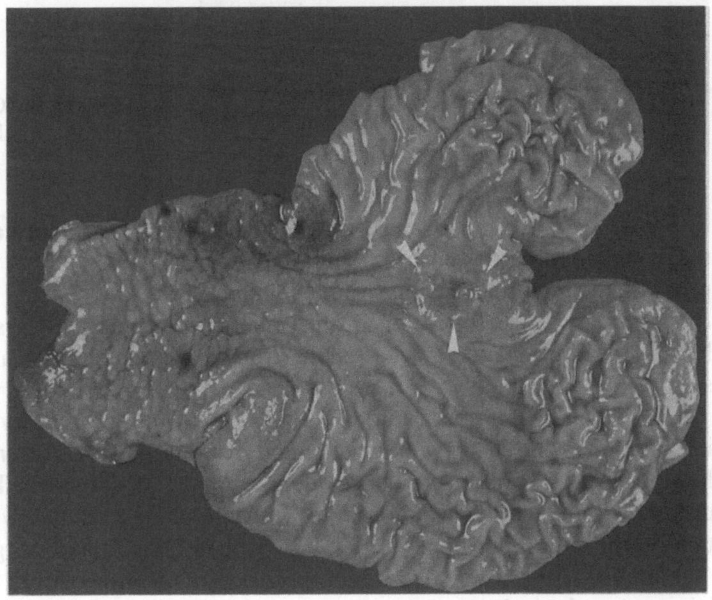

Fig. 12.13. Case 4. Resected specimen of a gastric cancer (Borrmann type II)

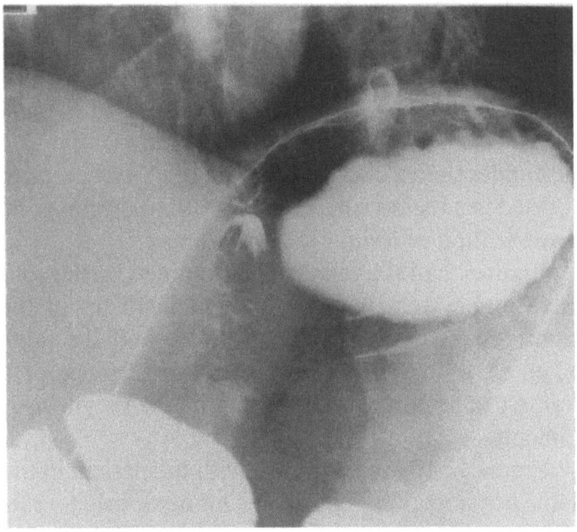

Fig. 12.14. Case 4. Conventional system; radiograph with the patient head up in the RPO with table tilted 40°–60°

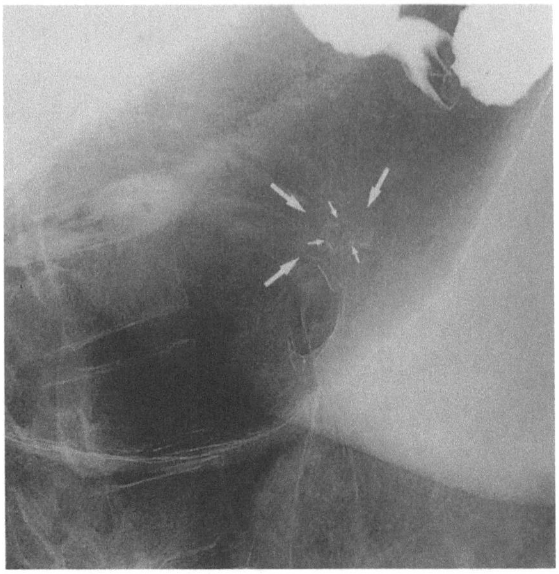

Fig. 12.15. Case 4. CR; radiograph with the patient in the same position as in Fig. 12.14

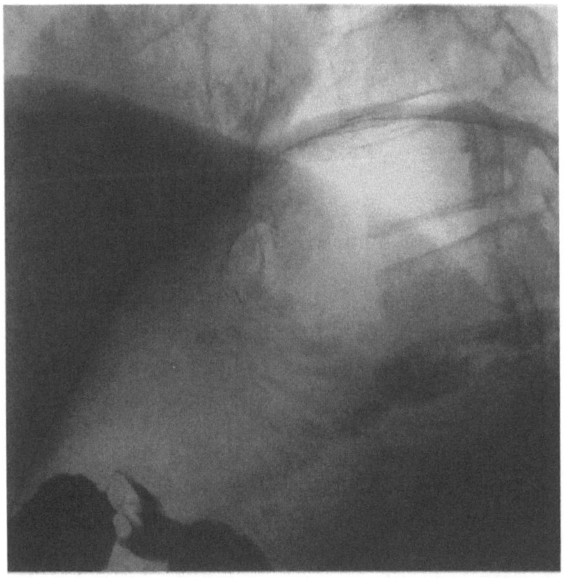

Fig. 12.16. Case 4. CR (with reversal processing); radiograph with the patient in the same position as in Fig. 12.14

Figure 12.18 is a double-contrast CR image with the patient in the lateral position. This film obviously provides distinguishable details of the lobulation of the lesion and ulcer formation in the center of the lesion. In addition, the mucosal pattern is clearly seen.

CR, which has a wide latitude and high sensitivity, minimizes the effects of geometric and dynamic lack of clarity. Therefore, it can provide clear images even by double-contrast radiography with the patient in the lateral position, such as in Fig. 12.18.

Figure 12.19 shows a resected specimen of a leiomyoma along the greater curvature in the lower part of the corpus ventriculi.

Exposure dose and radiographs

The CR system has a wide range of sensitivity and, therefore, allows a great decrease in exposure dose as compared with the film/screen system. However, granularity becomes more marked with a decrease in the exposure dose, resulting in greater difficulty with diagnosis. The relationship between the dose and the granularity is shown in Fig. 12.20, which presents the rms value (granularity) with the exposure dose varying within the range of ten magnifications. With regard to the limits to the image quality subject to the effects of noise in clinical radiography, it is impos-

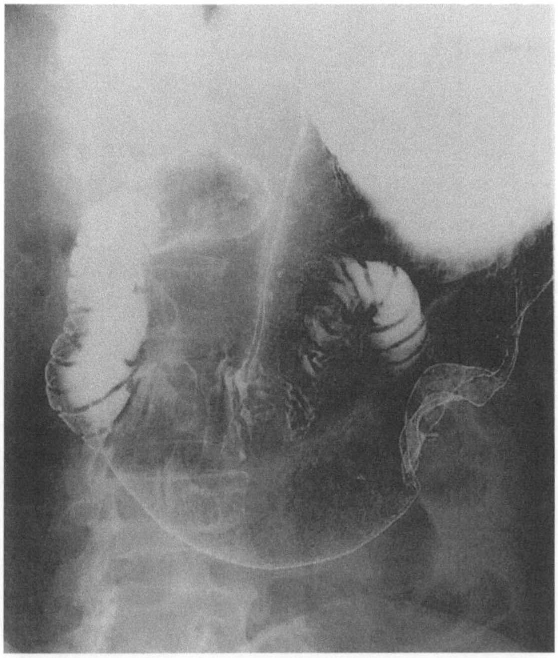

Fig. 12.17. Case 5. CR; double-contrast radiograph with the patient in the supine position

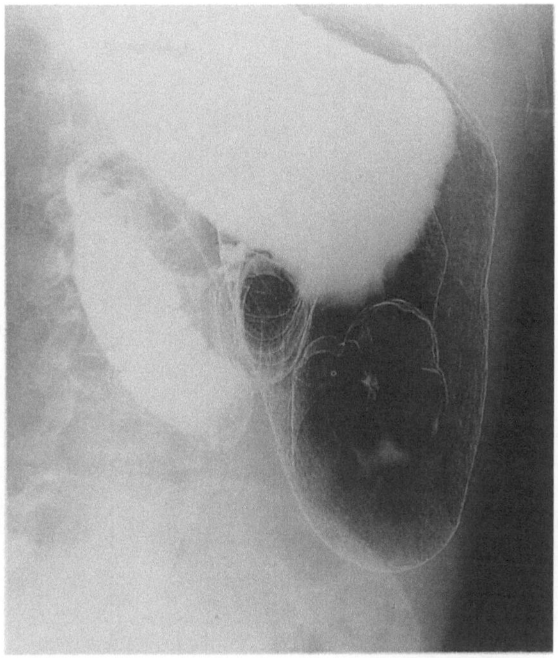

Fig. 12.18. Case 5. CR; double-contrast radiograph with the patient in the lateral position

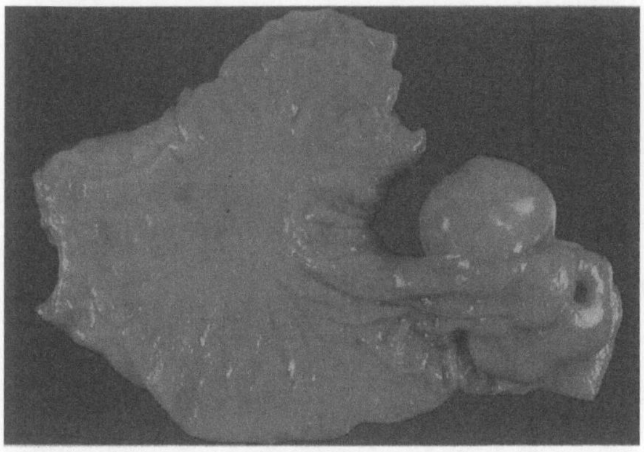

Fig. 12.19. Case 5. Resected specimen of a leiomyoma

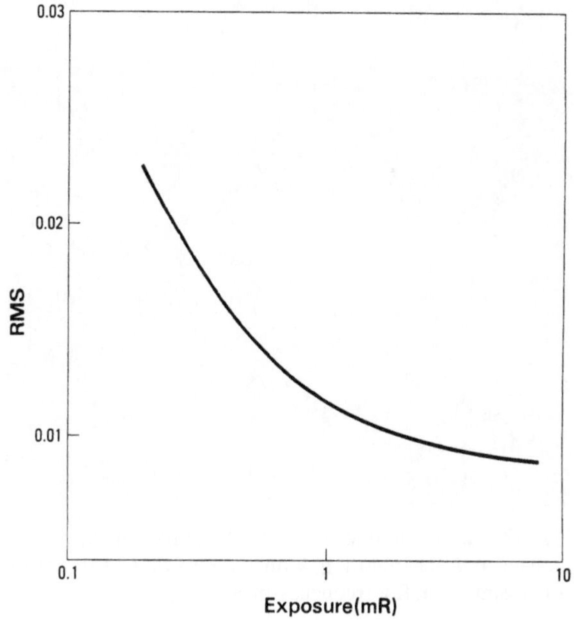

Fig. 12.20. Exposure dose (mR) vs. RMS

sible to give definition *to the GI tract.* As for the balance in radiography of the stomach, the clarity of the picture varies according to the adhesion of the contrast medium and the air content.

Figure 12.21 shows double-contrast CR images at decreased exposure doses. This subject was intentionally chosen as a representative case of cicatrized gastric ulcer with poor adhesion of the contrast medium. The doses used were one-half, one-quarter, one-eighth, and one-twelfth (Fig. 12.21a–h) that of the standard dose required for the film/screen system. With the gradation processing alone, the visualization of the scar was possible at as low a dose as one-eighth (Fig. 12.21e), whereas it was impossible at that dose by the frequency processing (Fig. 12.21f) because of the great effect of noise. Taking these results into account, a decrease in exposure dose to one-tenth seems to be the limit. How-

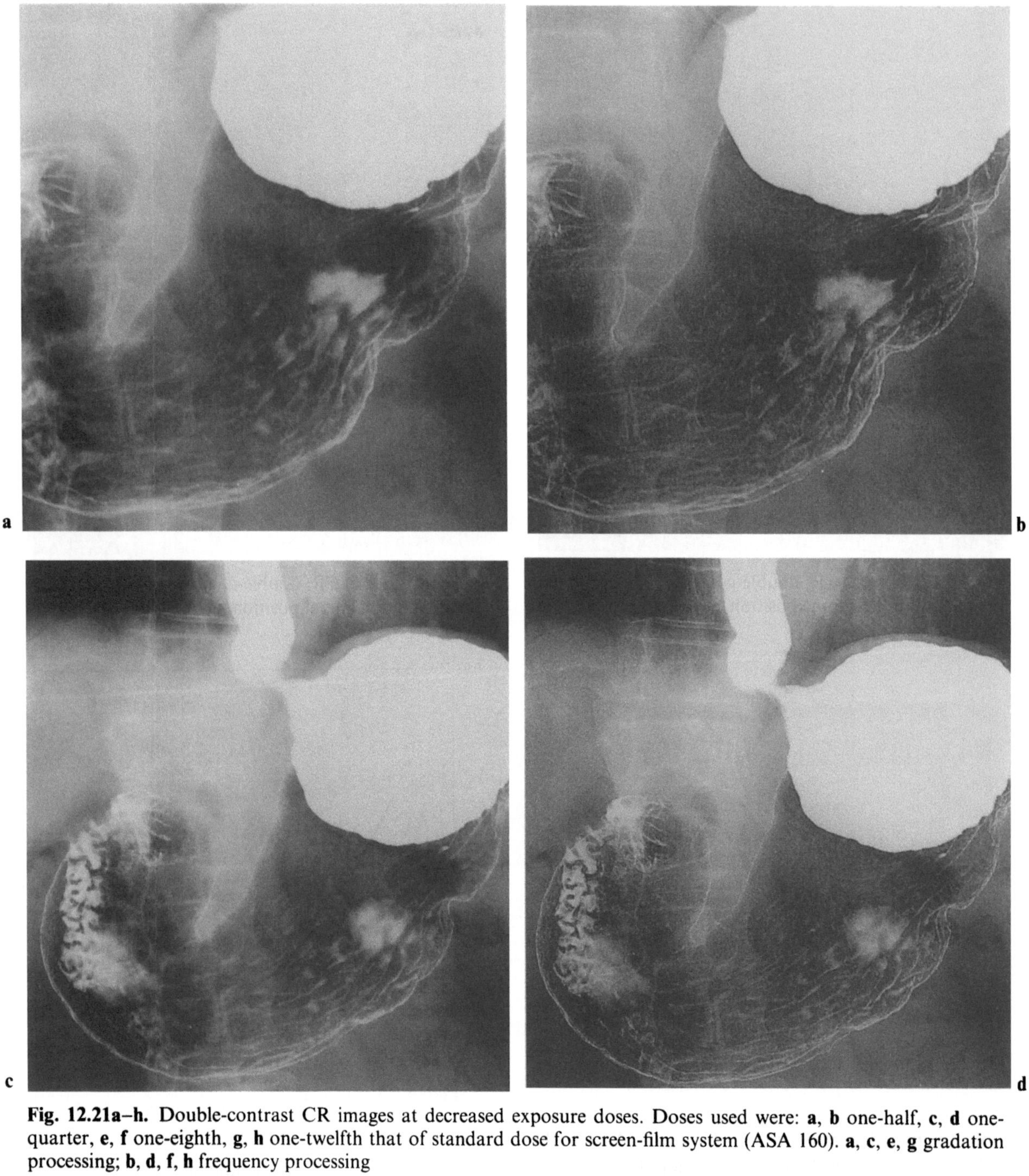

Fig. 12.21a–h. Double-contrast CR images at decreased exposure doses. Doses used were: **a**, **b** one-half, **c**, **d** one-quarter, **e**, **f** one-eighth, **g**, **h** one-twelfth that of standard dose for screen-film system (ASA 160). **a**, **c**, **e**, **g** gradation processing; **b**, **d**, **f**, **h** frequency processing

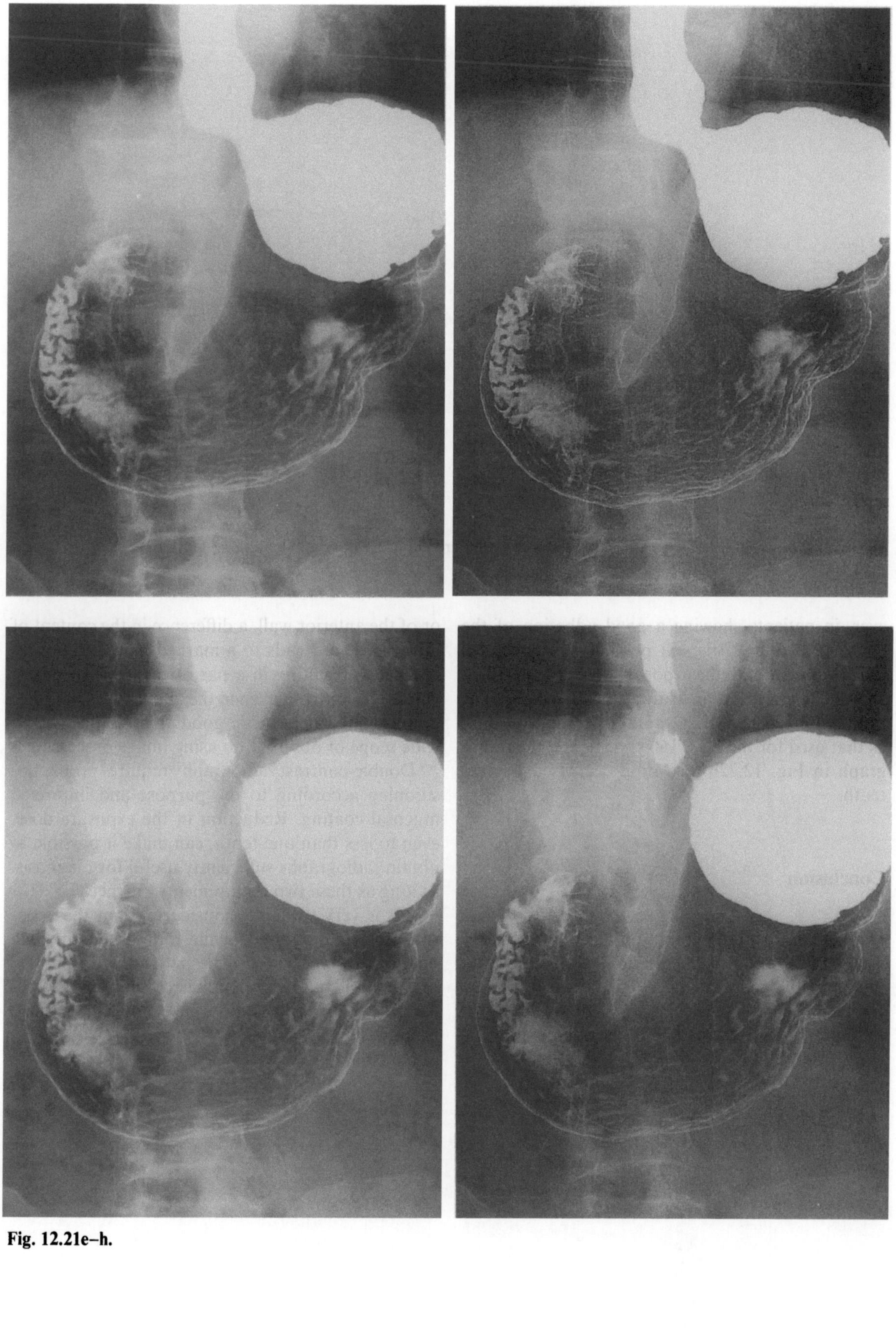

e f g h

Fig. 12.21e–h.

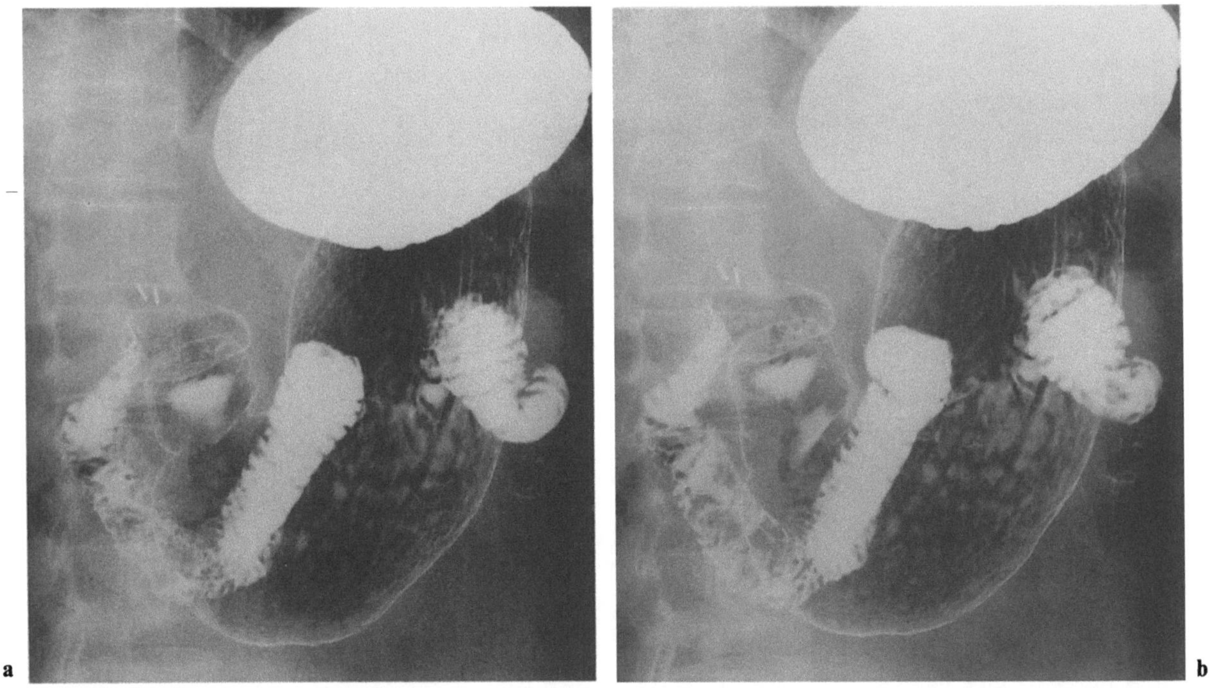

Fig. 12.22a, b. Double-contrast CR images at **a** the standard exposure dose and **b** at a one-tenth dose

ever, in patients showing a good adhesion of the contrast medium, it is still possible to distinguish lesions at one-tenth the exposure dose. Figure 12.22 shows CR images as an example of such cases. The radiograph in Fig. 12.22a was taken at the same dose as that used for the ASA 160 system, and the radiograph in Fig. 12.22b at a dose decreased to one-tenth.

Conclusion

Double-contrast radiographs obtained by CR allow clear visualization of microlesions and provide an image quality guaranteeing a high diagnostic ability. In the film system, particularly with double-contrast radiography with the patient in the oblique position or of the anterior wall, a difference in the content of intragastric air leads to a marked manifestation of blackness, resulting in a narrowed scope of observation. In contrast, even in the above types of radiography, CR can provide good radiographs with a wide scope of observation using image processing.

Double-contrast radiography requires proper positioning according to the purpose and improved mucosal coating. Reduction in the exposure dose, even to less than one-tenth, can make it possible to obtain radiographs sufficiently useful for diagnosis, as long as these two requirements are met.

CR is very valuable, considering that the exposure dose is decreased, while radiographs of high quality can be secured. The patients are thus free of the disadvantageous influence of a high exposure dose by the upper gastrointestinal (uGI) examination.

Chapter 13

Excretory urography

Hiromu Nishitani[1], Hideo Onitsuka[2], Kosaburo Kawahira[2], Hiroshi Honda[2], Keiichi Matsuura[2], Jyoichi Kumazawa[3], and Hideki Yuzuriha[1]

This chapter reports our experience with computed radiography (CR) applied to excretory urography, where CR is useful in decreasing the exposure dose. This method has also been found to allow analysis of renal parenchymatous lesions in excretory urography when used for tomography in the nephrogram phase.

Introduction

CR uses a photo-stimulable phosphor for a wide range of applications [1]. CR is more advantageous than screen/film systems in the following aspects: (1) it provides images with a wide latitude through image processing; (2) it permits a wide range of radiographic conditions and can continually provide high-quality pictures; and (3) it allows low-dose exposure [2].

Excretory urography is a representative method of X-ray examination of the urinary system. However, this method seems to be less advantageous, despite its simplicity, than computed tomography (CT), which has recently made great progress [3]. In addition, excretory urography, in most cases, inevitably includes the gonads in the exposure field, thus raising the problem of how to decrease the exposure dose.

CR, which has a wide dynamic range, excellent contrast resolution and image-processing functions, and the capability for low-dose exposure, may compensate for the defects of current excretory urography. This chapter reports our recent early experience in using CR for excretory urography.

Routine image processing in CR

Although details of the image-processing function in CR are described elsewhere (Chap. 4), mention of this will also be made here as the determination of a routine processing program is important for excretory urography from the viewpoint of reproducibility during follow-up observation.

In urography, components of high spatial frequency, such as in the thoracic region or skeletal system, are relatively rare. Thus, it seems better to use image processing for enhancing the low-frequency components. The enhancement of the high spatial-frequency components makes it difficult to see slight changes in contrast in the kidney and increases the obstructing effect of intestinal gas, which hinders the kidney shadow. The tonal conversion processing by CR in our study was not particularly superior to other methods and showed almost the same qualities.

In our facility at present, relatively low spatial-frequency areas are enhanced in routine image processing, and a set of two pictures, one undergoing almost no enhancement and the other moderately enhanced, are produced on each film. A straight line with a gradient of 1 is used as the gradation curve (characteristic curve).

Excretory urography by CR

In our experience, CR provides pictures that are not diagnostically inferior to those obtained by screen/film systems. In particular, images of the stable density can be obtained as a characteristic feature. In abdominal X-ray examinations prior to the injection of the contrast medium, calcification becomes more distinct through image processing. In cases of renal calculus, the internal structure is often more clearly visible than with screen/film systems (Fig. 13.1). Changes in gradation, which are visualized as completely white by screen/film systems, can be shown by CR. Such visualization, observable not only in the abdominal region but also in the thoracic region, seems to be based on the wide dynamic range of CR. On the other hand, intestinal gas may be more distinct than with screen/film systems depending on the procedure of image processing. This issue of obstructive shadows remains to be solved in the future.

Division of Radiology[1], Department of Radiology[2], and Department of Urology[3], Faculty of Medicine, Kyushu University, 1-1, Maedashi 3-chome, Higashi-ku, Fukuoka, 812 Japan

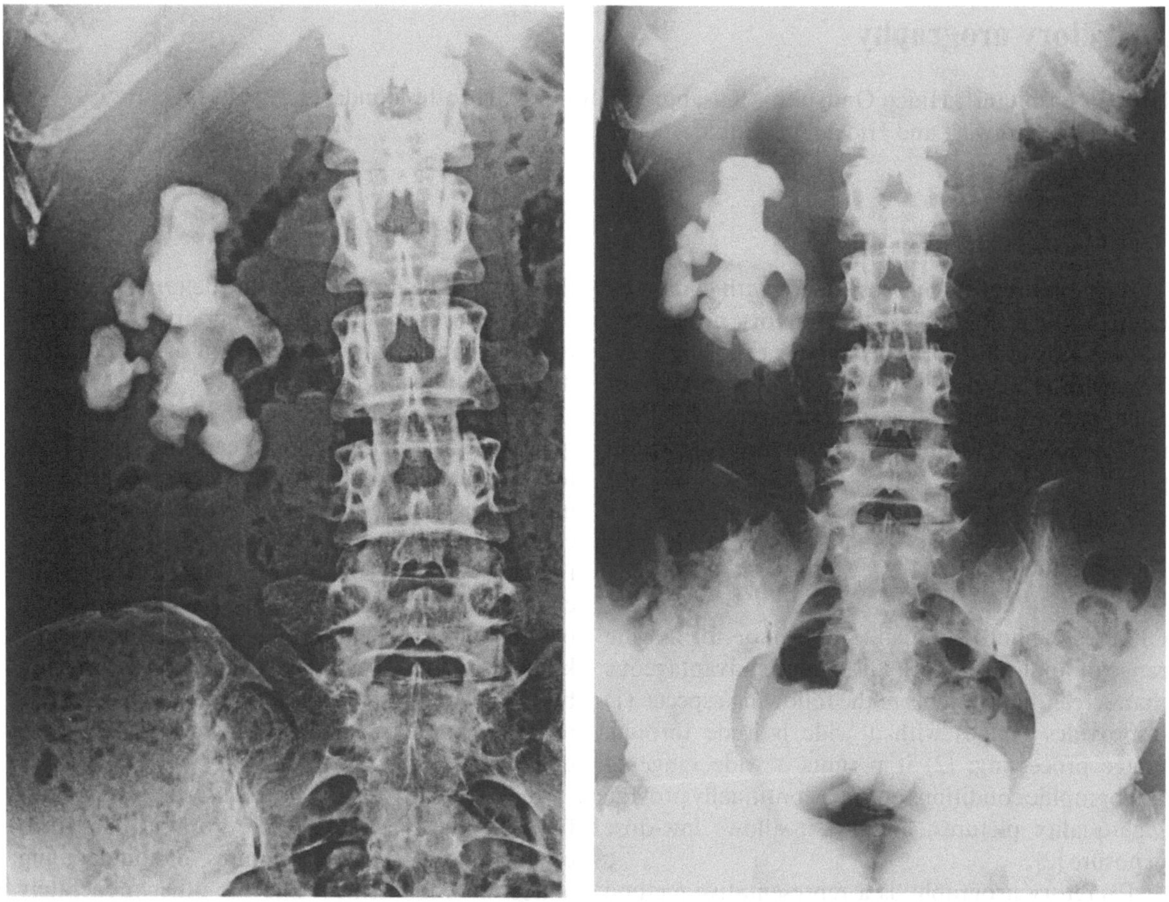

a b

Fig. 13.1a, b. Staghorn calculus in the right kidney. **a** After "conservative" processing by CR. **b** After "radical" processing by CR, showing the internal structure of the calculus

Images obtained after the injection of a contrast medium were stable and provided all the information necessary for diagnosis, despite the fact that they were reduced and thus smaller than those with screen/film systems (14 × 17 in.).

Possibility of a decrease in exposure dose

In our facility, a rare-earth screen/film system has been used for excretory urography in order to decrease the exposure dose. The currently used intensifying screen is KH (Konishiroku Medical), and the film is of AO type (Konishiroku Medical). With these materials, the exposure dose is about one-quarter less than that with the screen/film system of the LTII-A type (Kyokko).

In this study with CR, we also investigated the possibility of a further decrease of the exposure dose. Although the number of images was small, it was found that the quality of the images obtained with half the usual exposure dose was almost the same as that obtained with the current screen/film system. Although granularity became worse at one-eighth of the usual dose, sharp images sufficient for diagnosis were obtained, indicating that the introduction of CR can present a decrease of exposure dose (Fig. 13.2).

CR allows exposure using three imaging plates in a cassette. The addition of these three pictures presents a decrease in noise, so that a further decrease in exposure dose can be expected without deterioration of the image quality.

Fig. 13.2a–d. **a** CR image obtained under the same exposure conditions as screen/film system. **b** CR image at half the exposure dose in **a**. The image quality is almost the same as in **a**. **c** CR image at one-quarter the exposure dose in **a**. The granularity is somewhat worse. **d** CR image at one-eighth the exposure dose in **a**. Although the granularity is worse, diagnosis is possible with a distinguishable renal pelvis and calyx. Analysis of bone trabeculation is rather difficult ▷

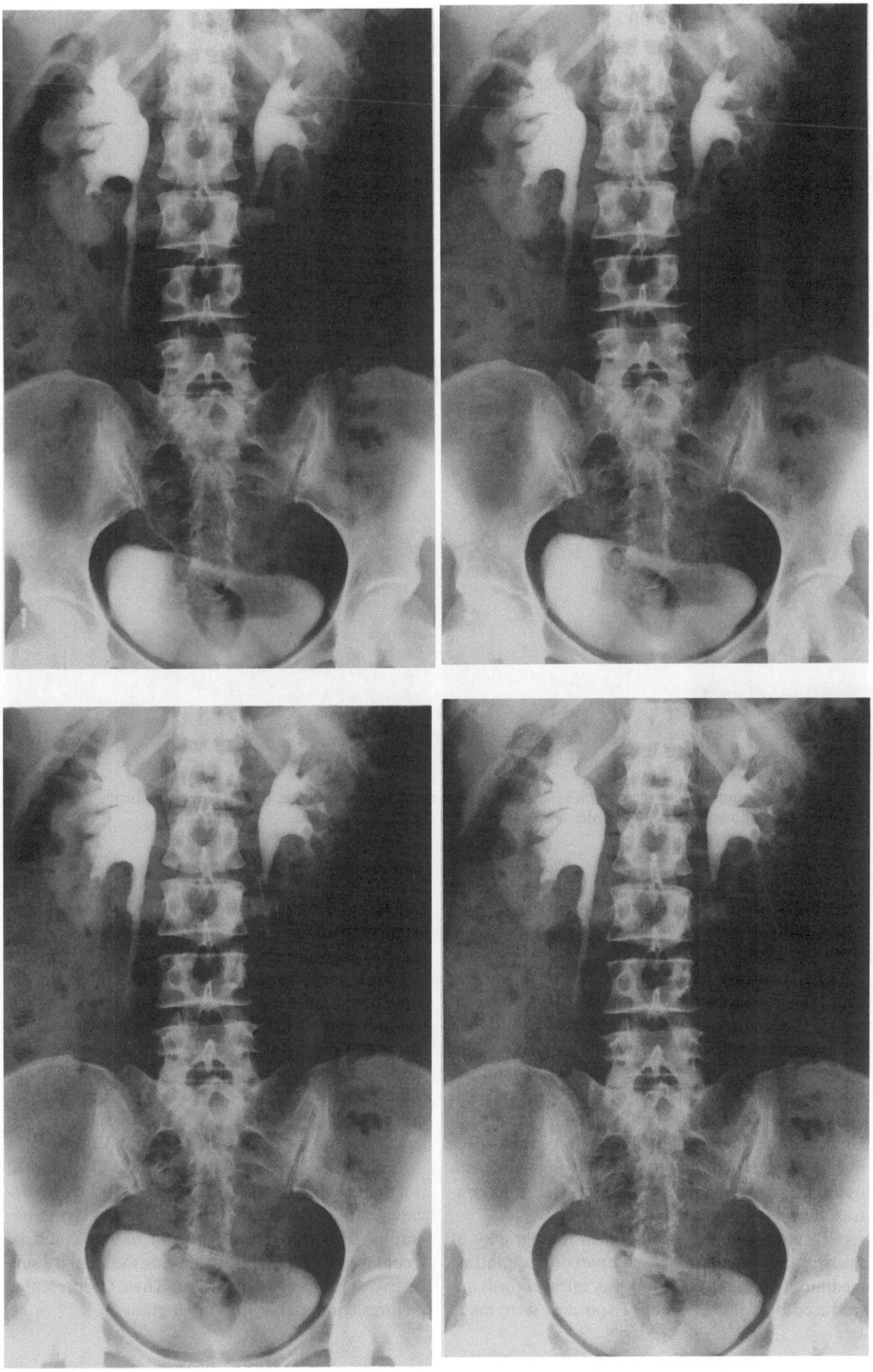

a b c d

Fig. 13.2a–d.

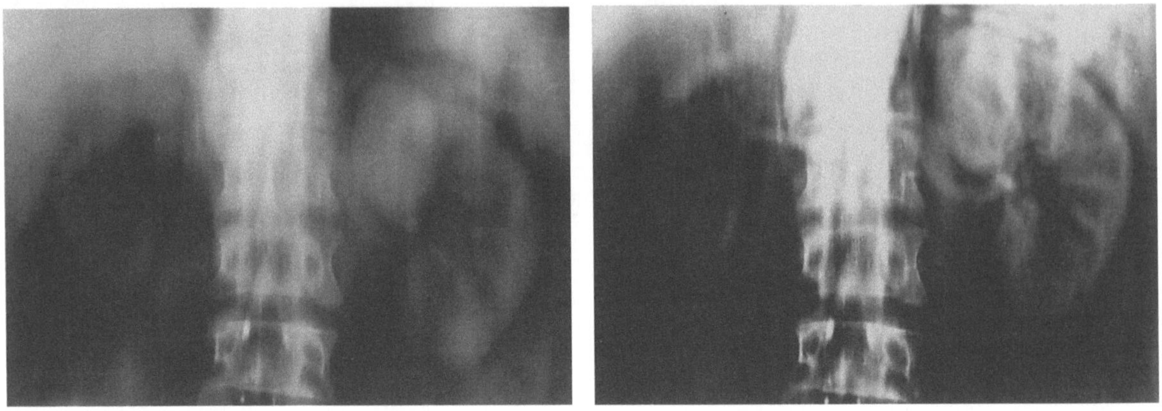
a b

Fig. 13.3. **a** Tomogram with screen/film system during the nephrographic phase. **b** CR tomogram of the same subject as **a** during the nephrographic phase. The blood vessels in the left kidney and a clear border between the renal cortex and the renal medulla can be seen

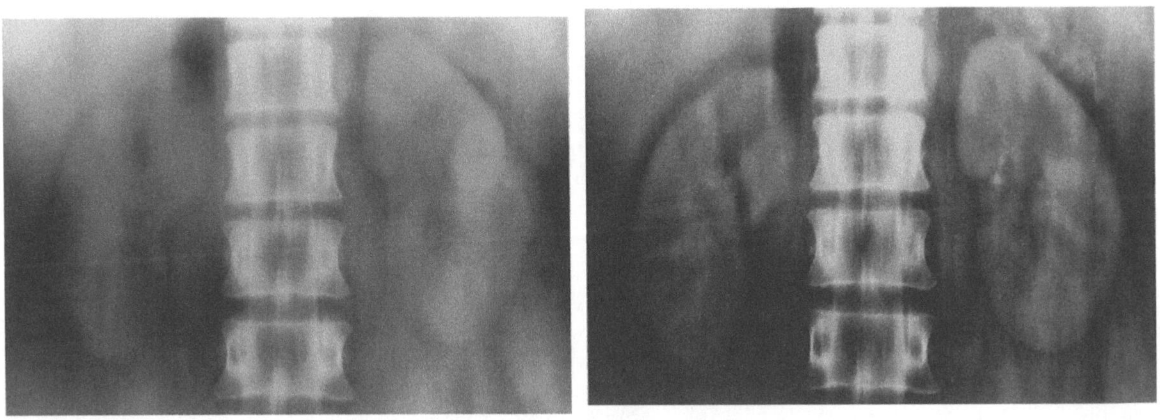
a b

Fig. 13.4. **a** Tomogram with screen/film system during the nephrographic phase. **b** CR tomogram during the nephrographic phase, showing a distinct border between the renal cortex and medulla

Possibility of improvement in contrast detectability

Excellent contrast detectability is one of the features guaranteeing the usefulness of X-ray CT. Compared with CT, excretory urography provides an outline of the information for the diagnosis easily but is inferior in terms of contrast detectability. Therefore, this method is less useful for the diagnosis of renal cysts and lesions localized in the renal parenchyma. However, the simplicity of excretory urography cannot be disregarded, and it could be reinstated as a useful method for screening renal diseases if its contrast detectability could be improved.

CR and a screen/film system in tomography of the nephrogram phase were compared. The contrast medium, Conray 400 (40 ml) was intravenously introduced by bolus injection. Exposures were made by CR and the screen/film system during the nephrographic phase, immediately after injection. Since the order of exposure, i.e., whether CR or the screen/film system is performed first, may affect imaging, the exposures were roughly divided in half as to which technique was used first. The investigation focused on the contrast between the renal cortex and medulla. Although the results will be described in detail in another paper, CR was found to be generally better in distinguishing the renal cortex and medulla, regardless of the order of exposure (Figs. 13.3, 13.4).

Improved contrast detectability was also marked in a patient with renal cancer (Fig. 13.5). This hypervascular renal cancer was hardly identified by tomography with the screen/film system but could be recognized easily by CR as a clear radiopacity suggestive of a hypervascular mass. Thus, CR presented pictures comparable with angiograms.

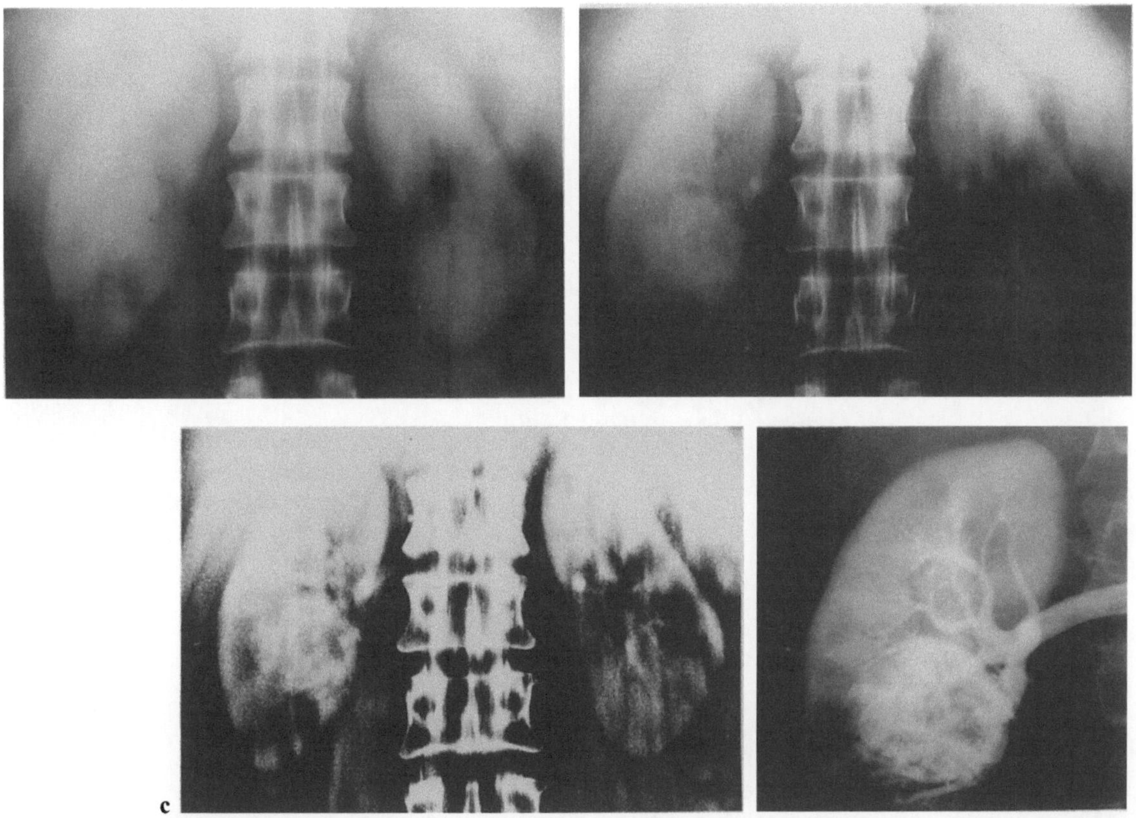

a b c d

Fig. 13.5. **a** Tomogram with screen/film system during the nephrographic phase. **b** CR tomogram during the nephrographic phase (with "conservative" processing) shows a radiopaque mass in the central area of the right kidney. **c** CR tomogram during the nephrographic phase (with "radical" processing) presents a clearer distinction of the mass. **d** Selective right renal angiography by screen/film system

It is likely that improvements in contrast detectability can be expected from excretory urography when used with tomography by CR during the nephrographic phase.

Conclusion

In a trial using excretory urography with CR, the exposure dose of radiography was successfully decreased without lessening the information necessary for diagnosis and an improvement in contrast detectability, which is required for diagnosis, was indicated. On the other hand, obstructing intestinal gas tends to be enhanced, depending on the image processing. This problem remains to be solved by further research, e.g., by use of the dual-energy subtraction method.

As a whole, CR is suitable for excretory urography and is a useful method of examination, apart from the problem of its high cost.

References

1. Takano M (1983) New computed radiography using stimulable phosphy. Japanese Jornal of Clinical Medicine 41 : 1395–1403
2. Hachiya J, Korenaga T, Miyasaka Y, Nitatori T, Wakasa K, Sakurai K (1983) Computed radiography utilizing photostimulable phosphor. Igakynoayumi 127: 1199–1206
3. Nishitani H (1984) Renal imaging diagnosis by computed tomography. Nippon Acta Radiologca 44: 1044–1061

Chapter 14

Biliary tract system

Yoshiro Furuya[1], Katsuhide Wakasa, Tuneaki Seki, Yasuo Miyasaka, and Junichi Hachiya

Data obtained on the biliary tract in initial clinical experience with the computed radiography (CR) system are presented in this chapter. In drip infusion cholecystocholangiography (DIC) combined with tomography, CR produced better images than the screen film; even in the case of low-dose exposure, CR was thought to be sufficiently usable for clinical diagnosis. In simultaneous multisection tomography, CR could probably present even better images if the exposure conditions were improved.

Introduction

The diagnosis of biliary tract disease has made remarkable progress following the recent development of medical imaging techniques such as ultrasound and X-ray CT. Above all, diagnosis by ultrasound, which is a noninvasive and simple procedure without discomfort to patients and involves no exposure to radiation, has become the method of first choice in clinical examinations. DIC, which is a common radiological diagnostic method, unlike ultrasound images the entire biliary tract and is excellent in demonstrating the relative positional relationships of various regions with high resolution; it is thus an important method.

This chapter describes initial clinical experience in the area of the biliary tract system with CR, a digital radiographic system using photostimulable phosphor.

Subjects and methods

Tomography was performed on patients with suspected abnormality of the biliary tract who had undergone DIC; the conventional screen/film system and the CR system were used under practically identical conditions, and the images were compared. Normal tomography was performed on 30 patients, and simultaneous multisection tomography on 21. In addition, images were taken of 15 patients who underwent normal tomography, with the CR system at a low exposure dose. Biliscopin DIC 50 was injected by intravenous drip infusion over a period of 30–45 min, and exposures were made 30 and 60 min after completion of the infusion. In normal tomography using a Fuji Hi-Screen STD screen and Fuji RX film, the exposure conditions were FFD 115 cm, 70–85 kV, 80mA, and 1.25 s. In simultaneous multisection tomography with five planes at intervals of 1 cm, using KYOKKO, STF 5–10 screens and Fuji RX film, the conditions were 115 cm, 77–85 kV, 160 mA, and 1.25 s. In the case of the CR system, imaging plates were used and exposures were made for tomography as far as possible under identical conditions. We obtained two images which were set as appropriate in advance (left image close to the conventional tomographic image and right image with intensive image processing). Images at low exposure doses by CR were made at half and one-fifth of the usual dose.

Eight items were used in the evaluations: three from the gallbladder—base, body, and neck; three from the biliary tract—common bile duct, common hepatic duct, and intrahepatic bile duct; and the others were the gallstone and lateroabdominal wall. The image of each region was compared with the conventional tomographic image and rated on a three-point scale (+, ±, and −). An FCR image superior to the conventional tomographic image was rated (+), an equivalent image (±), and an inferior image (−). With simultaneous multisection tomography, evaluation was made of five sections together, separate from the individual assessment of each section

Comparison with normal tomography

At the same exposure dose as the normal tomographic image, CR presented no (−) ratings, but gave a (+) rating in 39% of cases, although this differed from region to region (Table 14.1). In par-

[1] Department of Radiology, Kyorin University School of Medicine, 20-2, Shinkawa 6-chome, Mitaka, Tokyo, 181 Japan

Table 14.1. Evaluation of CR images at normal and low exposure doses in normal tomography

	Rating	Same exposure dose	Half exposure dose	One-fifth exposure dose
Gallbladder				
Base	+	12/25 (48.0%)	0/4	2/8 (25.0%)
	±	13/25 (52.0%)	4/4	6/8 (75.0%)
	−	0/25 (0%)	0/4	0/8 (0%)
Body	+	6/25 (24.0%)	0/4	1/8 (12.5%)
	±	19/25 (76.0%)	4/4	7/8 (87.5%)
	−	0/25 (0%)	0/4	0/8 (0%)
Neck	+	5/25 (20.0%)	0/4	1/8 (12.5%)
	±	20/25 (80.0%)	4/4	7/8 (87.5%)
	−	0/25 (0%)	0/4	0/8 (0%)
Bile duct				
Common bile duct	+	8/27 (29.6%)	0/4	1/9 (11.1%)
	±	19/27 (70.4%)	4/4	5/9 (77.7%)
	−	0/27 (0%)	0/4	3/9 (33.3%)
Common hepatic duct	+	4/28 (14.3%)	0/4	2/9 (22.2%)
	±	24/28 (85.7%)	4/4	7/9 (77.7%)
	−	0/28 (0%)	0/4	0/9 (0%)
Intrahepatic bile duct	+	7/30 (23.3%)	0/4	2/9 (22.2%)
	±	23/30 (76.7%)	4/4	7/9 (77.7%)
	−	0/30 (0%)	0/4	0/9 (0%)
Gallstone				
	+	6/10 (60.0%)	0/1	0/3 (0%)
	±	4/10 (40.0%)	1/1	3/3 (100%)
	−	0/10 (0%)	0/1	0/3 (0%)
Lateral abdominal wall	+	30/30 (100%)	4/4 (100%)	9/9 (100%)

Table 14.2. Evaluation of CR images in simultaneous multisection tomography

	Rating	Simultaneous multisection	First section	Second section	Third section	Fourth section	Fifth section
Gallbladder							
Base	+	8/17 (47.0%)	71.0%	57.1%	14.3%	6.7%	7.1%
	±	9/17 (53.0%)	29.0%	42.9%	78.6%	66.7%	42.9%
	−	0/17 (0%)	0%	0%	7.1%	26.7%	50.0%
Body	+	4/17 (23.5%)	57.0%	57.1%	14.3%	0%	7.1%
	±	13/17 (76.5%)	43.0%	42.9%	78.6%	78.6%	50.0%
	−	0/17 (0%)	0%	0%	7.1%	21.4%	50.0%
Neck	+	4/17 (23.5%)	57.0%	57.1%	14.3%	0%	7.1%
	±	13/17 (76.5%)	43.0%	42.9%	85.7%	78.6%	42.9%
	−	0/17 (0%)	0%	0%	0%	21.4%	50.0%
Bile duct							
Common bile duct	+	4/21 (19.0%)	81.8%	41.7%	28.6%	12.5%	7.1%
	±	17/21 (81.0%)	18.2%	58.3%	71.4%	87.5%	50.0%
	−	0/21 (0%)	0%	0%	0%	0%	42.9%
Common hepatic duct	+	4/21 (19.0%)	69.2%	46.7%	33.3%	6.2%	6.6%
	±	17/21 (81.0%)	30.8%	53.3%	66.6%	93.8%	53.3%
	−	0/21 (0%)	0%	0%	0%	0%	40.0%
Intrahepatic bile duct	+	7/21 (33.3%)	100%	62.5%	43.8%	5.4%	7.1%
	±	14/21 (66.6%)	0%	37.5%	56.2%	82.3%	35.8%
	−	0/21 (0%)	0%	0%	0%	11.8%	57.1%
Gallstone							
	+	5/7 (71.4%)		0%	100%	60.0%	0%
	±	2/7 (28.6%)		100%	0%	20.0%	0%
	−	0/7 (0%)		0%	0%	20.0%	100%
Lateral abdominal wall	+	21/21 (100%)	100%	100%	100%	100%	100%

ticular, the lateral abdominal wall, gallstone (where present), and base of the gallbladder were imaged more clearly. Structures can thus be better visualized, as has been reported for other body regions [1,2,3]. In some of the patients with gallstones, the presence and shape of the stones were identified for the first time in the CR image. With the images at half the exposure dose, all the regions with the exception of the lateral abdominal wall were rated to be *equivalent to the conventional tomographic images*. At one-fifth the normal exposure dose, the (+) rating was found in 28.6% of cases, and the (−) rating was given to the common bile duct in three cases. A possible cause for the latter was a deviation of the planes due to respiratory movement. The relationship between the decrease of exposure dose and the degradation of the image quality is also controversial in other medical fields. However, it was thought that the images at half the normal exposure dose could definitely be utilized for clinical diagnosis, while the one-fifth exposure dose could be used for a few problems.

Comparison with simultaneous multisection tomography

No (−) rating was found in the overall evaluation of the five sections together, and the (+) rating, which differed among the regions, was observed in 40.1% of cases (Table 14.2). This method was excellent, particularly in imaging the lateral abdominal wall, gallstone, base of the gallbladder, and intrahepatic bile duct. In the simultaneous evaluation of the five sections, a (+) rating was frequent for the first and second sections, with no (−) rating; the first (−) rating appeared in the third section, with marked degradation in the fifth plane. As shown in Fig. 14.1, this can be attributed to deterioration in the granularity that occurs in the third to fifth sections in simultaneous multisection tomography. Experimental results simultaneously obtained indicate that the granularity of the third or fifth sections could be improved by raising the tube voltage.

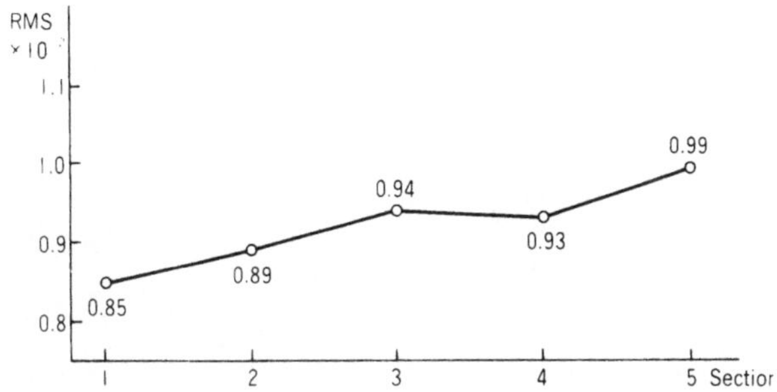

Fig. 14.1. Granularity of the imaging plate in multisection tomography

Case reports

Case 1: Normal case

Figures 14.2 and 14.3 show the conventional tomographic image, left and right CR images at the same exposure dose, and right CR image at one-fifth the exposure dose of the gallbladder and bile duct of the same patient. The gallbladder, especially the margin from the base to the body, is clearer in the right CR image. The intrahepatic bile duct and periphery of the common bile duct as well as the bile duct itself are more easily seen. The one-fifth dose CR image is practically the same as the conventional tomographic image.

Case 2: Normal case

Figures 14.4 and 14.5 show the images obtained by simultaneous multisection tomography using the screen/film system and the first to fourth sections of the right CR image, respectively. The margin of the gallbladder, particularly the base, is clearly demonstrated in the first and second sections of CR but is degraded in the fourth section.

Case 3: Multiple gallstones

Figure 14.6 shows the conventional tomographic image, right CR image at the same exposure dose, and the right CR image at one-fifth the exposure dose. In the right CR images of both the same and one-fifth the exposure dose, the contours of the stones are slightly clearer, especially the stones close to the lateral abdominal wall. The same applies to the bile duct.

Case 4: Stones in the gallbladder

Figure 14.7 shows the conventional tomographic image and right CR image at the same exposure dosc. Because the portion near the lateral abdominal wall is obscured due to the low image density of contrast medium in the conventional tomographic image, the presence of the stone cannot be determined. The CR image definitely indicates the contour of the gallbladder and the presence of a stone within it.

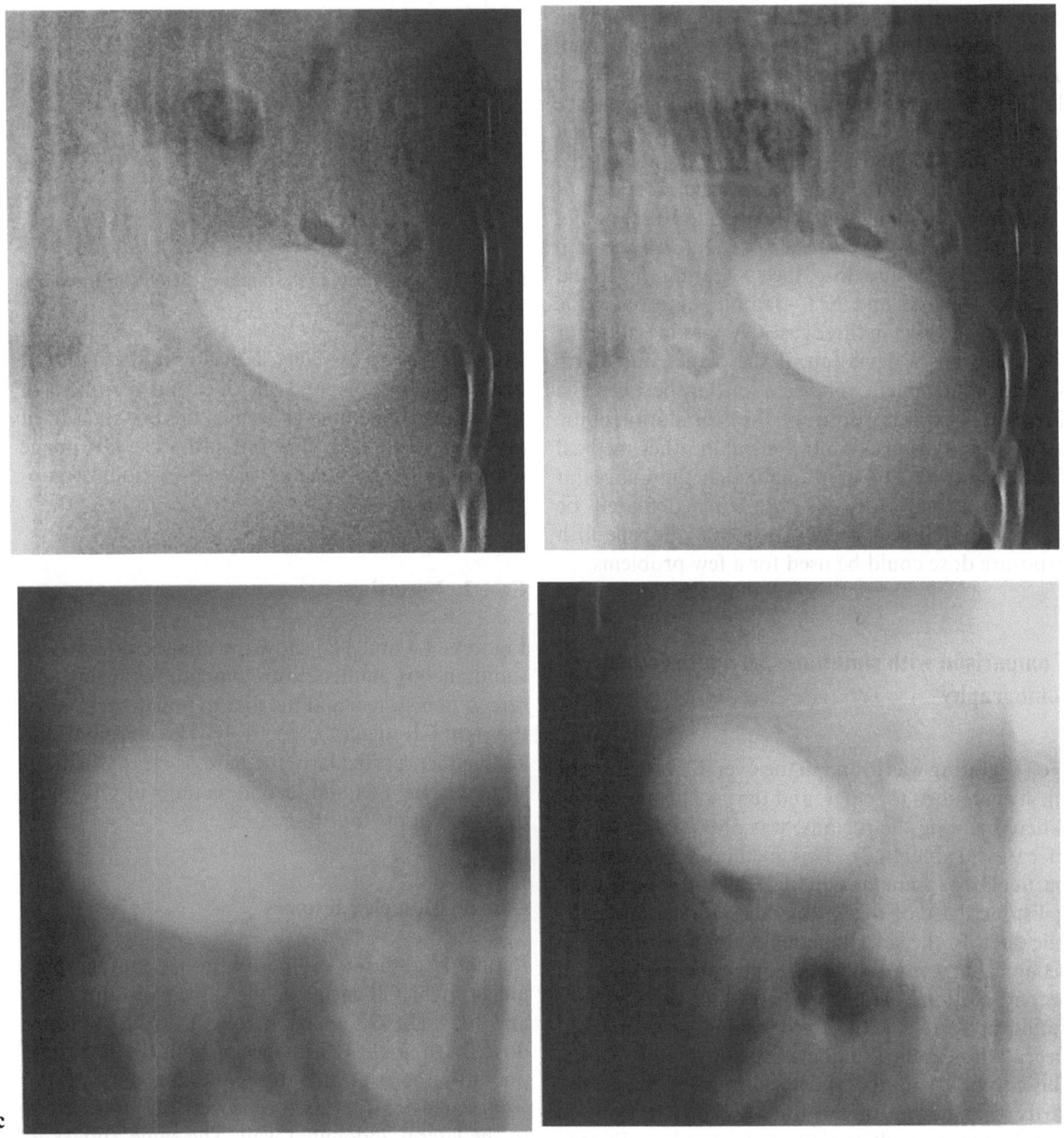

Fig. 14.2a–d. Case 1: Normal gallbladder. **a** Conventional tomogram; **b** left CR image; **c** right CR image; **d** right CR image at one-fifth exposure dose

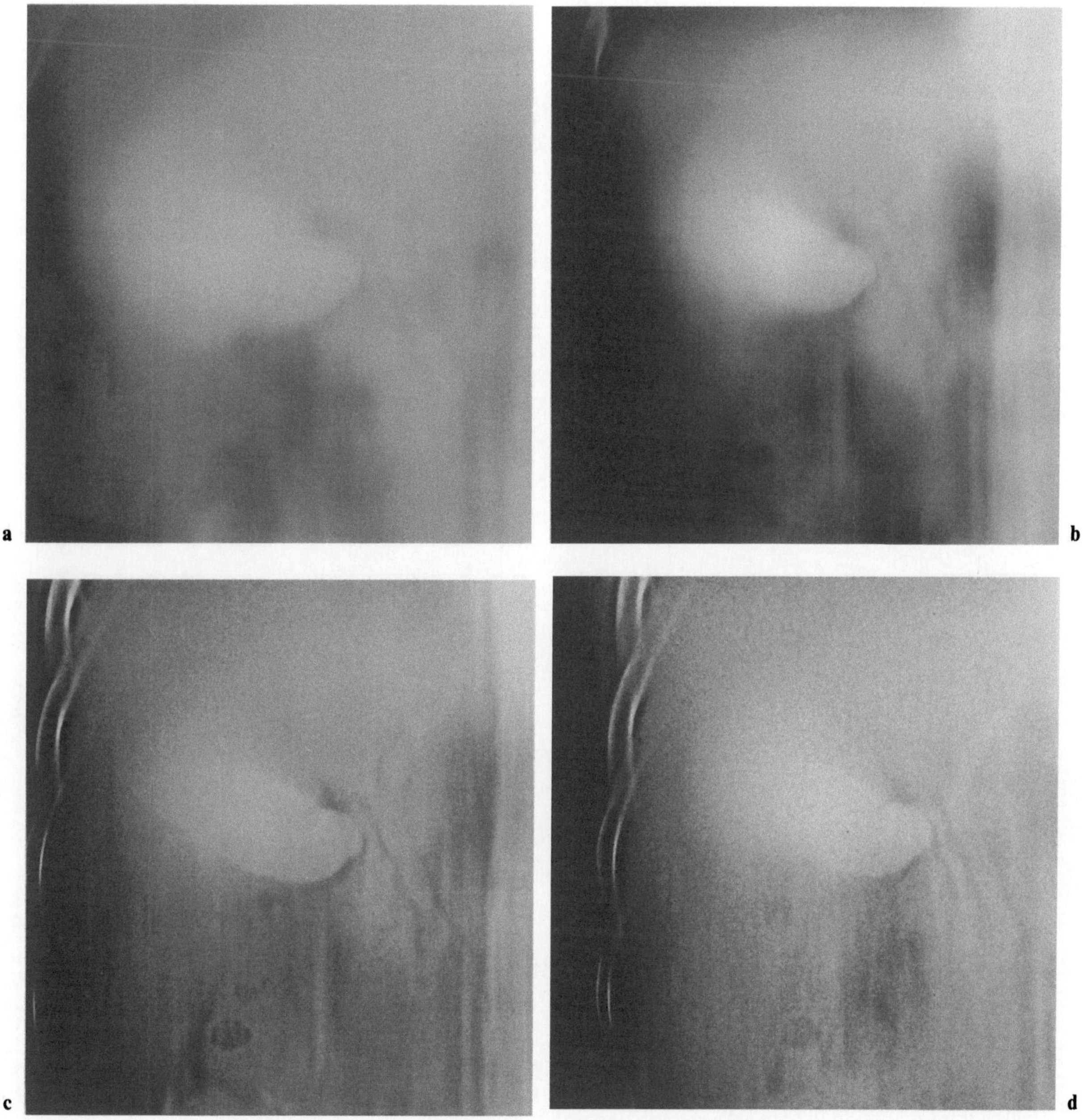

Fig. 14.3a–d. Case 1: Normal bile duct. **a** Conventional tomogram; **b** left CR image; **c** right CR image; **d** right CR image at one-fifth exposure dose

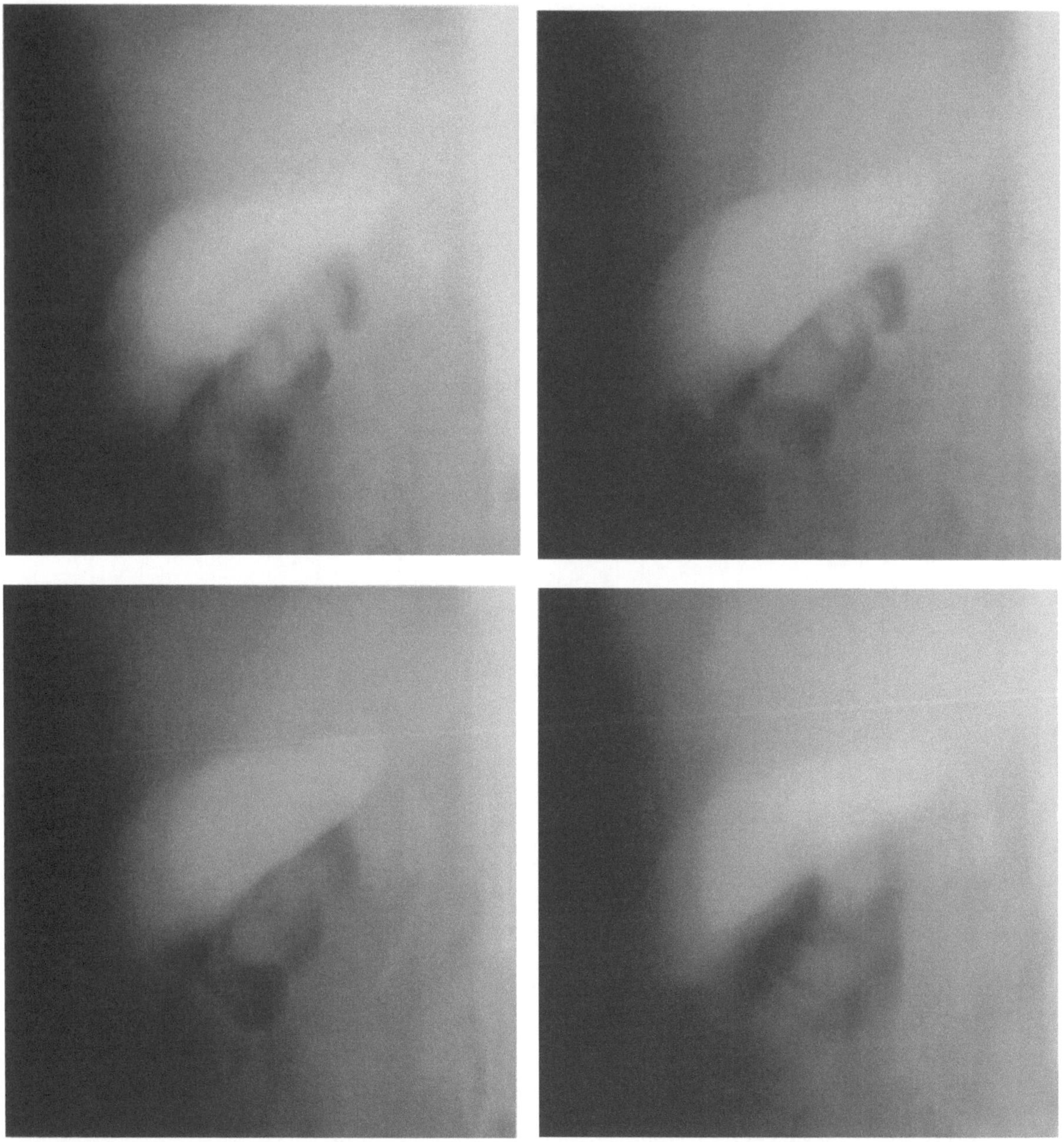

a b c d

Fig. 14.4a–d. Case 2: Simultaneous multisection tomographic image using the screen/film system. **a** First section; **b** second section; **c** third section; **d** fourth section

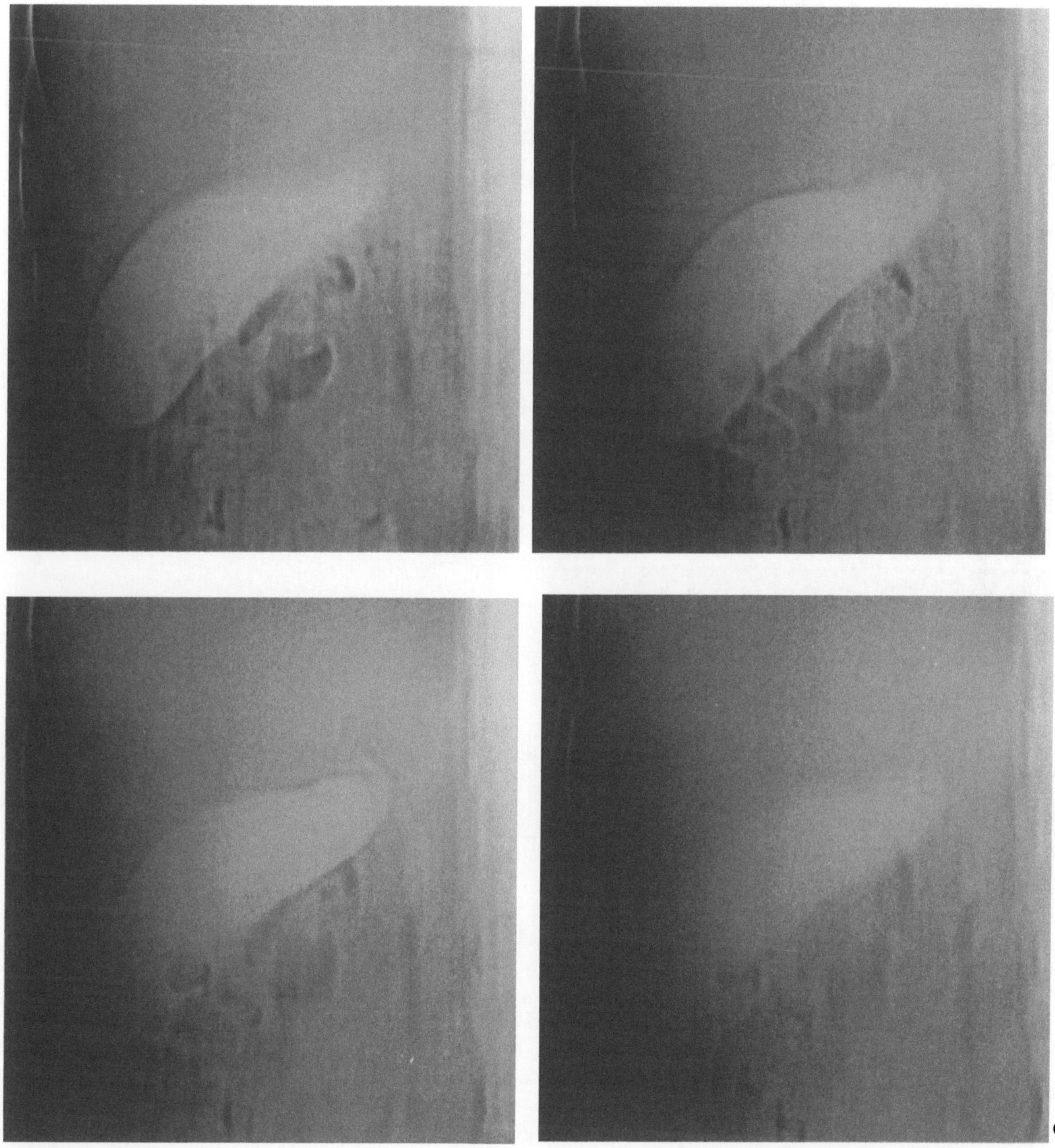

a b c d

Fig. 14.5a–d. Case 2: Simultaneous multisection tomographic image using CR. **a** First section; **b** second section; **c** third section; **d** fourth section a 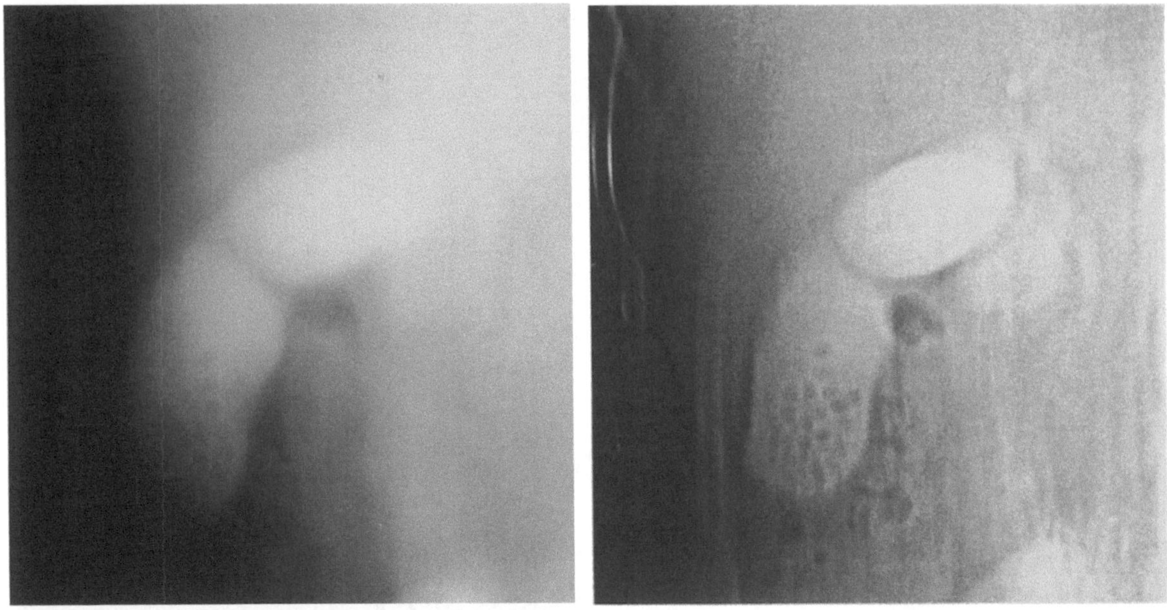 b

c 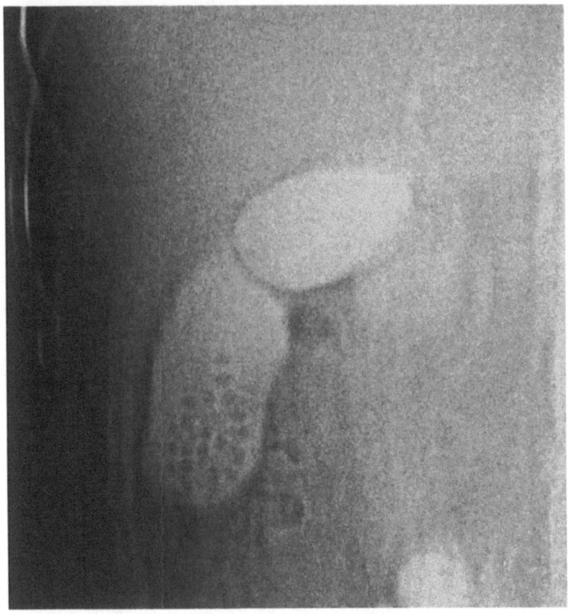

Fig. 14.6a–c. Case 3: Stone in the gallbladder. **a** Conventional tomogram; **b** right CR image at the same exposure dose; **c** right CR image at one-fifth exposure dose

Case 5: Stone in the periphery of the common bile duct

Figure 14.8 shows the conventional tomographic image and the right CR image at the same exposure dose. The stone and margin of the common bile duct are substantially equivalent in terms of visualization, but the intrahepatic bile duct is well visualized in the image.

Case 6: Gallstone

Figures 14.9 and 14.10 show the second to fourth sections of the simultaneous multisection tomographic image obtained by the screen/film system and the right CR image, respectively. The image of the stone itself is virtually equivalent in these two systems. However, the margin of the gallbladder is slightly clearer in the second and third sections of the CR image; the degraded margin of the gallbladder and common bile duct is inferior in the fourth section.

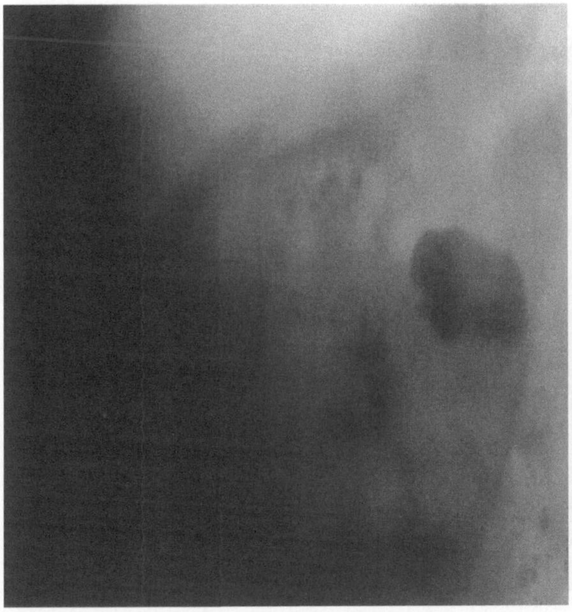
a
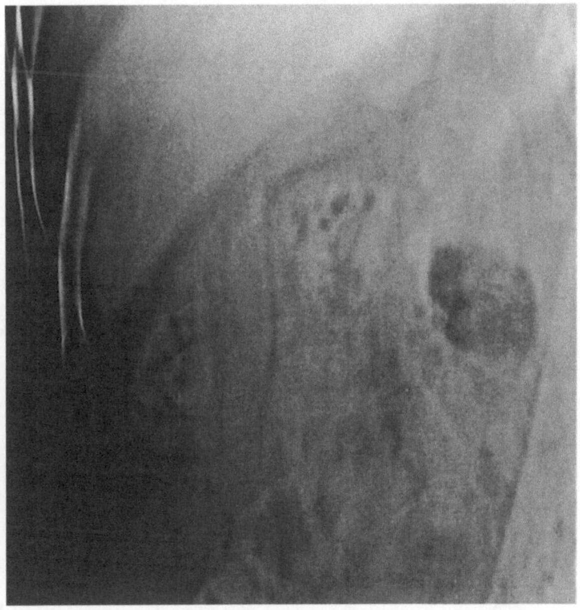
b

Fig. 14.7a, b. Case 4: Stone in the gallbladder. **a** Conventional tomogram; **b** right CR image

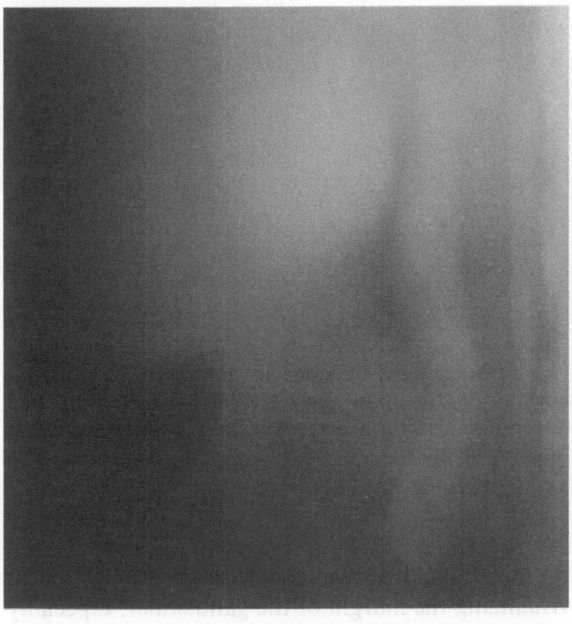
a
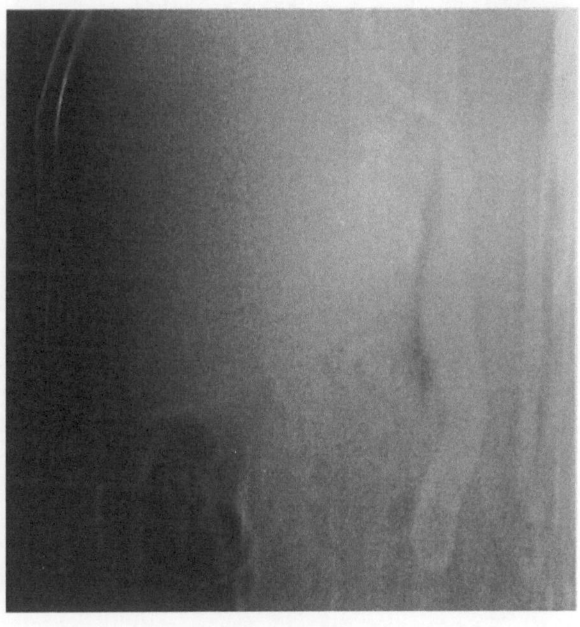
b

Fig. 14.8a, b. Case 5: Common bile duct stone. **a** Conventional tomogram; **b** right CR image

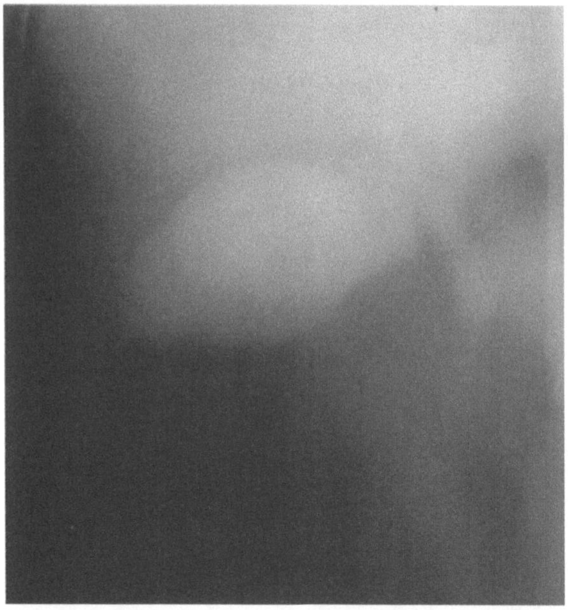

a

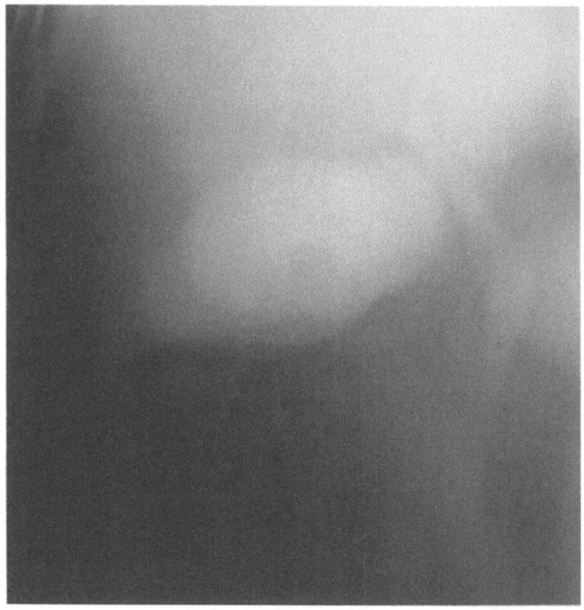

b

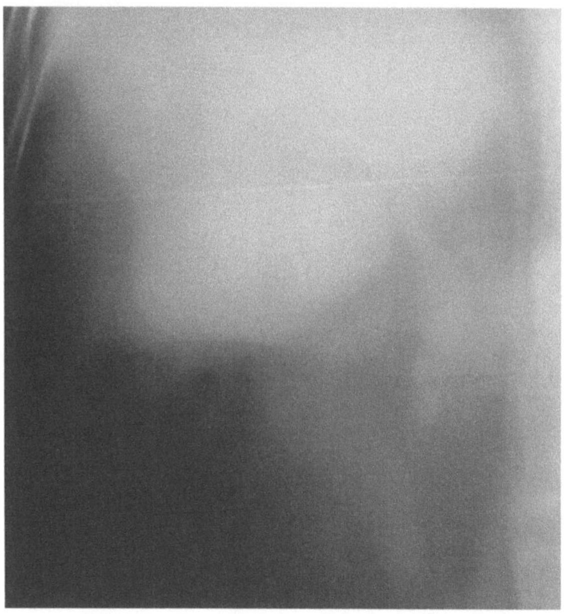

c

Fig. 14.9a–c. Case 6: Stone in the gallbladder. Simultaneous multisection tomographic image using the screen/film system. **a** Second section; **b** third section; **c** fourth section

Conclusion

The CR system can be used for the diagnosis of various anatomical regions [4–7]. Its major clinical merit is that it presents constantly clear images of high quality, and this was confirmed with respect to the area of the biliary tract in the present study. In both methods of tomography, the CR images of the lateral abdominal wall were better than the screen/film system in all cases. This is due to the wide dynamic range and the image processing of the CR system; this advantage was successfully utilized in visualizing the margin of the gallbladder, especially the base, and the intrahepatic bile duct, resulting in image improvement. In comparing the two CR images, improvement was more marked in the right image with more intensive image processing. Gas images and calcifications were also shown clearly. Such characteristics of CR are useful for presenting better image of gallstones, particularly those in the base of the gallbladder, and in cases where the density of the image of the contrast medium is low in the gallbladder and common bile duct.

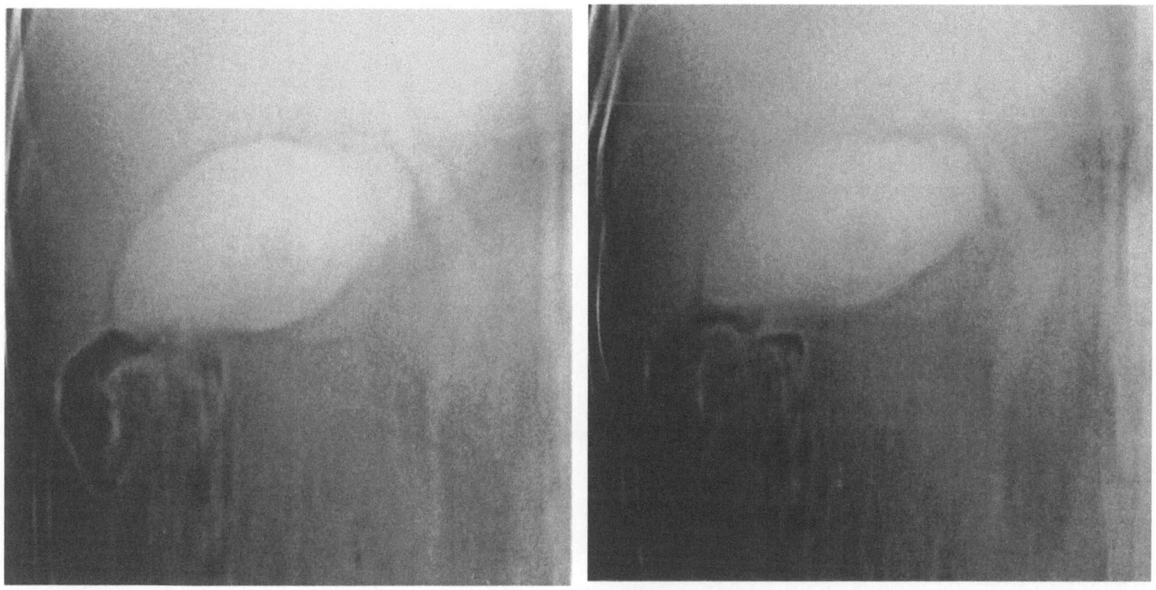

a b

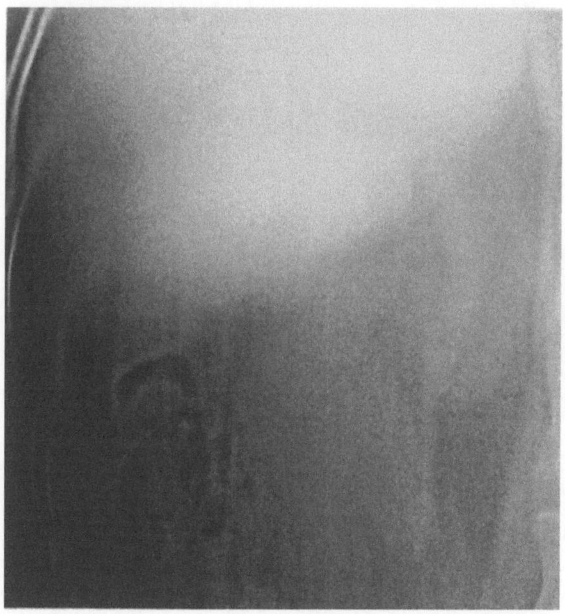

c

Fig. 14.10a–c. Stone in the gallbladder. Simultaneous multisection tomographic image by CR. **a** Second section; **b** third section; **c** fourth section

The images at half the normal exposure dose were practically equivalent to conventional screen/film images as a result of the high sensitivity, and the images at one-fifth the exposure dose were fully usable clinically, despite slight degradation observed in some regions. The present results, which were from a rather small number of cases, are significant in view of the reduction of the exposure dose.

It seems likely that simultaneous multisection tomography by means of CR, which is still partly inferior to the screen/film method, would present even better images if optimal exposure conditions could be established, and this is presently being investigated at our department.

References

1. Hachiya J, Korenaga T, Miyasaka Y, Nitatori T, Wakasa K, Sakurai K (1983) Computed radiography utilizing photostimulable phosphor. Igakunoayumi 127:1199–1206
2. Sakurai K, Hachiya J, Korenaga T, Nitatori T, Miyasaka Y, Furuya Y (1984) Digital radiography of the chest—Evaluation of normal anatomical structure and low dose radiography Nippon Acta Radiologica 44:11–22
3. Korenaga T, Hachiya J, Sakurai K, Nitatori T, Miyasaka Y, Furuya Y (1983) Digital subtraction angiography (DSA) by Fuji computed radiography (FCR) system. Radiolgycal Medicen 2:61–70

Chapter 15

Intravenous digital subtraction angiography

Toshiaki Nitatori[1], Junichi Hachiya, Tateo Korenaga, and Yoshiro Furuya

Introduction

Computed radiography (CR) is a general-purpose diagnostic system developed for application to various kinds of radiography by means of unique imaging plates. We have so far used this system for angiography, taking advantage of one of the features of CR, namely that it allows easy subtracton because it performs analog to digital conversion in the course of processing [1, 2]. Another type of digital angiography is digital fluorography (DF) using an imaging intensifier (II). Since the DF system was originally developed for angiography, there are naturally fundamental differences between the two systems. In routine angiography, however, the two systems have already been widely used, currently for the same purposes in most cases. This chapter includes a comparison of the two systems.

Examination procedure

To perform serial radiography, imaging plates, 10 × 12 in., are inserted into a vinyl pack (Fig. 15.1), which is then placed in a rapid changer that allows a maximum of 12 serial exposures at a rate of three per second (Fig. 15.2). With the use of a specially designed program selector (Fig. 15.3), optional conditions, i.e., X-ray exposure interval and operation of the contrast medium injector, are programmed by means of a keyboard.

Prior to exposure, a 4-F arteriographic catheter (with side holes) is inserted into the superior vena cava by cubital vein puncture under fluoroscopic guidance. If cubital vein puncture is difficult to carry out, femoral vein puncture is used. However, we make it a routine procedure to use cubital vein puncture so as to reduce pain and perform the examination on an outpatient basis. Now that we are fully trained in this technique, cubital vein puncture is possible in almost all adults, and catheter insertion requires about 0.5 min.

In most cases, a contrast medium 40 ml Iopamidol is injected at a rate of 15 ml/s for one examination. The standard exposure techniques are almost the same as those for conventional angiography: 90 kV, 200 mA, and 0.1 s for the head; 80 kV, 300 mA, and 0.1 s for the neck and chest; 80 kV, 200 mA, and 0.1 s for the abdomen; and 70 kV, 200 mA, and 0.1s for the extremities. The low dose exposure, which is one of the characteristic features of CR, is not performed for digital subtraction angiography (DSA).

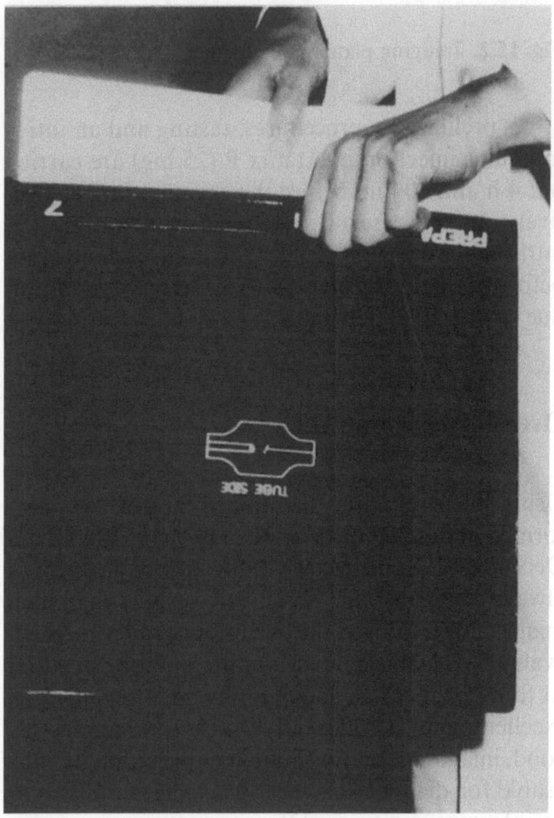

Fig. 15.1. Imaging plates inserted into the vinyl pack for serial radiography

[1] Department of Radiology, Kyorin University, School of Medicine, .20-2, Shinkawa 6-chome, Mitaka, Tokyo, 181 Japan

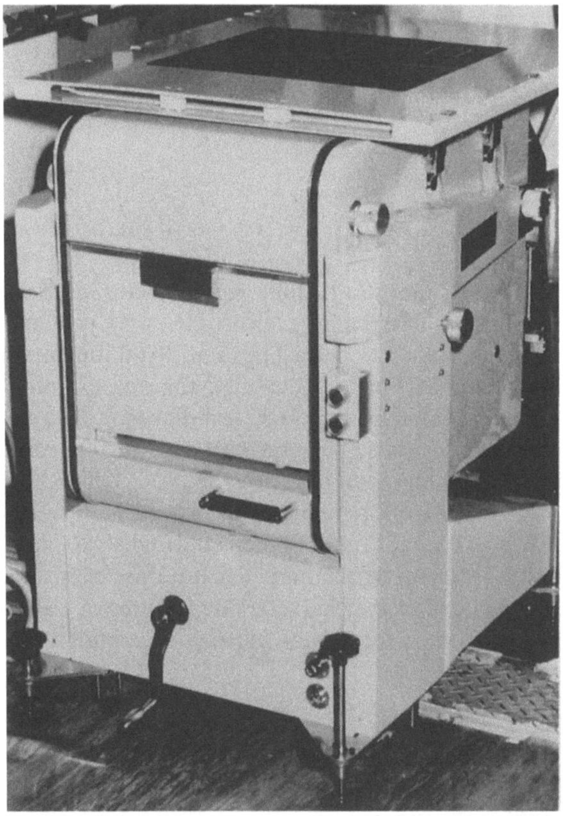

Fig. 15.2. Imaging plate rapid changer

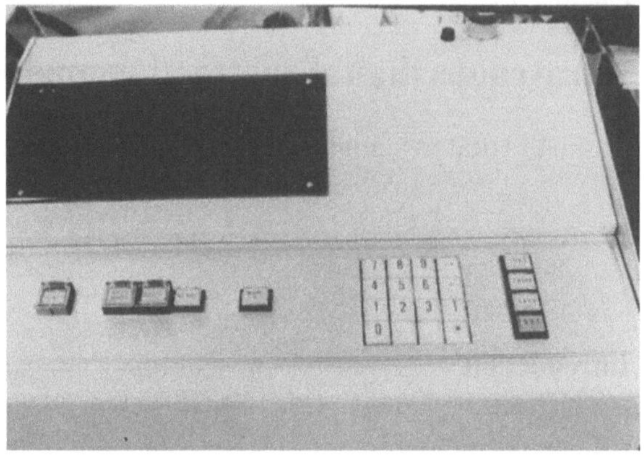

Fig. 15.3. Program selector

As preliminary procedures, fasting and an intramuscular injection of Atarax P (25 mg) are carried out 4 h and 30 min before the examination, respectively. In the case of abdominal radiography, an intravenous injection of glucagon, 1 mg, is done 1 min before radiography in order to avoid misregistration due to intestinal peristalsis.

Overall image evaluation

Using the technique mentioned above, we performed about 300 intravenous DSA (IVDSA) from December 1982 to March 1984. Although there is always difficulty in objectively evaluating such images, we employed the usual four grades for such evaluation. Namely, images approximating or equal to those of invasive angiography were regarded as excellent, images sufficiently usable for diagnosis as good, images presenting limited information but not usable for diagnosis as fair, and images presenting no information as poor [3].

Table 15.1 shows the results of the four-grade image evaluation in relation to the site of examination. Of the 305 cases, 16 were rated as poor, accounting for about 5%. Considering the success rate of the examination, it is appropriate that fair cases are included in the failure category. In this sense, the success rate was about 89% [4, 5].

Causes for failure in the examinations are shown in Table 15.2. Misregistration due to patient body motion was the most frequent, followed by technical failure, including poor system functioning and poor exposure conditions. Technical failure was found only during the early period of our use of IVDSA. The third cause, misprogramming in serial exposures, has also markedly decreased since the circulation time was determined by the dye dilution method (see below). Misregistration due to body motion is the greatest problem at present. We have been trying to decrease failures by performing tape fixation of the examination site, instruction just before the examination, dilution of the contrast medium, etc. Although the number of such cases is decreasing, this factor is still the most probable cause of degradation in image quality. This is particularly serious in old patients.

The success rate for the last 6 months exceeds 95%.

Table 15.1. Image evaluation in relation to the site (● represents five cases and ○ represents one case)

	Abdomen	Pelvis	Neck	Extremities	Thorax	Head	Total
Excellent	○○○○○●	○○○	○○○○○○ ○○●●●	○○○○○○ ●●	○○○●●	○○○○○○ ●	160
Good	○○○○●● ●	○○●●	○○○●	○○○●	○○●●●●	○○○○○● ●●●	110
Fair	●●●		○●	●●	●	○●●	19
Poor	●●●●	●●●	●●	●●●		●●●●	16

Table 15.2. Causes of failed radiography (● represents one case)

	Abdomen	Pelvis	Neck	Extremities	Thorax	Head	Total
Misregistration motion	●●	●	●●●●● ●●●	●●	●	●●●●● ●	20
Gas, swallowing	●●						2
Technical failure	●			●		●●●●●	7
Inadequate timing exposure	●●	●●	●	●			6

Evaluation indications according to site

With regard to image evaluation and indications for IVDSA according to site, representative cases are given below.

Head

The 10 × 12-in. imaging plate can include a full view of the intracranial vessels. Because all blood vessels are outlined on both the right and left sides, it is inevitable that IVDSA will produce difficulties in reading due to overlapping in some areas. However, this examination allows the visualization of vessels as small as the anterior and middle cerebral arteries, so that it is useful in screening for arterial occlusion, arteriovenous malformation (AVM), and aneurysm. In cases of brain tumor, this examination is far inferior to conventional arteriography in terms of diagnostic information. The use of this examination should be decided taking its limitations into account.

The IVDSA images shown are those of suprasellar meningioma in the arterial and venous phases from the frontal and lateral views (Fig. 15.4)—the meningioma showing as a deeply stained tumor (Fig. 15.5)—and right internal carotid artery occulusion found on screening in a patient complaining of a decrease in visual acuity inthe right eye (Fig. 15.6).

Neck

The neck seems to be one of the most suitable sites for IVDSA. In IVDSA, imaging of this site allows a wide view, covering the aortic arch, common carotid artery, internal and external carotid arteries, vertebral artery, and subclavian artery. In order to avoid overlapping of the vertebral and common carotid arteries, we generally perform two different exposures, in bilateral oblique positions, for each examination.

Figures 15.7 and 15.8 represent two cases of multiple arterial changes due to Takayasu's arteritis and Fig. 15.9 shows a case of stenosis of the right internal carotid artery due to occlusive arteriosclerosis.

Chest

With regard to the aorta in the thoracic region, the scope of IVDSA includes the aortic arch and its proximal major branches and the thoracic descending aorta. Abnormalities of the large artery such as thoracic aortic aneurysm, dissecting aortic aneurysm, and aortic coarctation are good indications for this examination. The pulmonary artery can also be visualized to its periphery, depending on the patient holding his breath. The heart is not considered suitable because its dynamics are better assessed in continuous exposures.

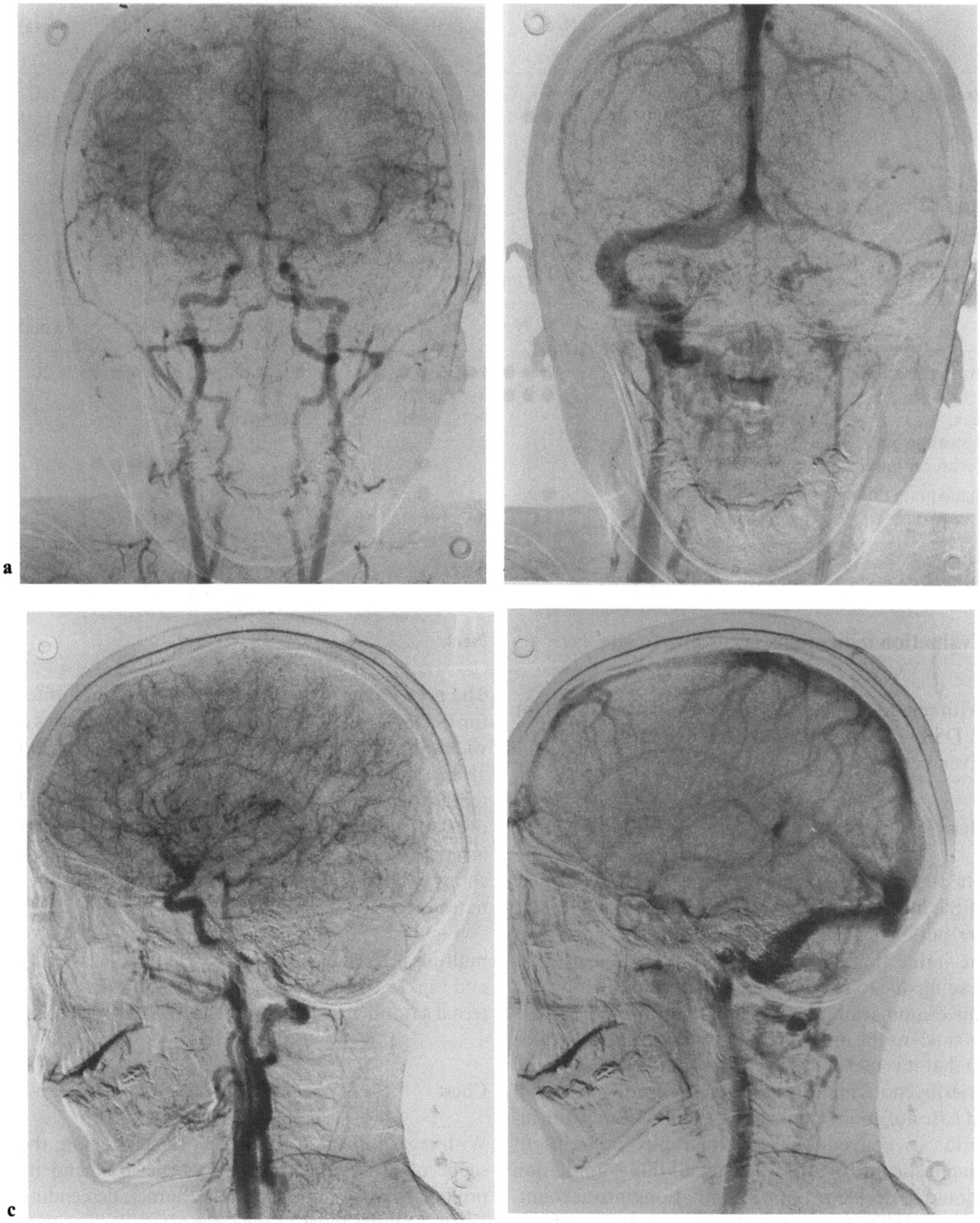

Fig. 15.4a–d. Normal case. **a** PA projection in the arterial phase. **b** PA projection in the venous phase. **c** Lateral projection in the arterial phase. **d** Lateral projection in the venous phase

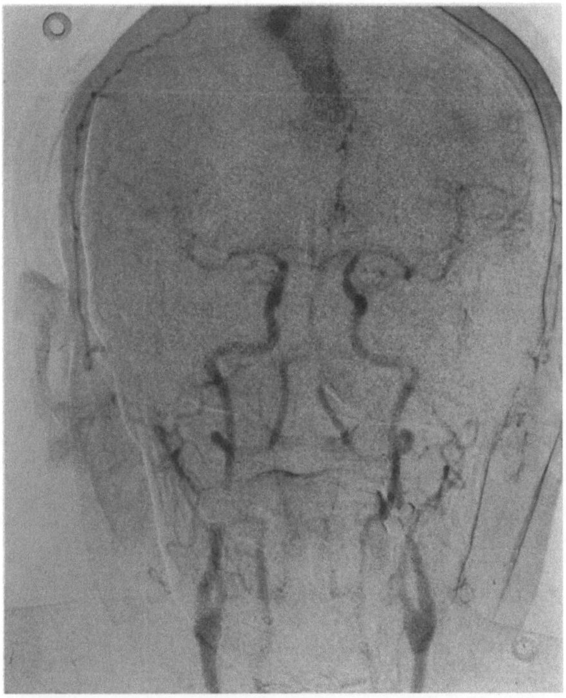

Fig. 15.5. After surgery for meningioma. The figure shows a deeply stained tumor, indicating that it could not be completely excised by surgery

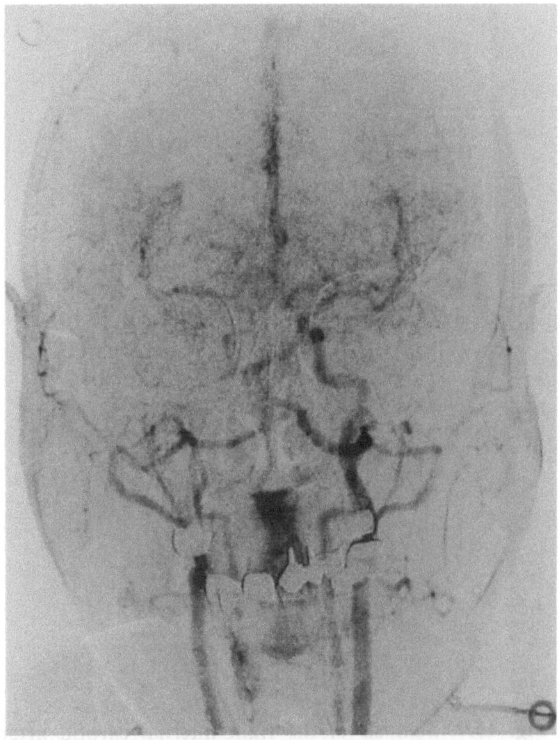

Fig. 15.6. Occlusion of the right internal carotid artery. This case was detected by IVDSA as a screening test in a patient complaining of decreased visual acuity in the right eye

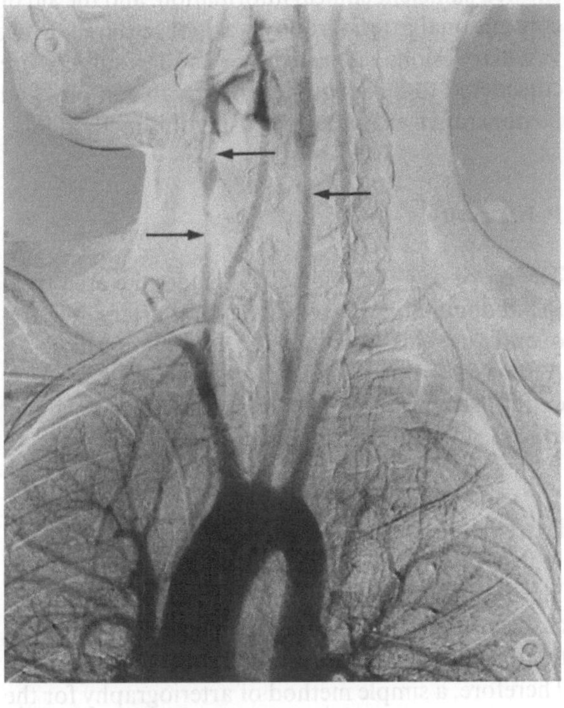

Fig. 15.7. Takayasu's arteritis. Multiple arterial changes in the neck are shown *(arrows)*

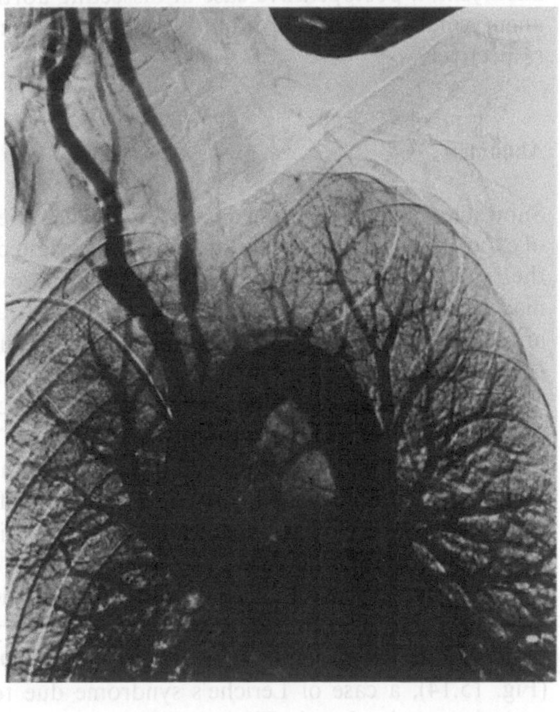

Fig. 15.8. Takayasu's arteritis. Occlusion of the bilateral subclavian arteries is visualized

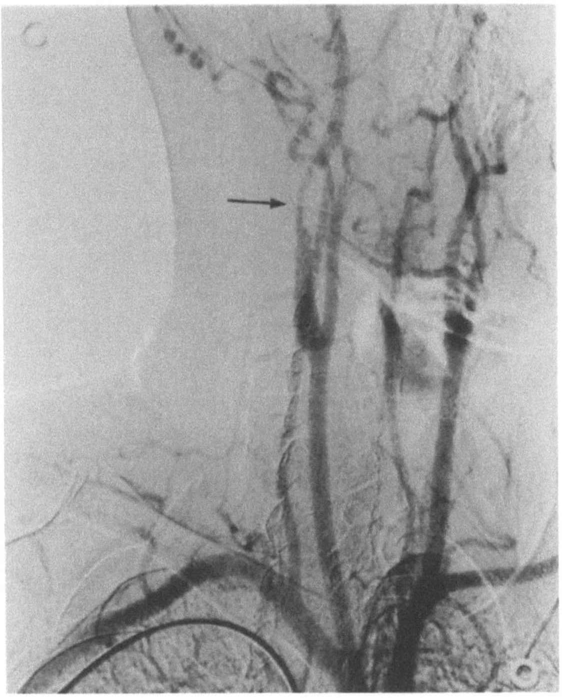

Fig. 15.9. Stenosis of the right internal carotid artery (*arrow*)

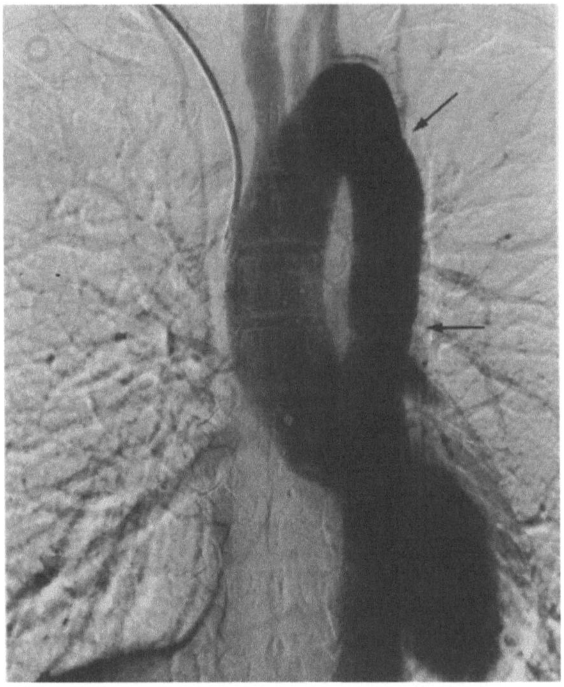

Fig. 15.10. After artificial vessel replacement for aortic aneurysm in the thoracic region. The anastomosed area can be easily observed (*arrow*)

Figures 15.10–15.12 show a postoperative case of artificial vessel replacement for thoracic aortic aneurysm, a postoperative case of dissecting aortic aneurysm, and a case of adult aortic coarctation, respectively.

Abdomen

Subtracted images performed with CR in one series of exposures can visualize the area from the origin of the renal artery to the bifurcation of the common iliac artery from the aorta, and further to the central area of the internal and external iliac arteries. The celiac artery and the superior mesenteric artery can also be well visualized. When registration was successful owing to reduced motion of the patient's body and intestinal gas, intrarenal branching of the renal artery could be seen clearly. The right and left hepatic arteries are often recognizable, although this may be difficult in some cases.

The figures represent a case of abdominal aortic aneurysm (Fig. 15.13), a case of right renal artery stenosis due to fibromuscular dysplasia (FMD) (Fig. 15.14), a case of Leriche's syndrome due to occlusive arteriosclerosis (Fig. 15.15), a postoperative case of artificial vessel replacement for abdominal aortic aneurysm (Fig. 15.16), and a case of adult aortic coarctation (Fig. 15.17). Each IVDSA image showed sufficient clinical information, and the aid of conventional arteriography was not required. Most vascular lesions subjected to vascular surgery are considered to be adequately observable in pre- and postoperative examinations using this method [6].

Pelvic cavity

The scope of IVDSA by CR includes the area from the abdominal aorta to the femoral artery. We frequently use this technique for gynecologic diseases. A case of trophoblastic disease and a case after extra-anatomical femorofemoral bypass are shown in Figs. 15.18 and 15.19, respectively.

Extremities

The incidence of occlusive arteriosclerosis has increased in Japan, and internal medical and surgical treatment of this disease is now frequently performed. Therefore, a simple method of arteriography for the lower extremities is particularly needed. IVDSA by CR with the wide visual field may be well suited to

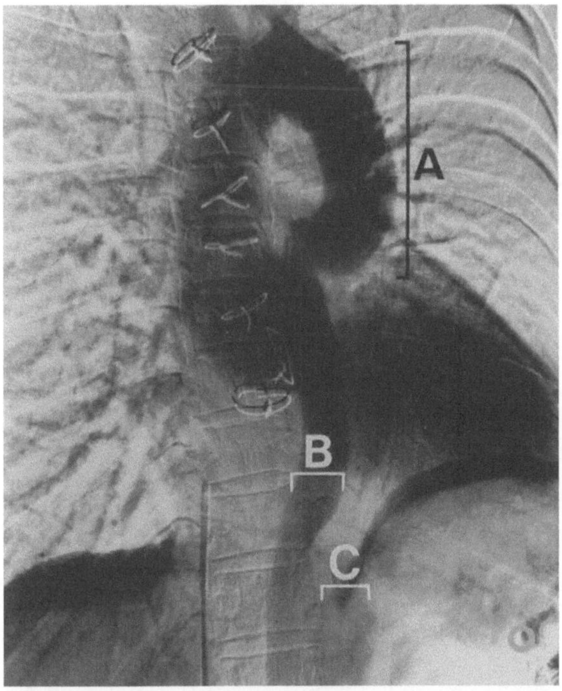

Fig. 15.11. After surgery for dissecting aortic aneurysm. *A* Artificial blood vessel, *B* true lumen, *C* false lumen

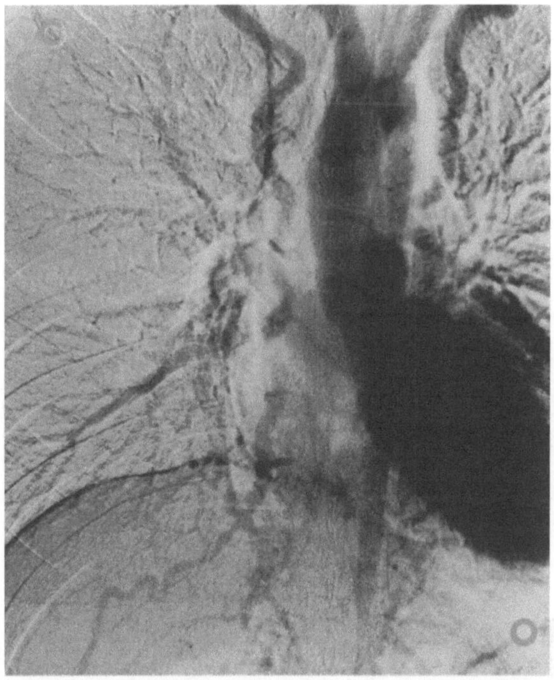

Fig. 15.12. Adult aortic coarctation. The patient was a 17-year-old boy who was admitted to our hospital with hypertension and cerebral hemorrhage. Diagnosis was performed by IVDSA

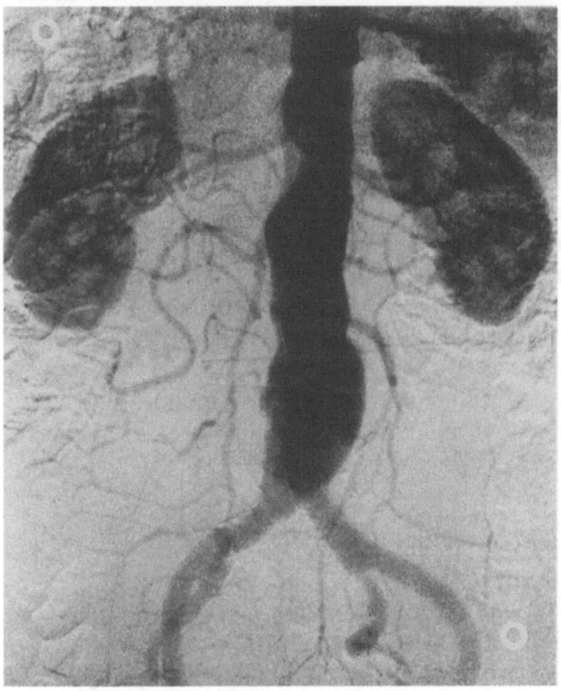

Fig. 15.13. Abdominal aortic aneurysm

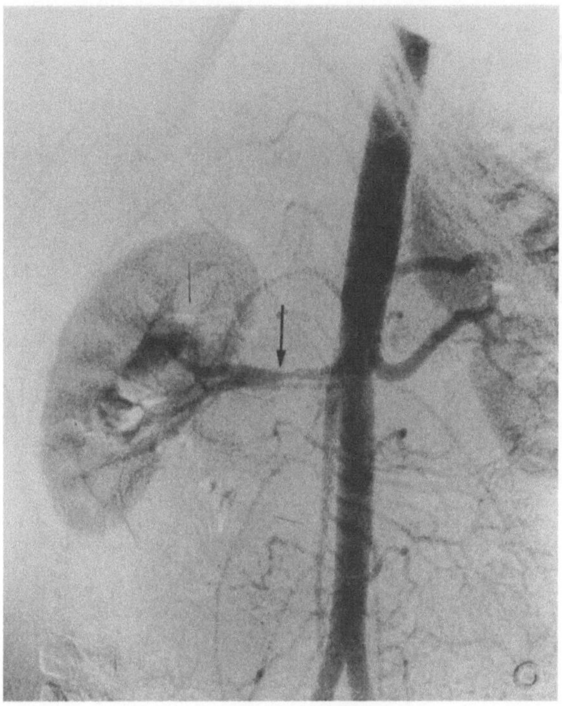

Fig. 15.14. Stenosis of the right renal artery. Narrowing due to fibromuscular dysplasia (*arrow*)

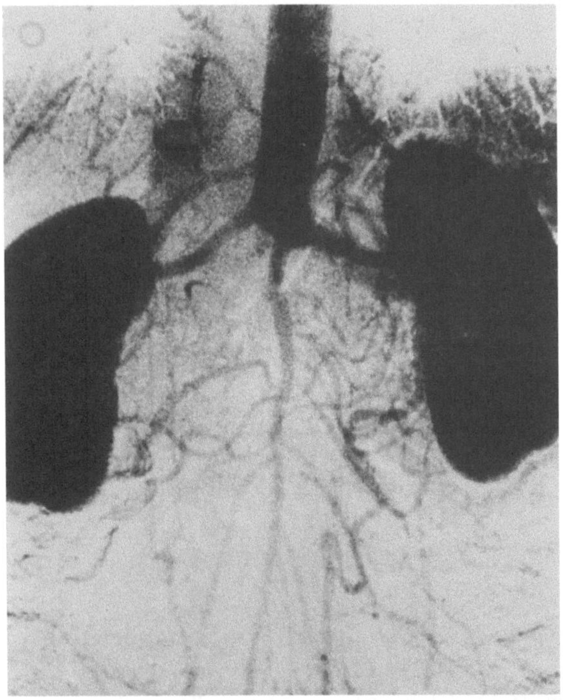

Fig. 15.15. Leriche's syndrome. Complete occlusion of the abdominal aorta in the areas under the origin of the renal artery

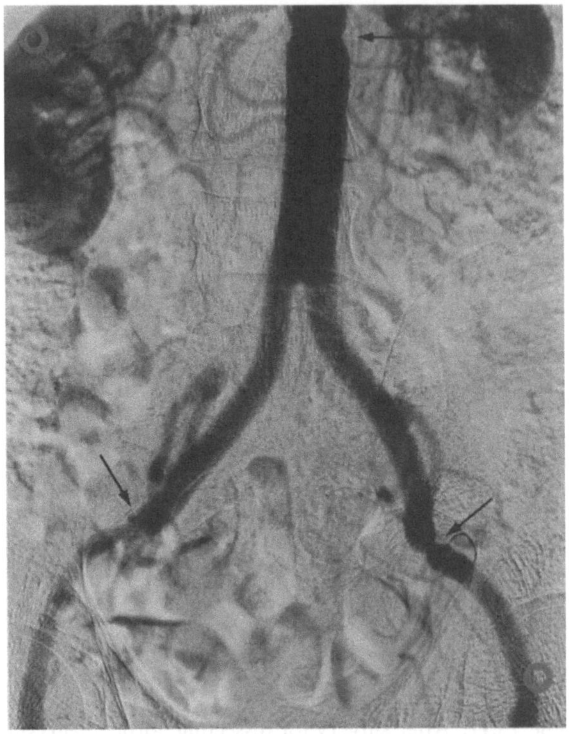

Fig. 15.16. After artificial vessel replacement for abdominal aortic aneurysm. The nature of the anastomosed area can be confirmed (*arrows*)

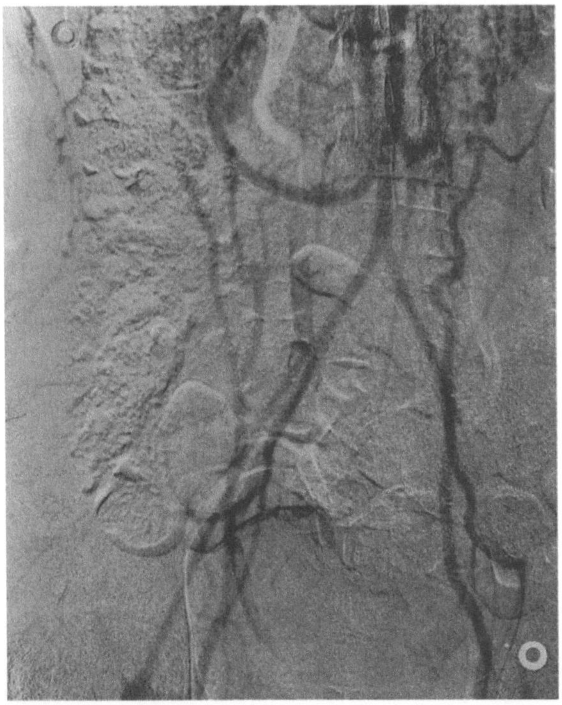

Fig. 15.17. Adult aortic coarctation (in the abdomen). Same patient as in Fig. 15.12

this purpose. We have frequently used this examination for follow-up observations in patients after vascular replacement or in those using thrombolytic agents and vasodilators. It seems that applications are likely to increase.

According to our experience, IVDSA can provide therapeutically sufficient clinical information about arterial changes in areas proximal to the trifurcation in the lower extremities and in areas down to the brachial artery in the upper extremities.

The IVDSA images shown in Figs. 15.20–15.23 are those of a case of occlusive arteriosclerosis, a case of the same disease after artificial vessel replacement, a case of Buerger's disease in the right leg, and a case of embolism in the left brachial artery, respectively.

Characteristics of IVDSA by CR

The greatest advantage of CR as compared with DF is that the decrease in spatial resolution due to the extension of the visual field is relatively slight. The spatial resolution of the DF method is inversely proportional to the size of the diameter of the II. The diameter of the II currently used is normally 5–9 in.

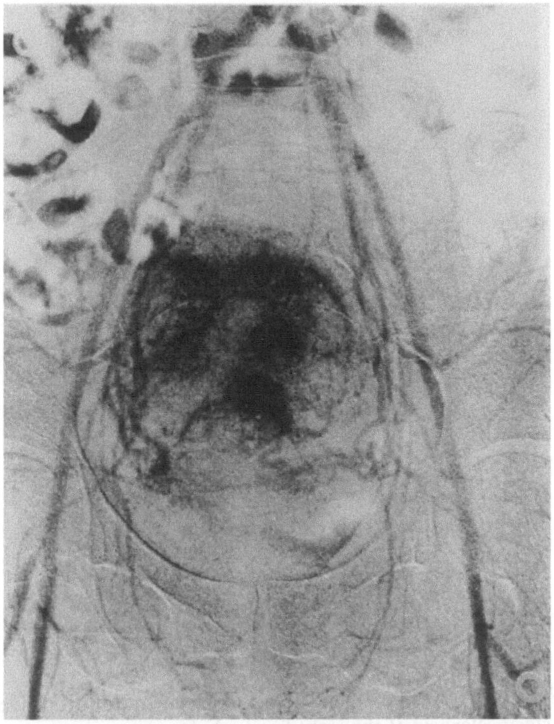

Fig. 15.18. Destructive hydatidiform mole

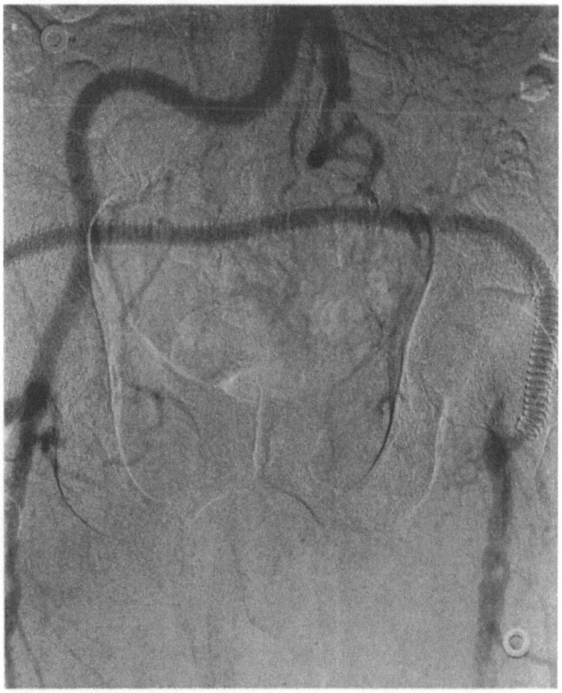

Fig. 15.19. After extra-anatomical femorofemoral bypass

(spatial resolution 1–2 cycles/mm). If the diameter is doubled, the spatial resolution decreases by half, moreover, noise due to scattered rays increases [7]. In contrast, the spatial resolution of CR varies only slightly according to the size of the imaging plate, showing a level of 3.4 lp/mm for 10 × 12-in. plates and 2.5 lp/mm for 14 × 14-in. plates. Whole-body IVDSA by CR, which will be described below, is one of the best uses of IVDSA by CR because it is unlikely to be subject to restriction of the visual field. IVDSA by CR, which allows a wide image field with ordinary 10 × 12-in. plates, as demonstrated by image evaluation according to site, is probably most suitable in the examination of diseases associated with extensive vascular lesions.

The weakest point of this system is that the processing takes a relatively long time. In this system, subtraction is carried out after all 12 exposed plates are processed, so that it takes 30 min from the beginning of processing to the end of subtraction for a set of 12 plates. This is critical when compared with the real-time processing in DF.

There is another restriction from the aspect of serial radiography, i.e., insufficiency in the dynamic analysis of angiography. This system is thus unsuitable for cardiac examination. Furthermore, because of the limited number of radiograms to a maximum of 12 sheets, optimal radiography is sometimes impossible within the IVDSA program.

Further technical considerations

We have developed a radiographic technique in order to take advantage of IVDSA by CR and to compensate for its deficiencies and this will be described briefly.

Whole-body IVDSA

This is a method by which a single injection of contrast medium and six exposures allows angiography of almost the entire body. A set of four 10 × 12-in. imaging plates is attached to each of the six faces of a hexagonal prism of an Angiorapid II (Siemens Fig. 15.24). As the hexagonal prism, to which a total of 24 plates are attached, is rotated, six exposures are produced from an X-ray tube set at a high position, providing angiograms of the whole body synthesized by optimal radiography (Figs. 15.25, 15.26). By this method, a full view can be obtained with a single injection of contrast medium for diseases involving extensive vascular lesions, such as occlusive arterio-

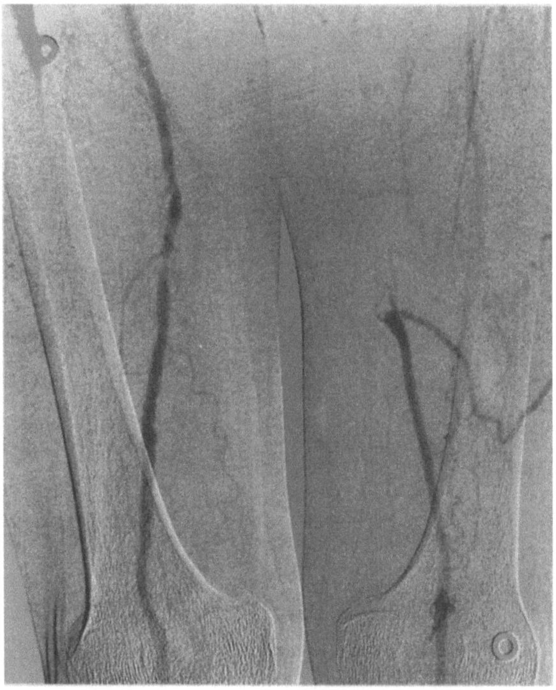

Fig. 15.20. Occlusive arteriosclerosis of the bilateral femoral arteries

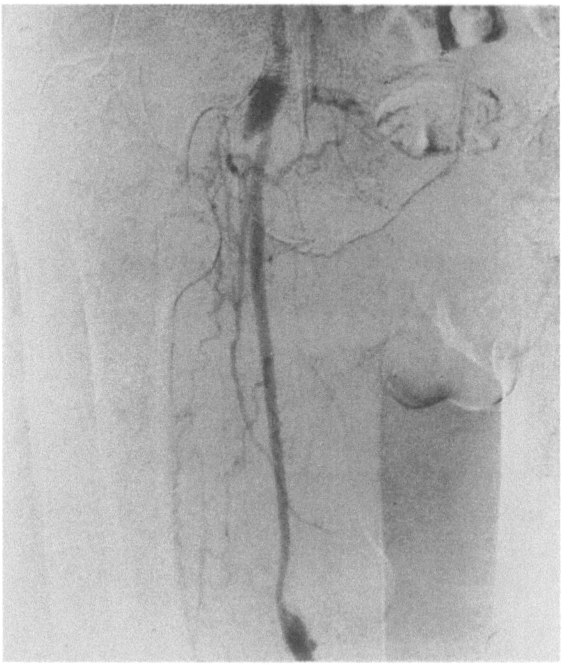

Fig. 15.21. After artificial vessel replacement for occlusive arteriosclerosis of the right femoral artery

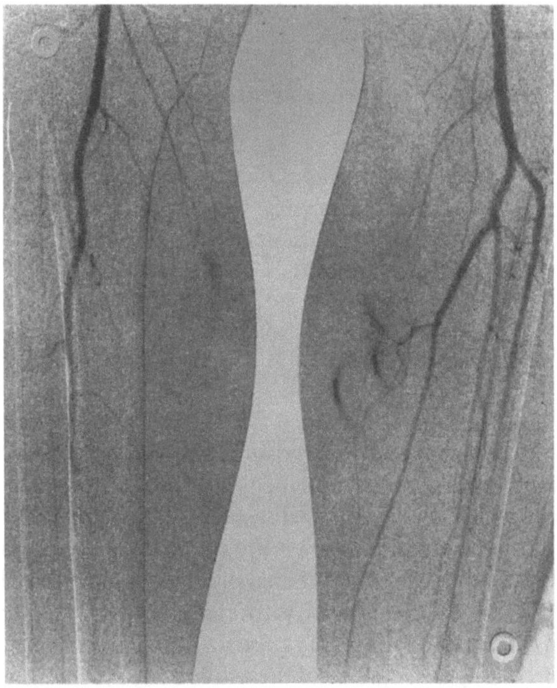

Fig. 15.22. Buerger's disease in the right crus. The right posterior tibial artery cannot be visualized

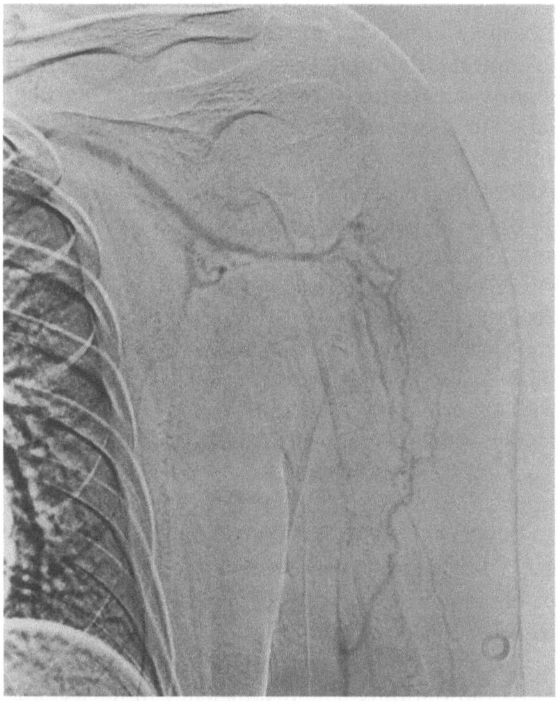

Fig. 15.23. Embolism of the left brachial artery. The patient had mitral stenosis with atrial fibrillation

sclerosis and Takayasu's arteritis. This method is particularly advantageous in older patients and those with renal hypofunction. This technique of IVDSA takes full advantage of CR, which is not subject to restriction of the visual field [8].

Elimination of misprogramming

It is sometimes difficult to perform optimal radiography of the site of interest at the best timing by means of 12 imaging plates. This is because there are individual differences in the circulation time from the central vein to the arterial system. According to our experience, the time required for the circulation of contrast medium from the superior vena cava to the renal artery was 7 s at the minimum and 23 s at the maximum. To predict such individual differences prior to the angiography, determination of the circulation time by the dye dilution method was carried out previously in clinical cases. With this procedure, it has been possible to predict at a high degree of probability the optimal exposure timing at various sites of the trunk, excluding the extremities [9].

This procedure uses Cardiac Output Computer MLC4200 manufactured by Nihon kohden Corp. and indocyanine green (ICG) at 5 mg (1 ml). ICG is infused from a catheter introduced into the superior vena cava, and the dye dilution curve is obtained by the earpiece method. The time of appearance (ta) and the time of peak concentration (tp) are obtained from the dye dilution curve, and then optimal exposure timings are programmed on the basis of the circulation time presumed for each site (Fig. 15.27). The time of approximately 1 min required for this technique has markedly decreased the incidence of misprogramming.

Table 15.3. Presumed time of optimal angiography and four-plate radiography program

Cephalic arterial phase	(1) Masked image (2) tp − 4 s (3) tp − 2 s (4) tp
Cervical artery and thoracic aortic artery	(1) Masked image (2) ta (3) ta + 2 s (4) tp + 4 s
Renal artery and upper abdominal aortic artery	(1) Masked image (2) tp − 4 s (3) tp − 2 s (4) tp
Common iliac artery, pelvic artery, and lower abdominal artery	(1) Masked image (2) tp − 2 s (3) tp s (4) tp + 2 s

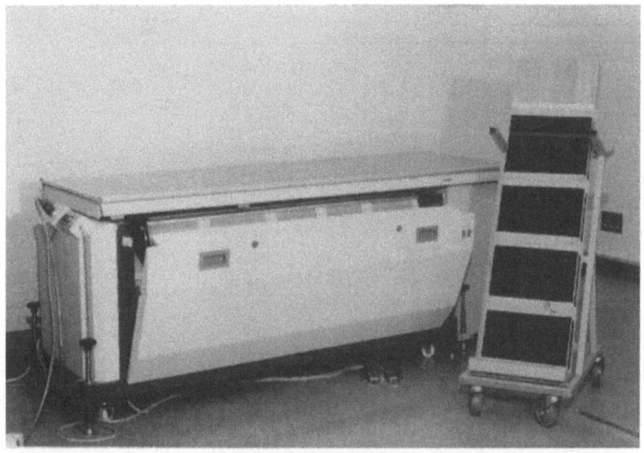

Fig. 15.24. Angiorapid II (Siemens) used for whole-body IVDSA. Four imaging plates are arranged in line and attached to a surface

Reduction of exposures in serial radiography

Since optimal exposure timings for each site could be predicted at a high probability, we tried to decrease the number of exposures in serial radiography. At present, we perform every IVDSA with four to six exposures. Table 15.3 shows the program for four exposures in serial radiography. Only two exposures for a mask image and a live image provide enough clinical information for diagnosis in the neck and chest, where the difference between true and presumed times is small.

This method, by which the number of exposures for angiography is decreased, has not only reduced the exposure dose for the patient but has also solved the problem of processing time to a large extent.

Conclusion

IVDSA using the CR system has been described. This method has some good features arising from its unique use of imaging plates. This method should be further applied, making good use of its advantages, while bearing in mind its indications and limits. As one of the several types of IVDSA, it may be used more widely as a simple angiographic method as it involves less invasion.

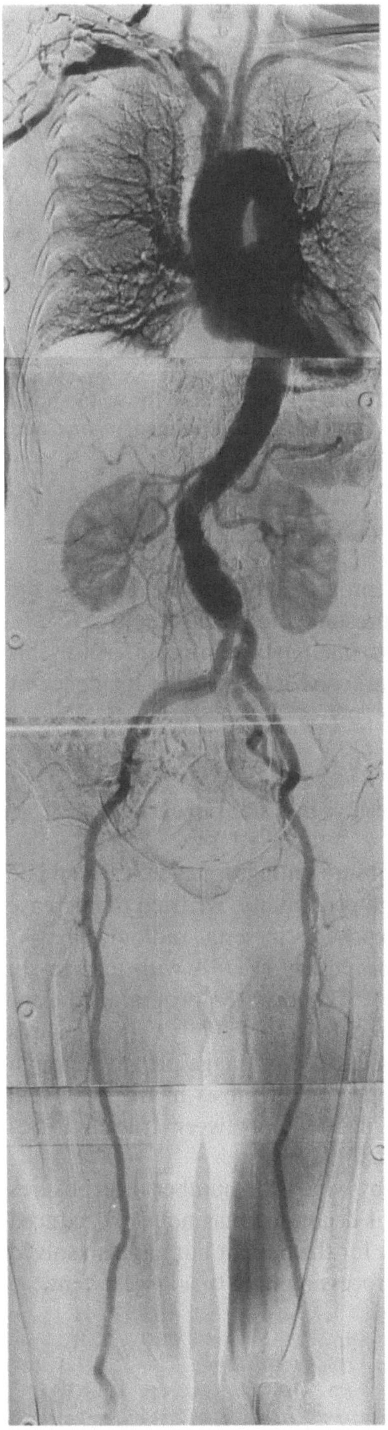

Fig. 15.25. Whole-body IVDSA. Abdominal aortic aneurysm

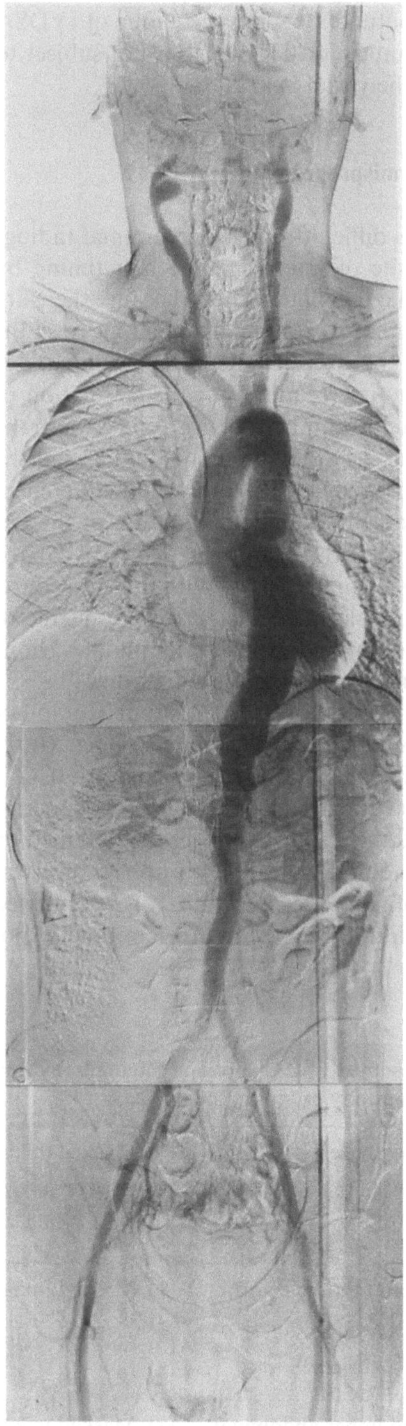

Fig. 15.26. Whole-body IVDSA. Takayasu's arteritis. Extensive arterial lesions from aneurysm of the right common artery to the distal areas of the renal artery can be seen

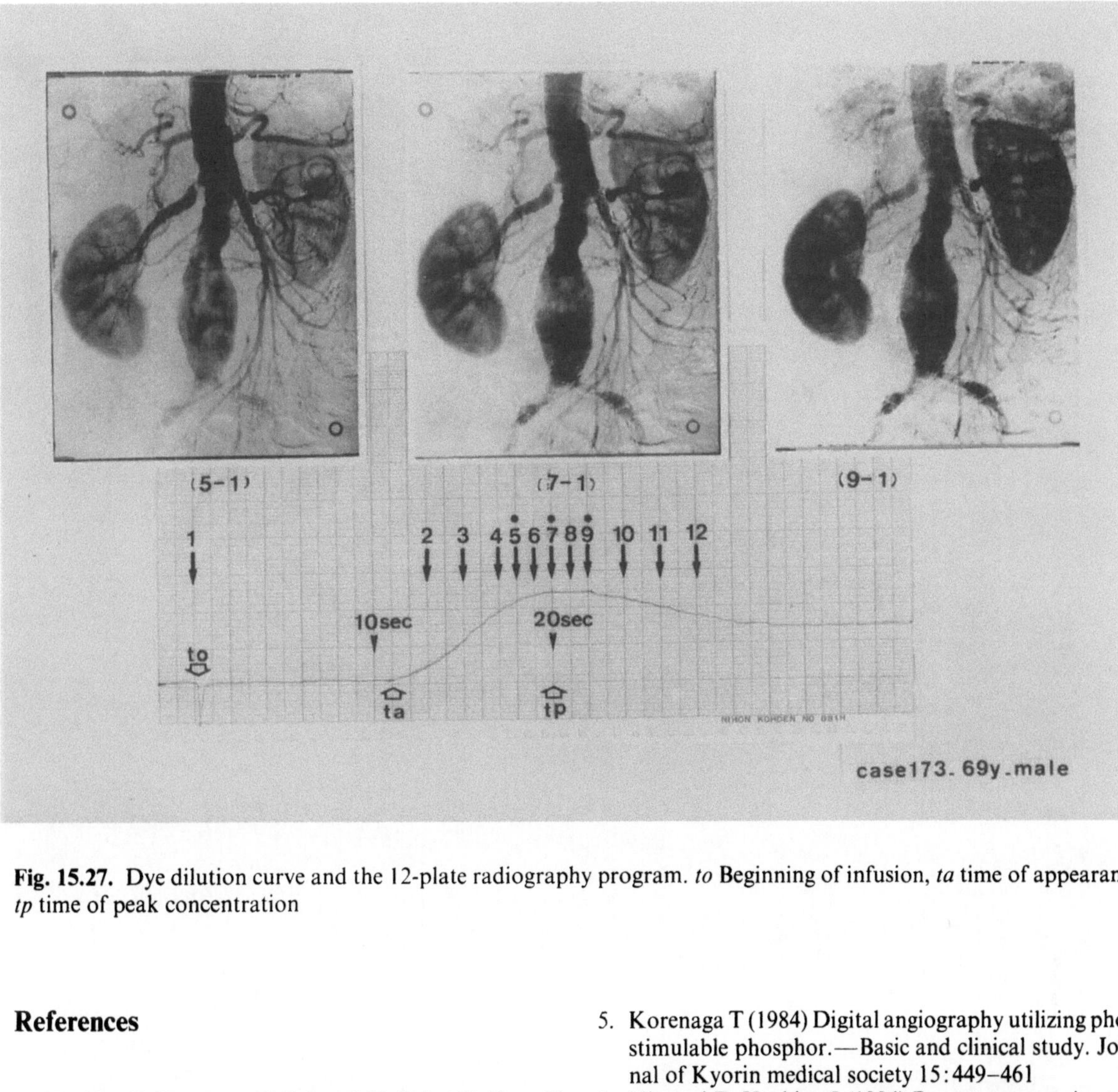

Fig. 15.27. Dye dilution curve and the 12-plate radiography program. *to* Beginning of infusion, *ta* time of appearance, *tp* time of peak concentration

References

1. Hachiya J, Korenaga T, Sakurai K, Sakai F, Kato K, Takano M (1982) New method of digital angiography. Japanese Jaurnal of Clinical Radiology 27:705–710
2. Hachiya J, Korenaga T, Miyasaka Y, Nitatori T, Wakasa K, Sakurai K (1983) Computed radiography utilizing photostimulable phosphor. Igakunoayumi 127:1199–1206
3. Nishimura G, Naitou H, Yamaguchi T, Takamiya M, Kozuka T, Ohno H, Yasuhara H (1982) Clinical aspects of digital video subtraction angiography (DSA) Journal of Medical Imagings 2:182
4. Korenaga T, Hachiya J, Sakurai K, Nitatori T, Miyasaka Y, Furuya Y (1983) Digital subtraction angiography (DSA) by Fuji computed radiography (FCR) system. Radiolgycal Medicen 2:61–70
5. Korenaga T (1984) Digital angiography utilizing photo stimulable phosphor.—Basic and clinical study. Journal of Kyorin medical society 15:449–461
6. Nitatori T, Hachiya J (1984) Recent progress in prosthetic vascular surgery—Comprehensive review for diagnostic radiologist. Journal of Medical Imagings 4: 66–74
7. Smith CW, Winfield AC, Price RR, Harding DR, Tucker SW, Witt WS, Hollifield JW (1982) Evaluation of digital venous angiography for the diagnosis of renovascular hypertension. Radiology 144:51
8. Nitatori T, Hachiya J, Korenaga T, Furuya Y (1985) Whole body intravenous digital subtraction angiography. Radiology 156:829–830
9. Nitatori T, Hachiya J, Korenaga T, Miyasaka Y, Wakasa K, Furuya Y, Sakurai K (1984) Dye dilution method for IVDSA. Journal of Medical Imagings 4: 438–445

Chapter 16

Intra-arterial digital subtraction angiography

Toshihide Sonoda,[1] Kouji Ikeda, Hisashi Kobayashi, and Shinji Shinohara

Computed radiography (CR) produces an image with frequency enhancement and a digital subtraction angiogram (DSA) as well as an image similar to that obtained with the conventional screen/film system. CR is thus recognized as a new system which can deal with many angiograms. This chapter presents clinical cases of intra-arterial DSA using CR and the advantages and problems of this system are discussed.

Introduction

Intravenous digital subtraction angiography (IVDSA) has been evaluated as an alternative noninvasive method of conventional arteriography [1–3], but it also has its limitations such that at present conventional arteriography has to be applied in many cases [4]. In the field of arteriography, however, a new procedure, intra-arterial digital subtraction angiography (IADSA), has recently been developed centering on digital fluorography (DF) [5–6]. We have carried out IADSA by CR, and the clinical results are discussed here.

Methods

In IADSA a procedure similar to conventional arteriography was employed, and a contrast medium diluted by a factor of two to four was used in most of the patients. For the exposure conditions, a phantom study with test charts revealed no degradation of image quality when the dose used was one-fourth of that in the conventional screen/film method (screen KS, film AO). Therefore, the low dose technique was applied to some of the examinations. By use of an imaging plate (IP) rapid changer, 12 serial exposures (three exposures per second) are available, and exposure intervals are programmable for each examination. All angiograms were displayed as two images on one film. Left images were usually treated by tonal conversion alone as live images before subtraction and resembled conventional angiograms. Right images were displayed as the processed images in which spatial frequency components near 0.5 cycles/mm were enhanced. Subsequently, digital subtraction was carried out using these live images as a reference.

Subjects and clinical evaluation

IADSA by CR was performed in 53 patients during the period between September 1983 and March 1984. A contrast medium at the usual concentration was used in 22 of these patients, and a contrast medium diluted by a factor of two to four was used in the remaining 31. According to the criteria of Nishimura et al. [2], these images were evaluated by being divided into the following four stages: images where the information was equal to or greater than that obtained by conventional angiography were classified as excellent; those whose image quality was poor despite the fact that the information obtained was sufficient for diagnosis were classified as good; those from which no diagnosis could be made but which yielded limited information were classified as fair; and those from which no information could be obtained were classified as poor.

As shown in Table 16.1, 46 of 53 examinations (87%) showed diagnostically effective images, evaluated as excellent or good. In seven examinations for the chest and abdomen, on the other hand, no effective DSAs could be obtained because of misregistration artifacts due to respiratory movement or intestinal gas. When these images were evaluated according to the concentration of contrast medium (Table 16.2), there was no difference in the image quality between the two groups of the usual concentration and the diluted contrast medium. Typical cases of IADSA are given below.

Case 1: Hepatoma

The DSA (Fig. 16.1) obtained by injection of 30 ml 76% Urografin, diluted twofold, into the proper

[1] Department of Radiology, Faculty of Medicine, Kagoshima University, 1208-1 Usuki-cho, Kagoshima, 890 Japan

Table 16.1. Evaluation of images

Site	Evaluation	Excellent	Good	Fair	Poor
Chest	9	5 (56%)	1 (11%)	3 (33%)	
Abdomen	38	29 (77%)	5 (13%)	2 (5%)	2 (5%)
Extremities	5	5 (100%)			
Head	1	1 (100%)			
Total	53	46 (87%)		7 (13%)	

Table 16.2. Evaluation of images according to concentration of contrast medium

Contrast medium	Evaluation	Excellent	Good	Fair	Poor
Usual concentration	22	17 (77%)	2 (9%)	2 (9%)	1 (5%)
Diluted concentration	31	23 (74%)	4 (13%)	3 (10%)	1 (3%)

hepatic artery revealed tumors of the right lobe, together with a small daughter lesion with a diameter of 1 cm in the left lobe, overlapped on the vertebral body. This daughter lesion in the left lobe had not been identified on the conventional angiogram.

Case 2: Carcinoma of the colon metastatic to the liver

The DSA (Fig. 16.2) obtained by injection of 40 ml 76% Urografin, diluted twofold, into the proper hepatic artery revealed a tumor of the left lobe overlapped on the vertebral body, together with multiple metastatic tumors in the hepatic right lobe.

Case 3: Regenerative nodule of the liver

The DSA (Fig. 16.3) obtained by injection of 45 ml 76% Urografin, diluted twofold, into the left hepatic artery revealed a favorable image of deeply stained parenchyma of the hepatic left lobe, together with a hypovascular lesion overlapped on the vertebral body. Identification of the latter had been difficult on the conventional angiogram.

Case 4: Lung cancer

The DSA (Fig. 16.4) obtained by injection of 8 ml 65% Angiografin into the right bronchial artery revealed marked staining of the tumor, facilitating easy determination of the range of tumor invasion toward the mediastinal side overlapped on the vertebral body.

Case 5: Uterine cancer

The live image by CR (Fig. 16.5a) obtained by injection of 28 ml 65% Angiografin, diluted fourfold, into the right internal iliac artery yielded no view of the lesion overlapped on the bladder because of residual contrast medium in the bladder despite the fact that a urethral catheter had been inserted into this organ. On the DSA (Fig. 16.5b), tumor vessels in the same region can be detected.

Case 6: Carcinoma of the kidney metastatic to the liver

A favorable image of the portal vein is observed on the DSA by arterial portography (Fig. 16.6) obtained by injecting 24 ml contrast medium, i.e., half the usual amount.

Case 7: Rhabdomyosarcoma of the femoral region

The DSA (Fig. 16.7) obtained by injection of 20 ml 65% Angiografin, diluted fourfold, into the femoral artery yielded an image of good quality, which was useful for determining the range of the tumor. The use of diluted contrast medium reduced pain on injection.

Case 8: Pleural mesothelioma

The DSA (Fig. 16.8a) of the right internal mammary artery obtained from a patient with pleural meso-

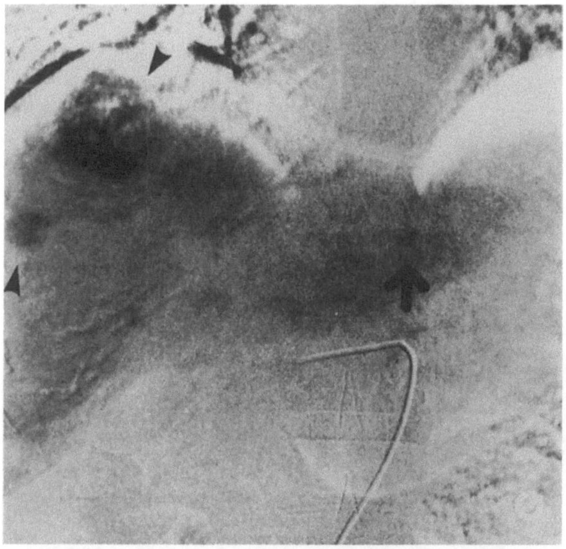

Fig. 16.1. Hepatoma. A small daughter lesion (*arrow*) in the left lobe overlapped on the vertebral body is shown, together with lesions (*arrowheads*) in the right lobe

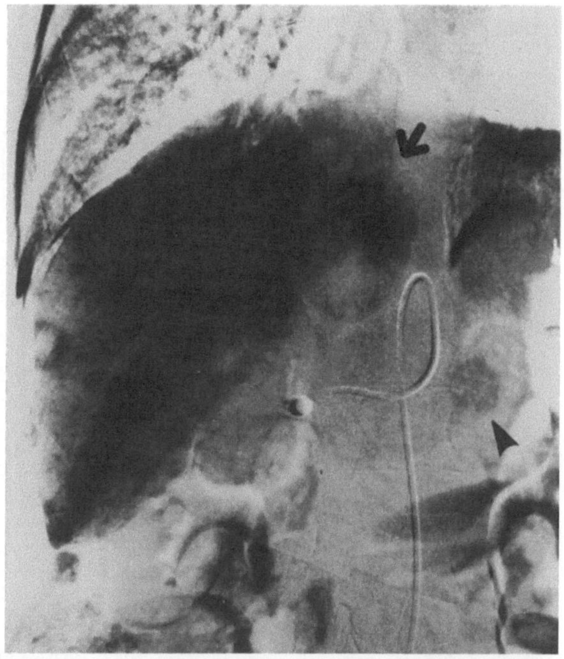

Fig. 16.2. Carcinoma of the colon metastatic to the liver. A tumor (*arrow*, *arrowhead*) in the left lobe overlapped on the vertebral body is favorably demonstrated, together with multiple metastatic tumors in the right lobe of the liver

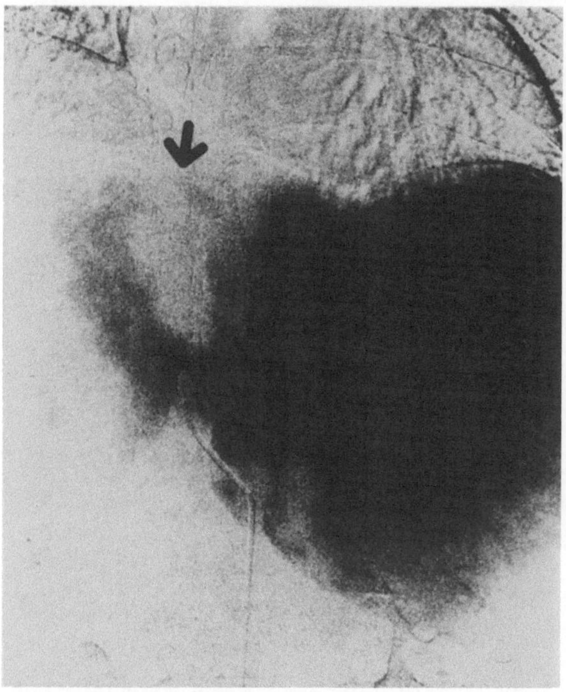

Fig. 16.3. Regenerative nodule of the liver. A deeply stained image of the hepatic parenchyma is distinctly seen, and a lesion overlapped on the vertebral body is pictured as a defect of the hepatogram (*arrow*)

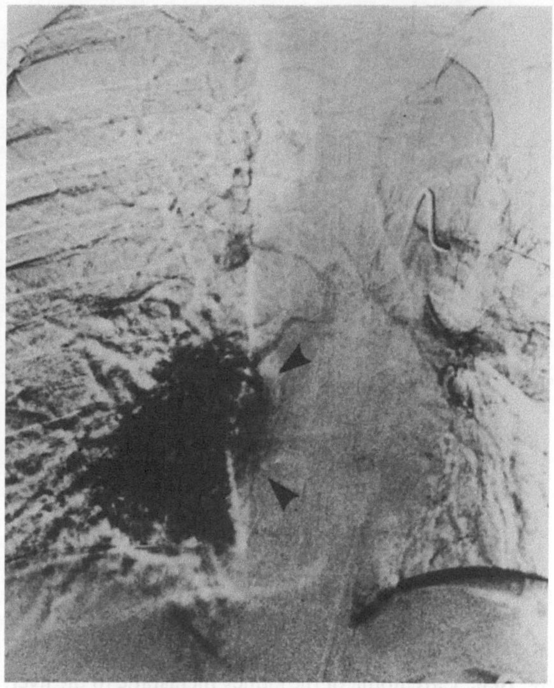

Fig. 16.4. Lung cancer. Tumor staining is remarkable, particularly facilitating easy determination of the range of advance (*arrowheads*) to the mediastinum overlapped on the vertebral body

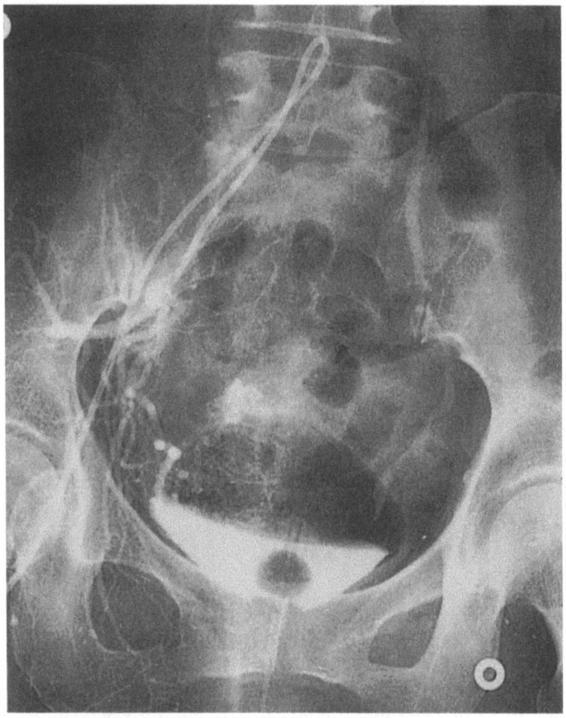

a

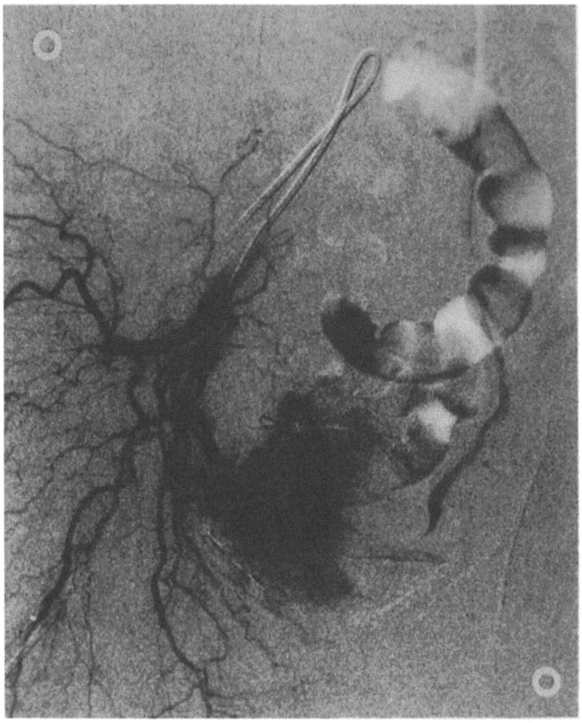

b

Fig. 16.5a, b. Uterine cancer. **a** Live image by CR. It is impossible to observe the lesion overlapped on the bladder. **b** DSA. Tumor vessels at the same site can be detected

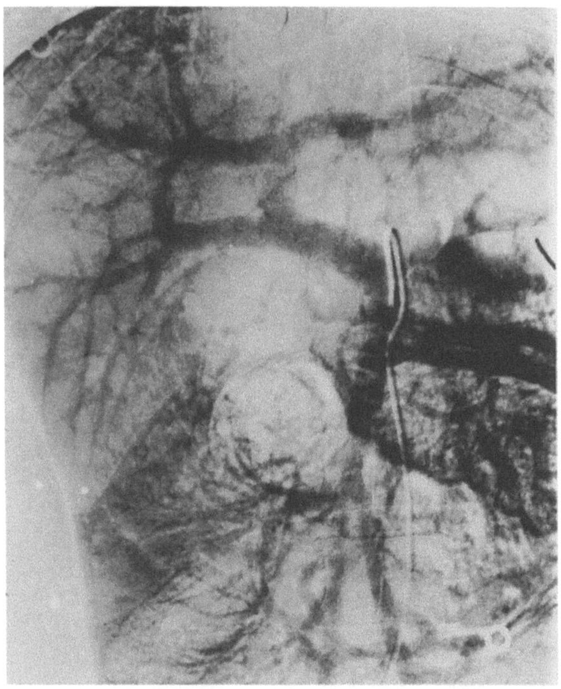

Fig. 16.6. Carcinoma of the kidney metastatic to the liver. DSA of arterial portography obtained by reducing the quantity of injected contrast medium by a factor of two

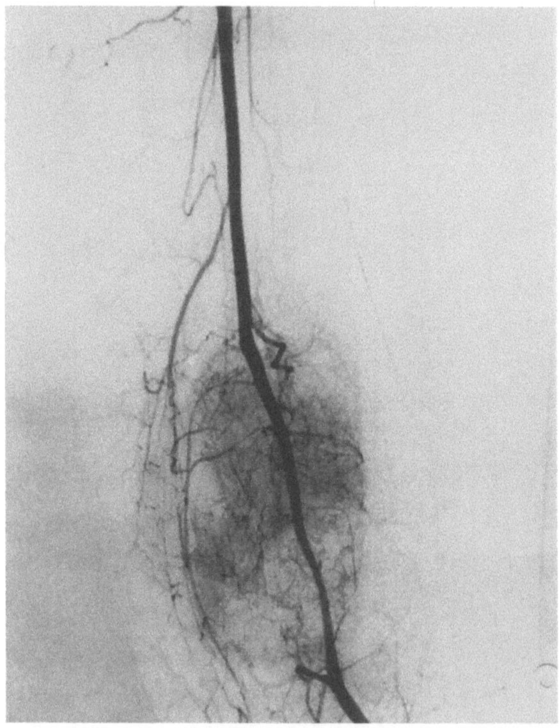

Fig. 16.7. Rhabdomyosarcoma of the femoral region. DSA obtained by injecting 20 ml 65% Angiografin, diluted four-fold, into the femoral artery. The range of advance of the tumor is well presented

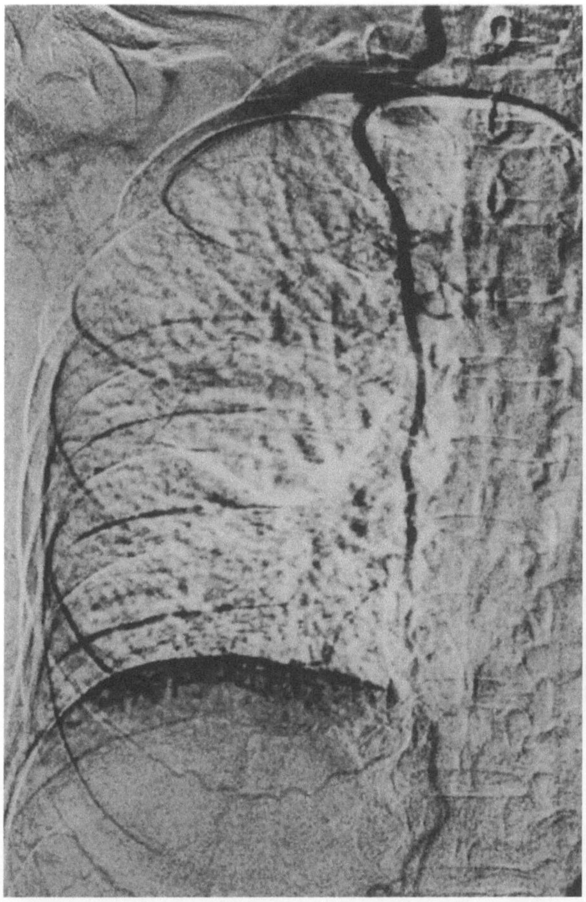
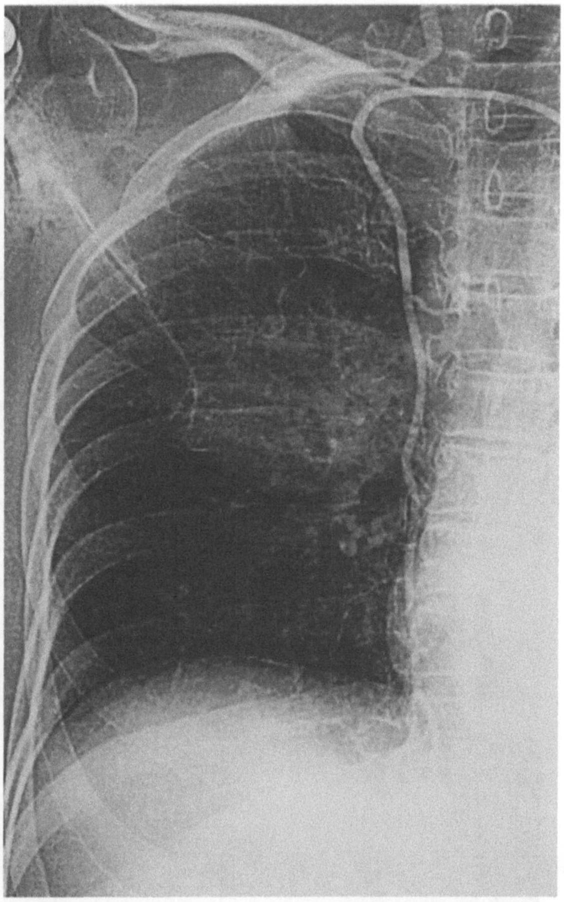

a b

Fig. 16.8a, b. Pleural mesothelioma. **a** DSA of the right internal mammary artery. It provided no diagnostic information because of motion artifacts. **b** Live image showing avascular tumor

thelioma provided no diagnostic information because of the poor image quality due to motion artifacts. However, the live image (Fig. 16.8b) obtained from the same patient facilitated diagnosis of the avascular characteristics of the tumor.

Case 9: Carcinoma of the ascending colon

The DSA (Fig. 16.9a) of the superior mesenteric artery obtained from a patient with carcinoma of the ascending colon provided no diagnostic information because of artifacts due to intestinal gas. On the live image by CR (Fig. 16.9b), which resembles the image obtained using the conventional screen/film system, the superior mesenteric artery and its major branches are clearly demonstrated. However, the vasa recta of the lesional area overlapped by the gas in the digestive tract cannot be observed. On the enhanced image (Fig. 16.9c), irregularity of the vasa recta of the lesion is clearly shown.

Case 10: Carcinoma of the lung metastatic to the adrenal gland

On the aortogram obtained from a case of carcinoma of the lung metastatic to the adrenal gland with one-fourth of the usual radiation dose, a lesion of the adrenal gland is presented as an avascular mass. There is no degradation of image quality due to the low dose reduction (Fig. 16.10).

Case 11: Carcinoma of the kidney

Figure 16.11 shows the live image of the angiogram of the renal artery obtained with one-fourth of the usual radiation dose. Tumor vessels of the upper pole of the right kidney are favorably demonstrated and show no degradation of image quality.

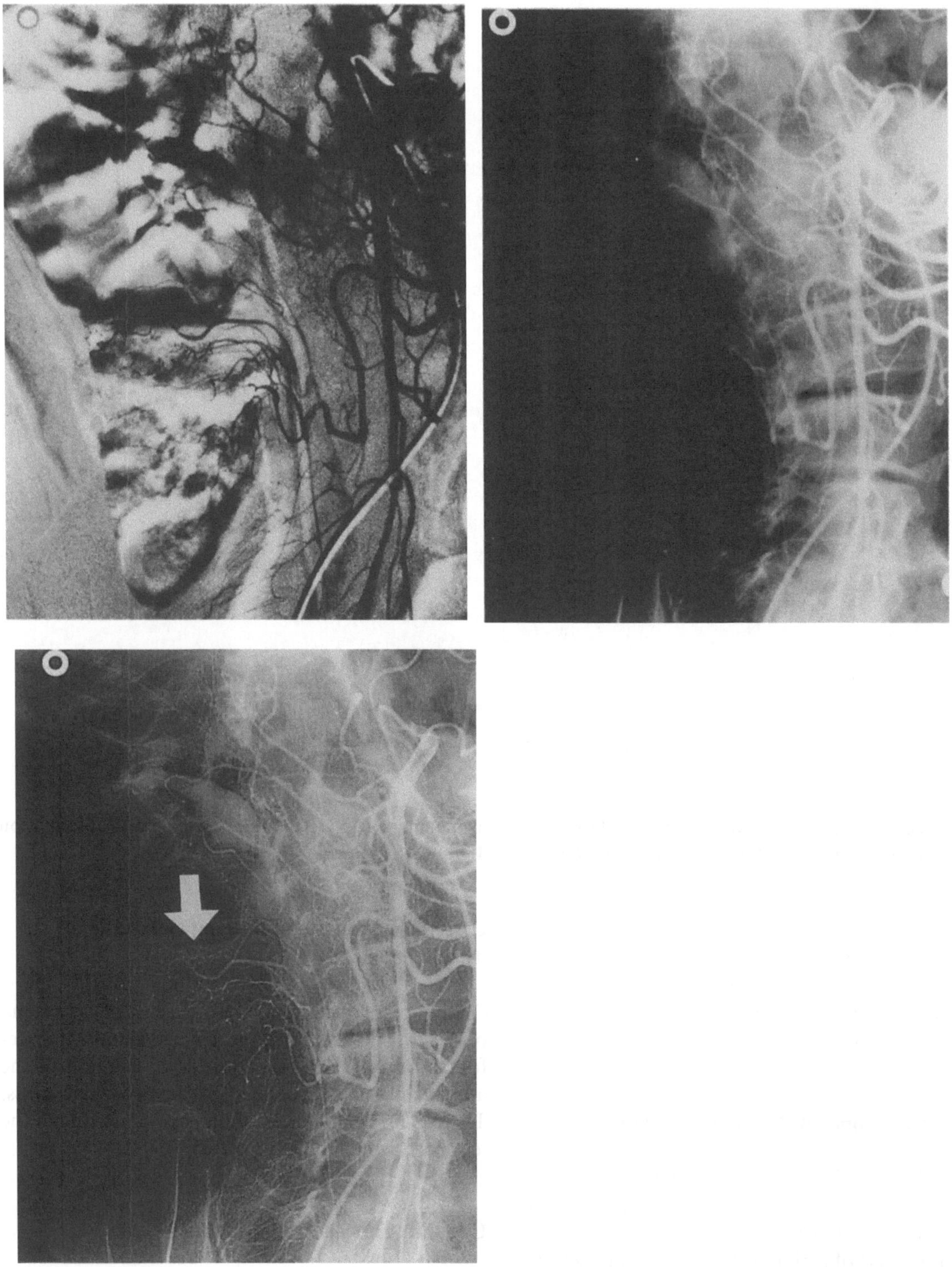

a b c

Fig. 16.9a–c. Carcinoma of the ascending colon. No diagnostic data on the lesional area are obtained on **a** DSA or **b** the live image, displayed as an image approximately similar to that obtained by the conventional screen/film method. **c** The live image whose spatial frequency components are enhanced, on the other hand, reveals irregularity of the vasa recta of the lesion (*arrow*)

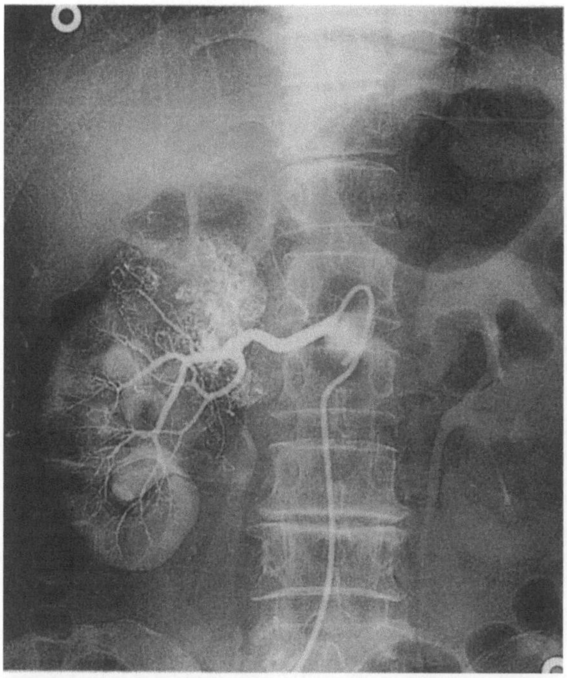

Fig. 16.10. Carcinoma of the lung metastatic to the adrenal gland. DSA obtained at one-fourth of the radiation dose used in conventional aortography. A metastatic lesion (*arrows*) of the adrenal gland is presented as an avascular mass

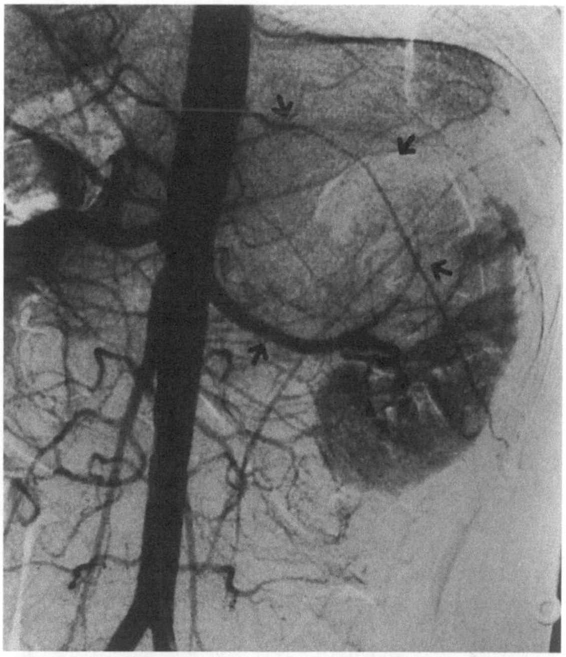

Fig. 16.11. Renal cancer. A live image by CR obtained at one-fourth of the radiation dose used in conventional angiography. Tumor vessels of the upper pole of the right kidney are favorably seen, showing no degradation of image quality

Case 12: Malignant histiocytoma of the gluteal region

In arteriography of the internal iliac artery in malignant histiocytoma of the right gluteal region, the DSA (Fig. 16.12a) obtained by twofold magnification clearly reveals tumor vessels in the low absorption area on the lateral side of the tumor as well as tumor vessels super-imposed on the high absorption area of the iliac bone, as compared with the coventional arteriogram (Fig. 16.12 b). The DSA is thus an excellent image for showing fine tumor vessels because of the twofold magnification.

Discussion

CR is a digital radiography system in which an imaging plate is used as an X-ray detector and the final image is displayed on the film rather than on CRT (cathode ray tube) [7]. This system, CR, can thus be siad to be a new system different from the DF system by use of the image intensifier (II)-TV camera. Angiograms performed with CR are displayed as two different types of image on a film. One is similar to the conventional angiogram and the other is an image in which the spatial frequency components are enhanced for easy analysis of fine blood vessels. With CR, in addition, even an over- or underexposed image can be corrected as a proper image by computer processing. Accordingly, it is held that a clear angiogram can always be obtained with CR. A subtraction processing function can also be added to CR, facilitating the application of DSA examination. CR with such advantages is a new method replacing the conventional screen/film method, and all angiograms obtained seem to be usable. We routinely use angiography by applying this system to clinical cases. As for examination by DSA, CR is employed both to IVDSA and IADSA. The characteristics of CR in IADSA are discussed below.

CR has high-contrast detectability like DF [8] and this makes it possible to perform IADSA, in which the quantity or density of injected contrast medium can be reduced. The reduction of the quantity of contrast medium used in IADSA is recognized as an advantage in arteriography for patients with hypofunction of the kidney or in arteriography of the nervous system from the viewpoint of preventing serious side effects. Dilution of contrast medium, on the other hand, is useful for decreasing the discomfort

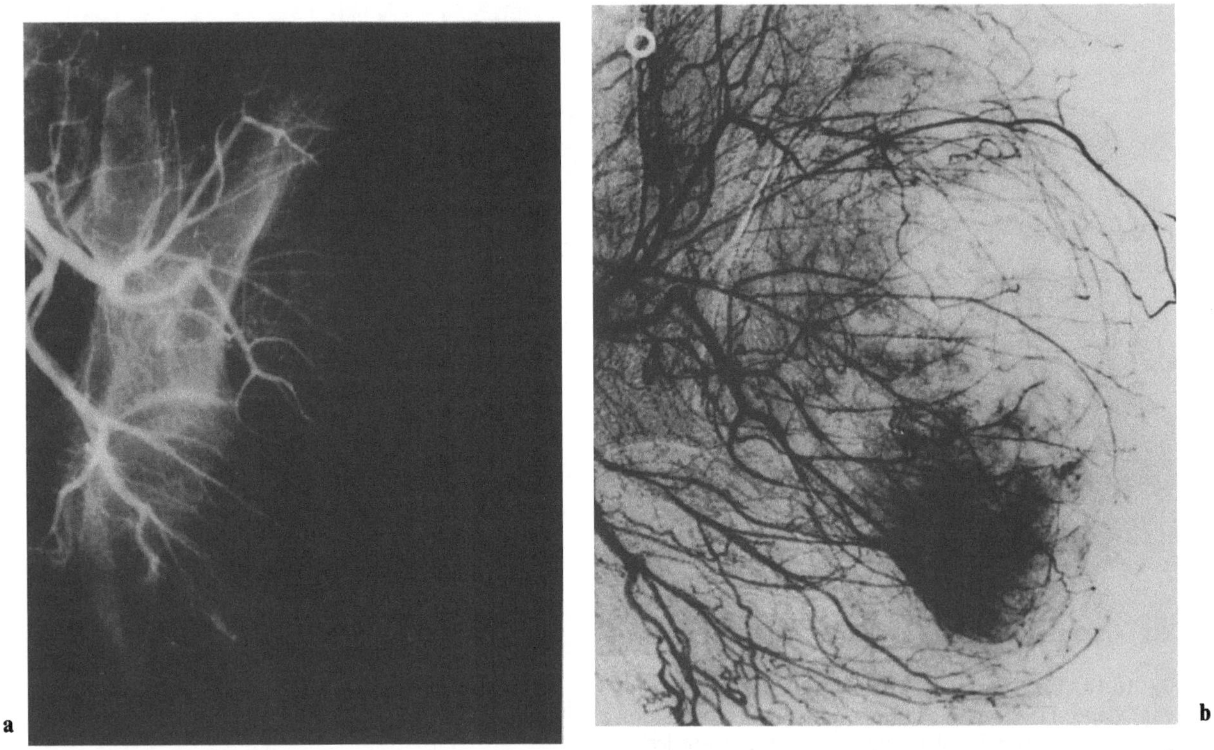

Fig. 16.12a, b. Malignant histiocytoma of the gluteal region. **a** DSA with twofold magnification clearly shows not only the tumor vessels superimposed on the high absorption area of the iliac bone but also those on the low absorption area on the lateral side of the tumor, as compared with **b** the conventional image. DSA is particularly good for the detection of fine tumor vessels because of the twofold magnification

experienced by a patient during its injection. In arteriography of the extremities particularly, the dilution of contrast medium is markedly effective for decreasing pain on injection. In arterial portography as well, images with favorable image quality can be obtained with use of CR, in which injection of the contrast medium is only half to one-third that of the conventional quantity. The high-contrast detectability of CR is, in addition, useful for presenting the staining of abdominal parenchymatous organs and faint tumor stains, and its subtracion is effective for detecting tumor stains overlapped by bony structure.

The spatial resolution of CR is 2.5 cycles/mm (14 × 14-in. imaging plate), being more favorable than the 1.8 cycles/mm of DF (512 × 512 Matrix, 5 in. II). Even this level, however, is inferior to the spatial resolution of the conventional screen/film system, 5.0 cycles/mm (intensifying screen KS, film AO). This point becomes a diagnostic issue in IADSA in which analysis of fine blood vessels is believed necessary. We, therefore, carried out magnified angiography by CR for the purpose of improving the spatial resolution. The spatial resolving power of CR as a result of a twofold magnification was increased to 4.0 cycles/mm, approximately equal to that of the conventional screen/film system. Its excellent effects were also obtained in clinical cases.

One of the characteristics of CR is that it provides an image with a wide latitude, which is very advantageous for angiography of areas with a large difference in contrast, such as the area between the lung field and the mediastinum. In CR, it was not necessary to use a compensating X-ray filter as in conventional angiography and DF. Three-dimensional radiography can also be easily carried out in CR by use of conventional apparatus, and this would probably lead to the prevention of a decrease in diagnostic ability due to the overlapping of vessels, which is one of the disadvantages of nonselective angiography.

The clinical maintenance of an adequate visual field has been possible since the development of an IP rapid changer for exclusive use in CR which facilitates angiography at the same visual field size as in conventional angiography. On the other hand, DF allows a wide visual field to be obtained by means of an II with a large diameter but results in a decrease of spatial resolution [6]. For CR, the decrease of the spatial resolution due to the wide visual field is slight [10]. From this point of view alone, CR may be more advantageous than DF. Such a wide visual field becomes a necessary condition for performance of the magnified angiography described above.

Misregistration artifacts, on the other hand, become an issue in IADSA by CR as well as by DF. Therefore, we explained to patients before examination that it was important for them to stop breathing during the exposure, and we carried out administration of glucagon for angiography of the abdomen. Of 53 patients in the present study, seven were evaluated as fair or poor because of misregistration artifacts. In such circumstances, CR has an advantage in that diagnostic information can be obtained by live images processed by tonal conversion and spatial frequency enhancement techniques.

CR is, in addition, useful for reduction of exposure dose. According to our phantom studies, CR had two to four times the speed of the conventional screen/film system (an intensified screen KS, film AO); angiography at a low dose—half to one-fourth that used in the conventional screen/film method—also provided a diagnostically adequate image in clinical examinations. From the viewpoint of dose reduction, CR can be said to be the best procedure for use in clinical examinations.

CR thus has many advantages except for the long processing time, which is its weakest point. It takes about 40 min to process 12 imaging plates and subtract about four or five images. This clearly indicates that CR is less advantageous than DF, which can provide real-time subtraction images. When considering standardization of angiographic examinations by the CR system in future, it will be necessary for the processing capacity to be upgraded.

Conclusion

IADSA performed by CR was clinically evaluated in 53 patients. The advantages of CR used in IADSA are that it allows angiography at a low dose, that side effects can be reduced by dilution and reduction of the contrast medium, and that it is diagnostically excellent in detecting lesions superimposed on high and low exposed areas. Its diagnostic ability can be expected to be further improved by combination with magnified angiography or three-dimensional radiography.

References

1. Mistretta CA, Crummy AB, Strother CM (1981) Digital angiography: a perspective. *Radiology* 139: 273–276
2. Nishimura G, Naitou H, Yamaguchi T, Takamiya M, Kozuka T, Ohno H, Yasuhara H (1982) Clinical aspects of digital video subtraction angiography (DSA) Journal of Medical Imagings 2:182–187
3. Meaney TF, et al. (1980) Digital subtraction angiography of the human cardiovascular system Am. J. Roentgenol 135:1153–1160
4. Takahashi M, et al. (1983) Usefullness of digital fluoroscopic angiography. Japanisch Deutsche Medizinische Berichte 27:303–331
5. Crummy AB, Stieghorst MF, Turki PA, Strother CM, Lieberman RP, Sackett JF, Turnipseed WD, Detmer DE, Mistretta CA (1982) Digital subtraction angiography—Current status and use of intra-arterial injection. *Radiology* 145:303–307
6. Bonnie DF, Antoinette SG, Edgar CS, Juan FL, S Osher Pais (1983) Intra-arterial digital subtraction angiography:Comparison with conventional hepatic arteiography. *Radiology* 148:17–21
7. Takano M (1982) New computed radiography—Fuji Intelligent Diagnostic X-ray system. Journal of Medical Imagings 2:202–207
8. Iinuma T (1983) Methods of digital X-ray imaging: Classification and definition. Japanese Journal of Clinical Medicine 41:1383–1387
9. Matsunaga N, et al. (1982) Clinical experiences of digital subtraction angiography by DVI (Philips) Japanisch Deutsche Medizinische Berichte 27:333–348
10. Korenaga T, Hachiya J, Sakurai K, Nitatori T, Miyasaka Y, Furuya Y (1983) Digital subtraction angiography (DSA) by Fuji computed radiography (FCR) system. Radiological Medicine 2:61–70

Chapter 17

Diseases of the bones and joints

Kazuhito Matsubara[1] and Sadayuki Sakuma[2]

Computed radiography (CR) has been applied to many kinds of radiographic examination in place of conventional screen/film techniques [1–3]. In most cases, the diagnostic capability of CR is evaluated as comparable or superior to that of screen/film techniques. However, due to its limited spatial resolution, this new technique does not have sufficient capability to visualize bone microstructures. In this chapter, the advantages and disadvantages of the application of CR to bone and joint diseases will be discussed.

Image processing and its clinical application

The image processor of CR allows several types of digital image processing. The major image-processing techniques of CR consist of gray-scale conversion and spatial frequency enhancement for the specified spatial frequency range. A computed radiograph processed by the image processor is printed as a hard copy on a single-emulsion transparent film by a high-resolution laser printer. For most examinations, two differently processed images obtained from one X-ray exposure are printed on a single 10 × 14-in. film. We think it appropriate to process one image to resemble the characteristic curve of a conventional screen/film radiograph ("normally" processed image), mainly by gradation conversion, and to process the complementary image like a zeroradiogram by gray-scale conversion to expand the image contrast and by strong spatial frequency enhancement so that it resembles a zeroradiogram (enhanced image). The normally processed image allows diagnosis mainly on the degree of X-ray attenuation of the tissue and the enhanced image allows diagnosis corresponding to the contour of the object. The type of processing required for a given examination is automatically selected and performed by the image processor. The CR system in current use can be programmed for image processing of up to 24 types of examination (programmed types). The type of image processing may also be manually controlled by using an image-processing console (post-processing mode), however this was seldom necessary in routine radiographic studies.

In the following, we describe lesion visualization with the various types of image processing, taking the clinical example of a 10-year-old girl who had osteosarcoma located mainly in the proximal region of the tibial diaphysis.

Figure 17.1a, b shows the two differently processed images printed on a single film after the programmed type for the extremity has been chosen. Figure 17.1a is the normally processed image with high contrast and weak frequency enhancement around the spatial frequency of 0.25 lp/mm. Figure 17.1b shows the enhanced image with expanded latitude and strong frequency enhancement at the same frequency as above. On the normally processed image, bone lesions can be well recognized, but the advancement of the tumor into the soft tissues is more difficult to assess. On the enhanced image, the density of the boundary areas of the bone lesion is too low, but the periosteal response and the shape of the tumors advancing into the soft tissues can be clearly recognized. Figure 17.1c represents a post-processed image obtained by strong frequency enhancement mainly around the lower frequency of 0.12 lp/mm. On this image, tumor advancement into the soft tissues can be recognized as clearly as in Fig. 17.1b, but no additional information can be obtained. Therefore, Fig. 17.1c seems to be unnecessary for general diagnosis. Figure 17.1d shows a post-processed image obtained by strong frequency enhancement mainly around the high frequency of 1.0 lp/mm. On this image, the bone crista is well visualized, but lesions in the soft tissues are difficult to assess compared with Fig. 17.1b. In addition, noise is enhanced, there is ircreased granularity, and tumorous changes of the bone itself are not as clear as in Fig. 17.1a. Figure 17.1d thus seems to be unnecessary for general diagnosis.

[1] Department of Radiology, Tosei Public Hospital, 160 Nishioimake-cho, Seto, Aichi, 489 Japan

[2] Department of Radiology, Nagoya University, School of Medicine, 65, Tsurumai-cho, Showa-ku, Nagoya, 466 Japan

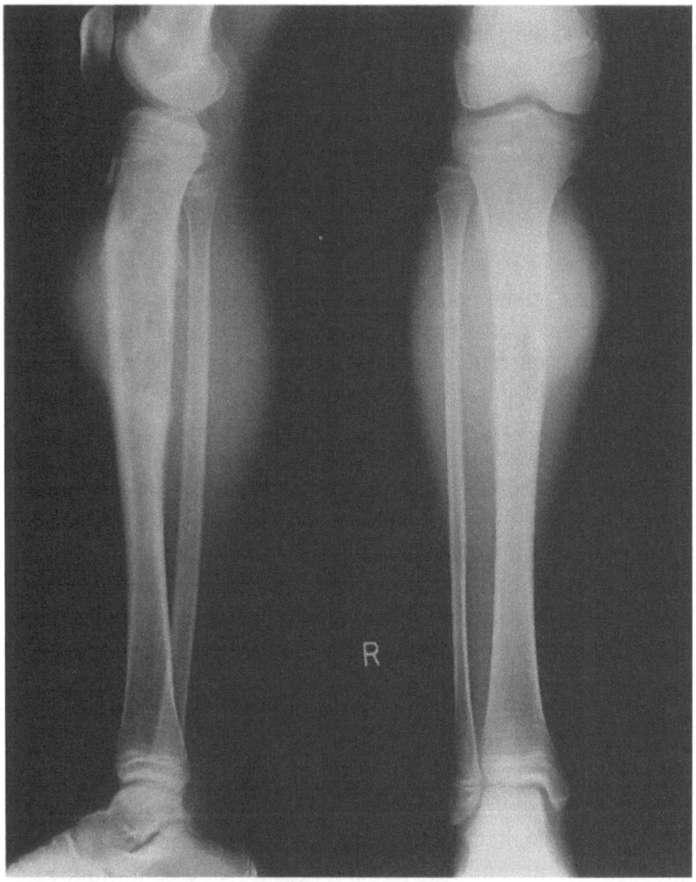

a

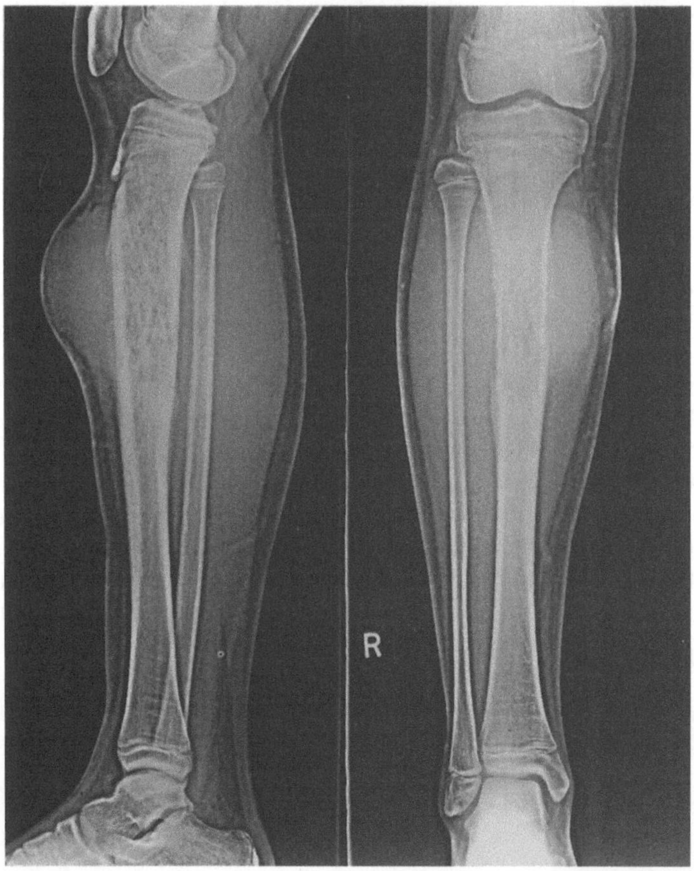

b

Fig. 17.1a–d. A 10-year-old girl with osteosarcoma. **a** The normally processed image with weak frequency enhancement, mainly at 0.25 lp/mm; osteosarcoma in and around the proximal part of the tibial diaphysis; bone lesions are well visualized. **b** The enhanced image with strong frequency enhancement, mainly at 0.25 lp/mm; periosteal response and tumor advancement in soft tissues are clear. **c** Image-processed low frequency-range enhancement (0.12 lp/mm; tumor advancement in soft tissues is as clear as in **b**. **d** Image with high frequency-range enhancement (1.0 lp/mm; bone cristae are well visualized, but granular appearance is marked)

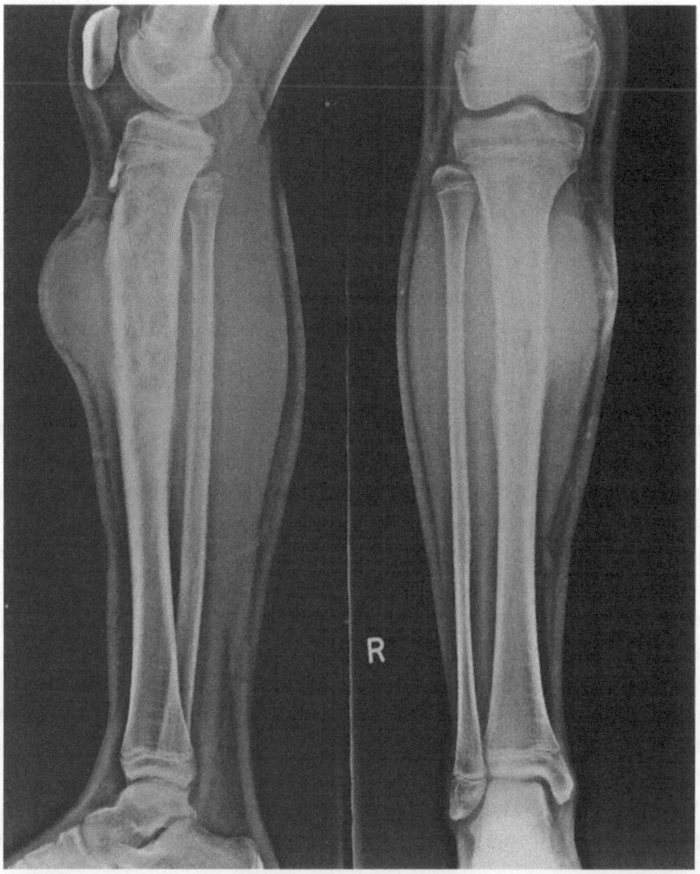

c

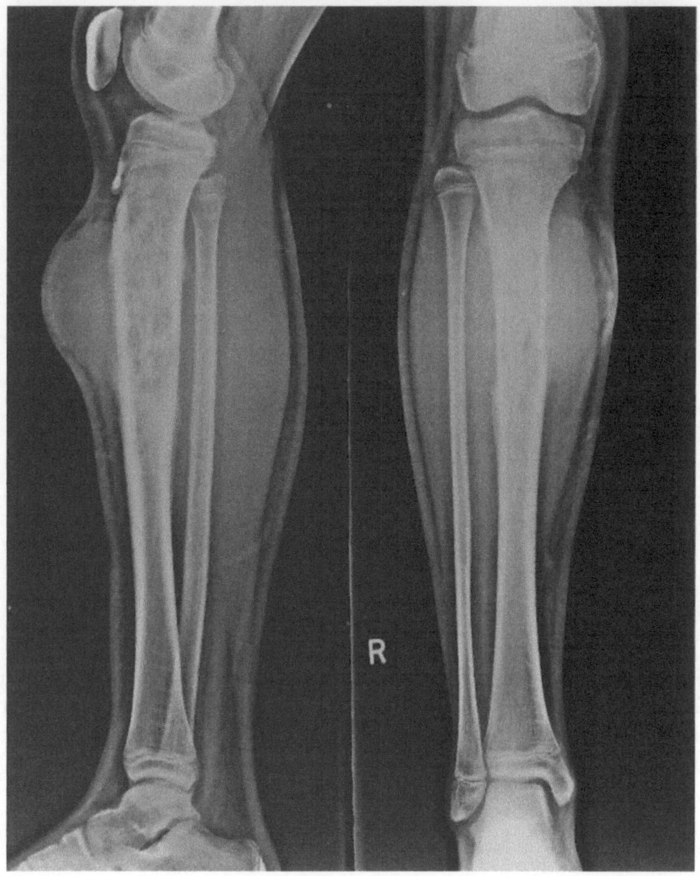

Fig. 17.1c, d. d

Application of magnification radiography

The spatial resolution of CR depends upon the size of the imaging plate and the number of sampling points. As the number of sampling points is limited to 2000 × 2000, the best spatial resolution is 2–2.5 lp/mm when the imaging plate is 12 × 10 in. or smaller. In the diagnosis of bone diseases, evaluation of the shadows of bone cristae is often required. However, on a radiogram taken using an imaging plate, bone cristae are less sharply visualized than on an image taken with conventional screen/film techniques. Magnification radiography is quite useful in the diagnosis of minute lesions [4, 5]. With magnification radiography, the contrast of the output image is decreased because of X-ray dispersion. However, such reduced contrast can be compensated for by using image processing. In our studies magnification radiography was conducted with a 50-μm focal spot X-ray tube; in ordinary cases, fourfold magnification was used at a distance from the focal spot to the imaging plate of 100 cm (100 FFD).

Figure 17.2 shows a 5.9-fold magnification radiogram of an 18-year-old female with tumors in the proximal metaphysis of the right fibula. This radiogram was taken under the exposure settings of 100 kV, 3mA, 0.14s, and 118 cm (FFD). Figure 17.2a demonstrates an image with gray-scale conversion. On this image, infiltrative bone destruction, bone sclerosis, periosteal response, and ossification in the soft tissues can be seen. However, the boundary of the fibula is recognized with difficulty. Figure 17.2b shows an image obtained by making the gradation a little flatter than in Fig. 17.2a by applying strong frequency enhancement mainly at around 0.12 lp/mm. In Fig. 17.2b, the fibular boundary is clear, but the difference in the X-ray absorption in the bone lesions has become less evident than in Fig. 17.2a.

Case reports

Primary bone tumor

Ewing's sarcoma in the iliac bone (17-year-old male)

This film was taken under the exposure settings of 90 kV and 16 mAs (Fig. 17.3). The enhanced image by strong frequency enhancement mainly at around 0.38 lp/mm is shown here. Bone sclerosis and partially osteolytic tumors, accompanied by periosteal response, are located at the center of the body of the right iliac bone; swelling is also present in the medial and lateral soft tissues. Bone lesions extend to the right sacroiliac articulation. Changes in soft tissues are clearly visualized.

a

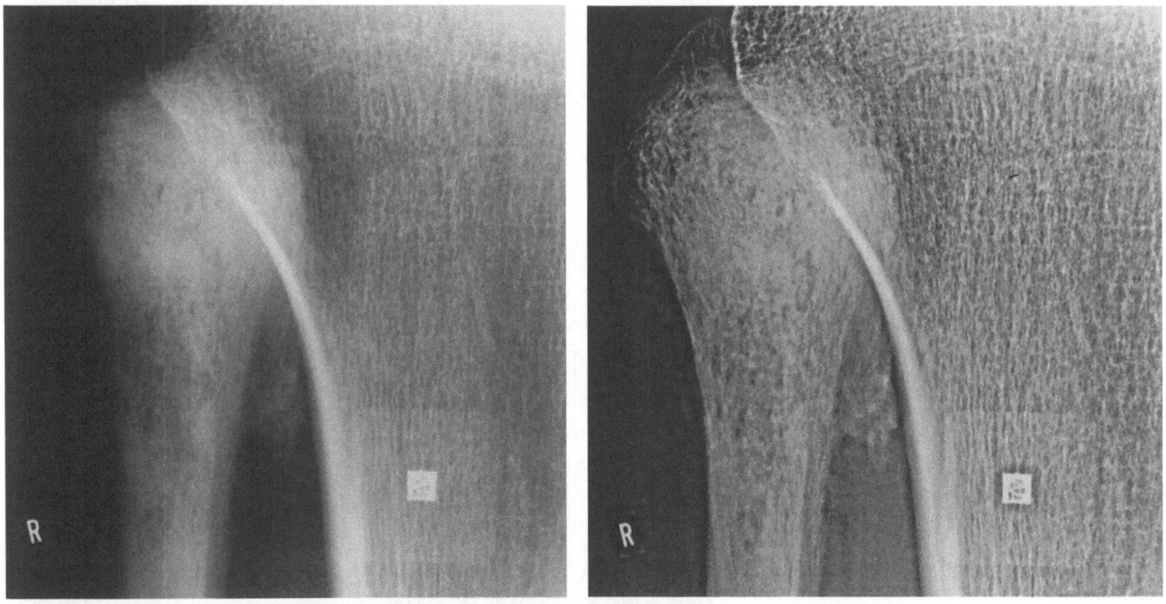

b

Fig. 17.2a, b. Magnification radiography of a bone tumor in the proximal metaphysis of the fibula in an 18-year-old female. **a** Image with tonal conversion only. Bone lesions and ossification in soft tissues are clear, but the contour of the fibula is less evident. **b** Image-processed strong frequency enhancement, mainly at around 0.12 lp/mm (the contour of the fibula is clear, but the difference in contrast in bone lesions is low)

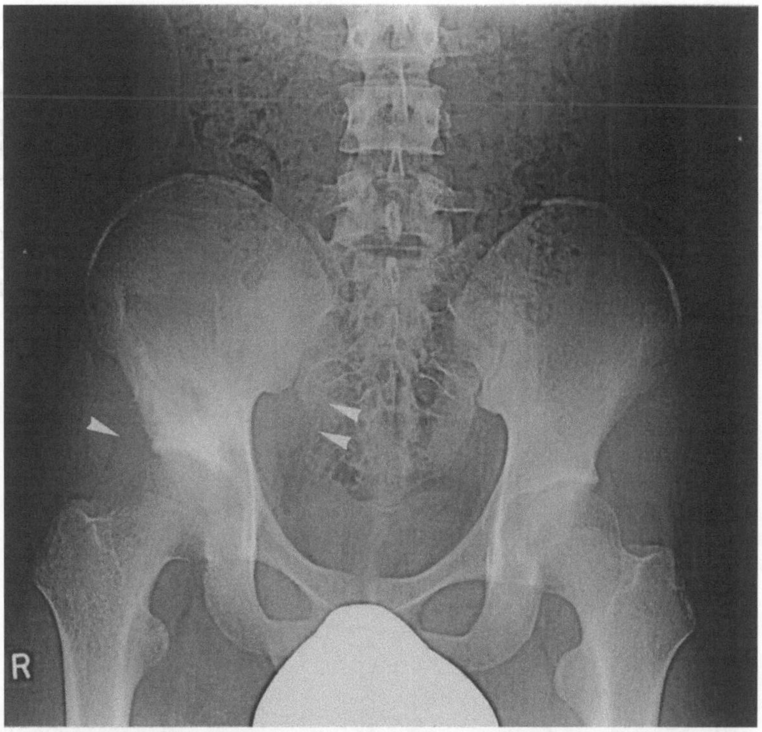

Fig. 17.3. Ewing's sarcoma in the iliac bone (a 17-year-old male). Image with strong frequency enhancement, mainly at 0.38 lp/mm. Mixed shadows centered at the body of the right iliac bone and extending to the right sacroiliac articulation are seen. Marked swelling of soft tissuesin the medial and lateral sides of the right iliac bone can be noted (*arrowheads*)

Giant cell tumor of the femur (30-year-old male)

The normally processed image (Fig. 17.4a) was obtained by weak frequency enhancement, mainly at 0.25 lp/mm. On this image, osteolytic lesions emerging at the distal metaphysis of the left femur are quite evident. The distal contour of the tumor is clear and a sclerosed edge is apparent, whereas the proximal contour is not clear.

The enhanced image (Fig. 17.4b) was obtained by making the gradation flat (expanding latitude) and applying strong frequency enhancement, mainly at 0.25 lp/mm. On this image, soft tissues surrounding the knee joint are well visualized. However, since this patient had no abnormality in the soft tissues, no additional information can be obtained from this image compared with Fig. 17.4a.

Postoperative recurrence of osteosarcoma in the femur (5-year-old girl)

The image was taken under the exposure settings of 50 kV, 200 mA, and 0.032 s, followed by strong frequency enhancement, mainly at 0.18 lp/mm (Fig. 17.5). After the osteosarcoma in and around the distal metaphysis of the femur had been surgically removed, tumors recurred in the soft tissues at the same location.

Metastatic bone tumor

Metastasis of lymphoepithelioma of the epipharynx to the ribs (19-year-old male)

Taken under the settings of 45 kV and 63 mAs, the normally processed image (Fig. 17.6a) was obtained by weak frequency enhancement, mainly at 0.5 lp/mm. On this image, the destroyed ribs are evident. The enhanced image (Fig. 17.6b) was obtained by making the gradation flat and applying strong frequency enhancement, mainly at 0.5 lp/mm. On this image, the relationship between the swelling of soft tissues at the lateral side of the ribs and the subcutaneous tissues is clear.

Metastasis of kidney cancer to the humerus (37-year-old male)

Figure 17.7a shows a normally processed image after strong frequency enhancement, mainly at 0.25 lp/mm. On this image, osteolytic metastasis of the cancer is found in a region from the cortex to the medulla of the diaphysis of the right humerus, but no advancement of the tumor into the soft tissues is seen. Details of the bone cristae and lesions are obscure.

Fig. 17.4a, b. Giant cell tumor (a 30-year-old male). **a** Image obtained by weak frequency enhancement, mainly at around 0.25 lp/mm. A typical case of giant cell tumor which developed at the distal metaphysis of the left femur. **b** Image obtained by making the gradation flat and applying strong frequency enhancement. Soft tissues surrounding the knee joint are easily recognized, but this image provides no more information than in **a**, since this patient had no abnormality in the soft tissues

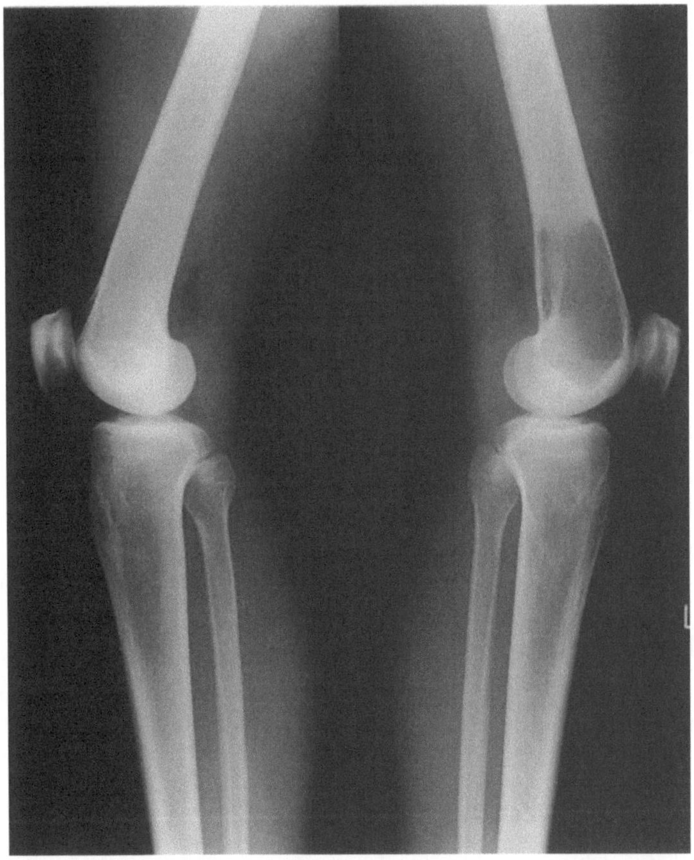

a

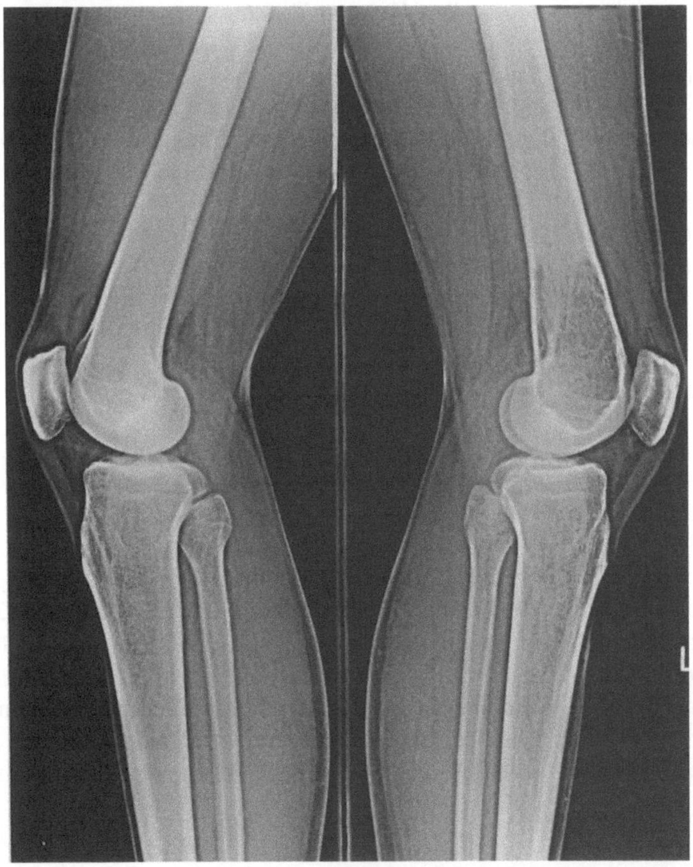

b

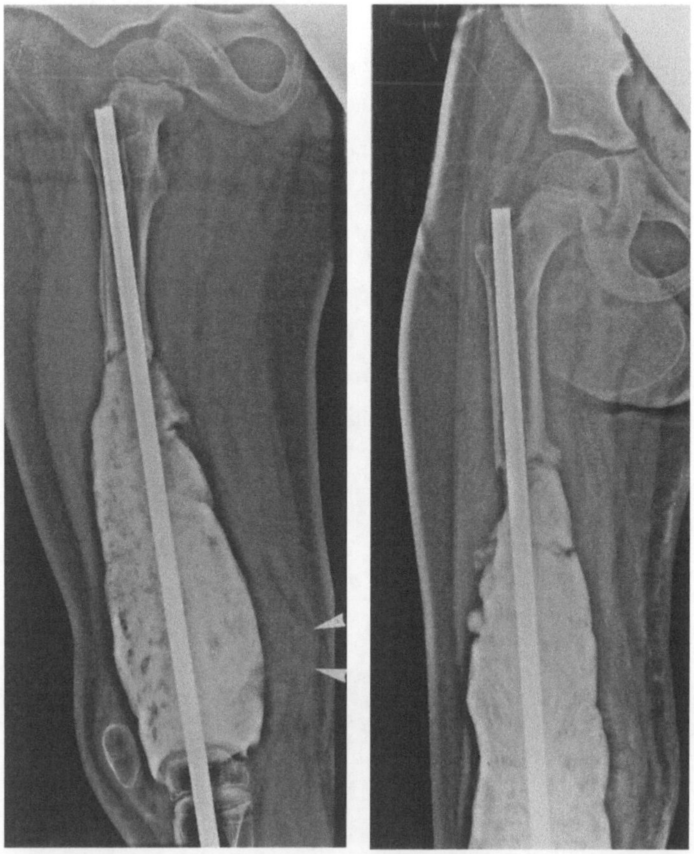

Fig. 17.5. Postoperative recurrence of osteosarcoma in the femur (a 5-year-old girl). Image with strong frequency enhancement, mainly at around 0.18 lp/mm. After removal of the osteosarcoma, a recurrent tumor is evident in the medial soft tissues (*arrowheads*)

a b

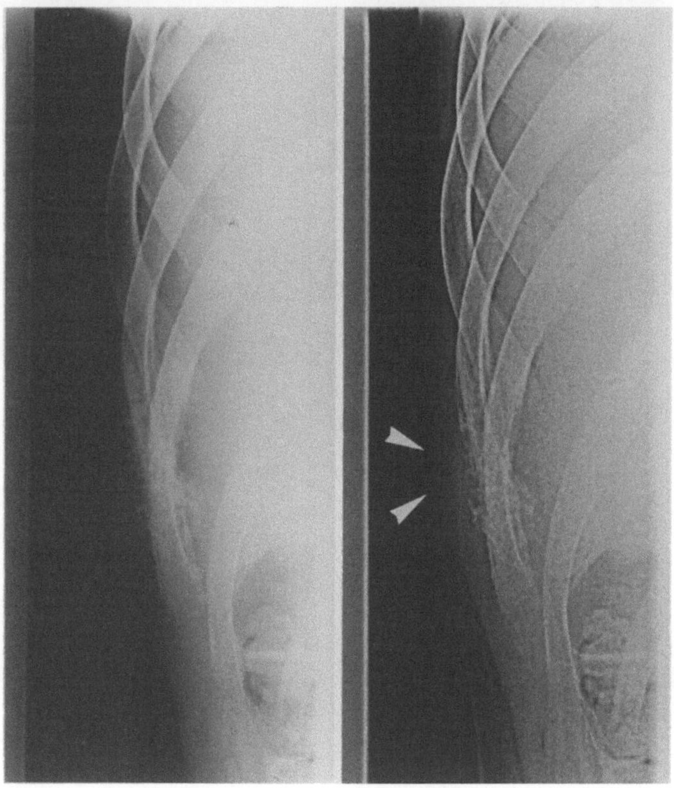

Fig. 17.6a, b. Metastasis of lymphoepithelioma of the epipharynx to the ribs (a 19-year-old male). **a** Image with weak frequency enhancement, mainly at around 0.5 lp/mm. Destroyed ribs are clear. **b** Image obtained by making the gradation flat and applying strong frequency enhancement. Swelling of soft tissues is clear (*arrowheads*)

a b

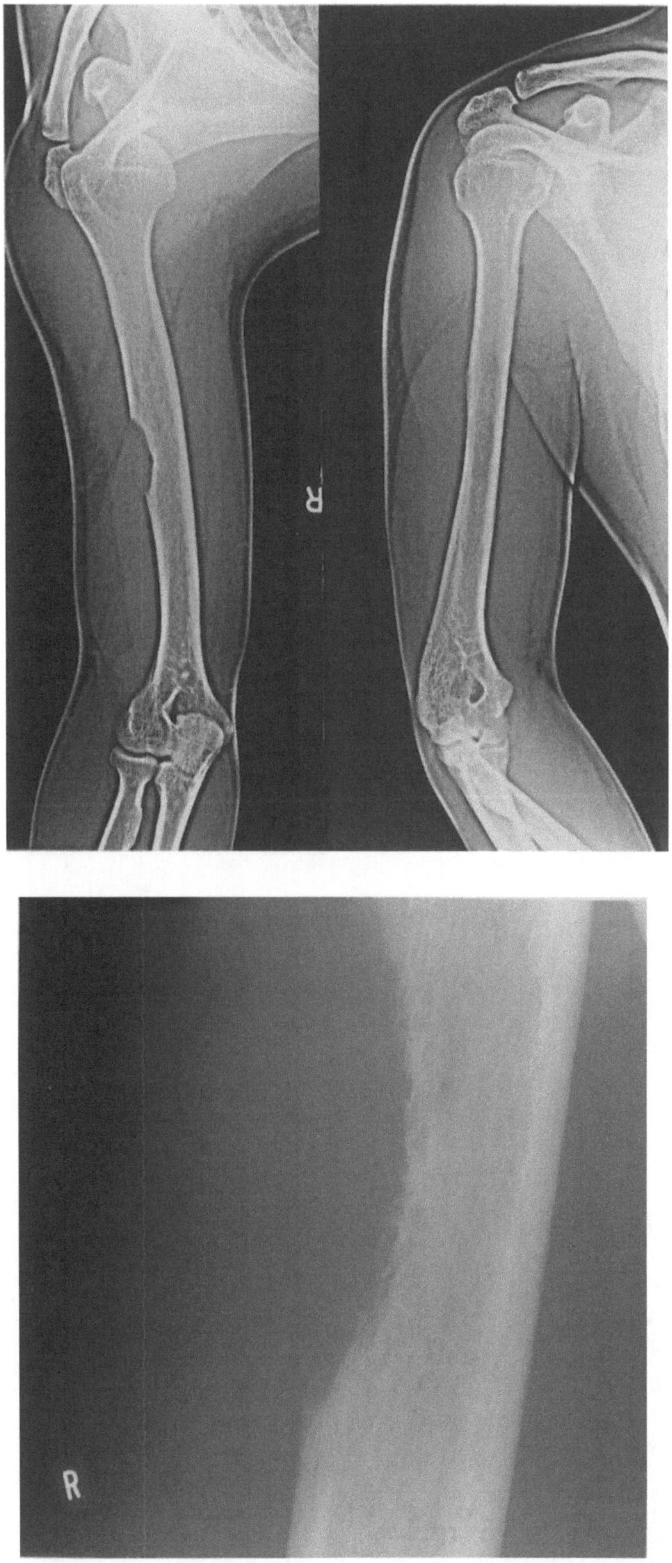

a

b

Fig. 17.7a, b. Metastasis of kidney cancer to the humerus (a 37-year-old male). **a** Image-processed strong frequency enhancement, mainly at around 0.25 lp/mm. Osteolytic metastasis of the cancer to the diaphysis of the right humerus is noted, but detailed changes are obscure. **b** Fourfold magnification. The boundary of the osteolytic metastasis and bone cristae is evident

Figure 17.7b presents a directly magnified image with a fourfold magnification taken at 80 kV, 0.05 s, and only processed for flat gradation. On this image, the boundaries of the osteolytic metastasis as well as the bone cristae are evident.

Metastasis of cervical cancer to the lumbar vertebren (63-year-old female)

Figure 17.8 is a lateral tomogram of the lumbar vertebrae after weak frequency enhancement, mainly at 0.38 lp/mm. Bone destruction is seen at the upper edge of lumbar vertebra IV.

Hyperparathyroidism

Secondary hyperparathyroidism (26-year-old male)

This patient developed secondary hyperparathyroidism in the course of hemodialysis which was started because of renal failure. A typical "salt-and-pepper" pattern is shown in the skull (Fig. 17.9a, b). Figure 17.9a presents an image after weak frequency enhancement, mainly at 0.5 lp/mm. Figure 17.9b shows an image obtained by making the gradation flat and applying strong frequency enhancement. The "salt-and-pepper" pattern is emphasized.

Figure 17.9c shows a CR image of the hand bones after weak frequency enhancement, mainly at 0.25 lp/mm. Figure 17.9d presents a radiogram of the same site taken with the screen/film combination.

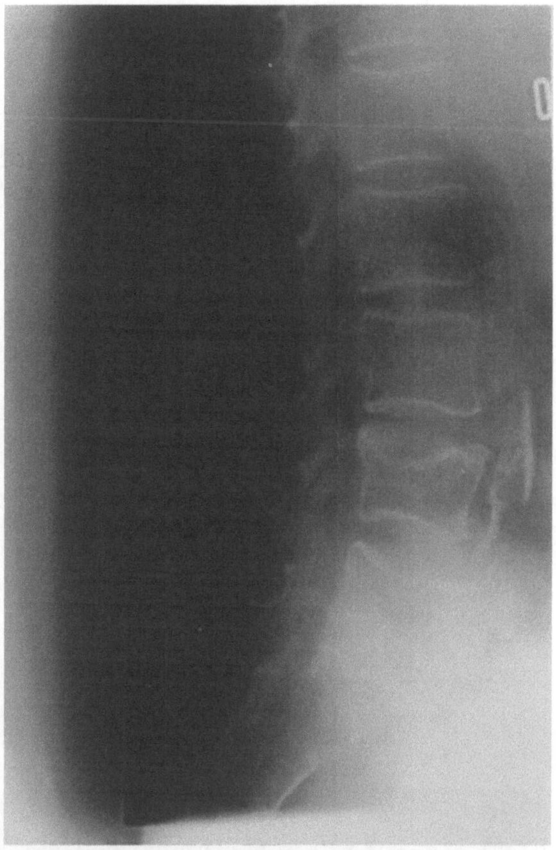

Fig. 17.8. Metastasis of cervical cancer to lumbar vertebrae (a 63-year-old female). Lateral tomogram of lumbar vertebrae with weak frequency enhancement, mainly at around 0.38 lp/mm. Bone destruction is seen in the upper end of lumbar vertebra IV

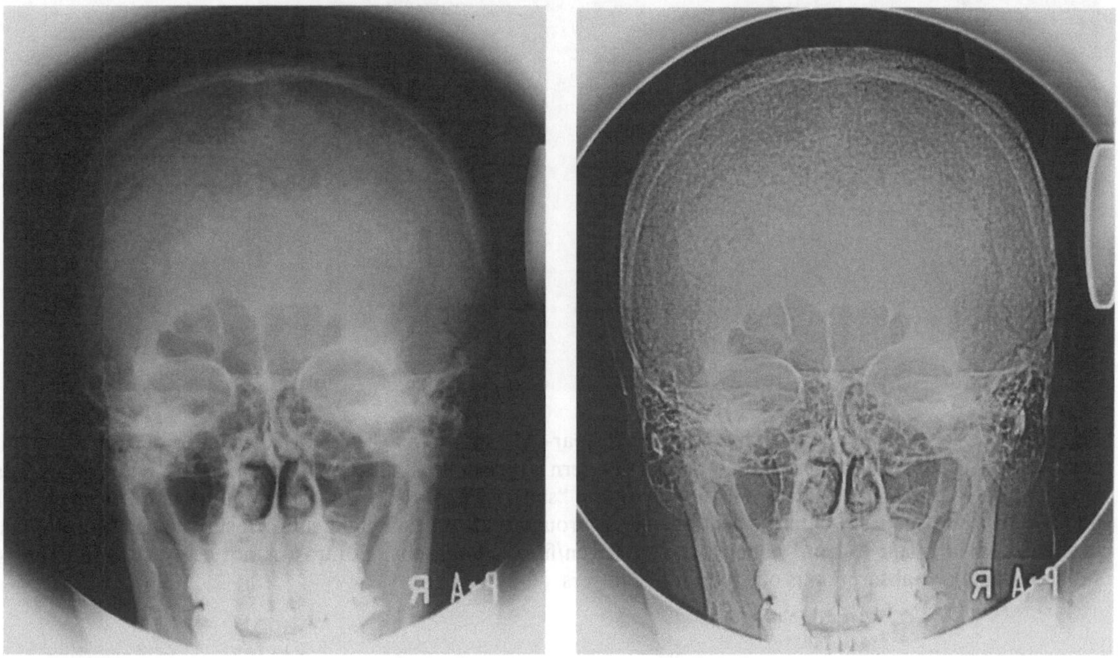

a b

Fig. 17.9a, b.

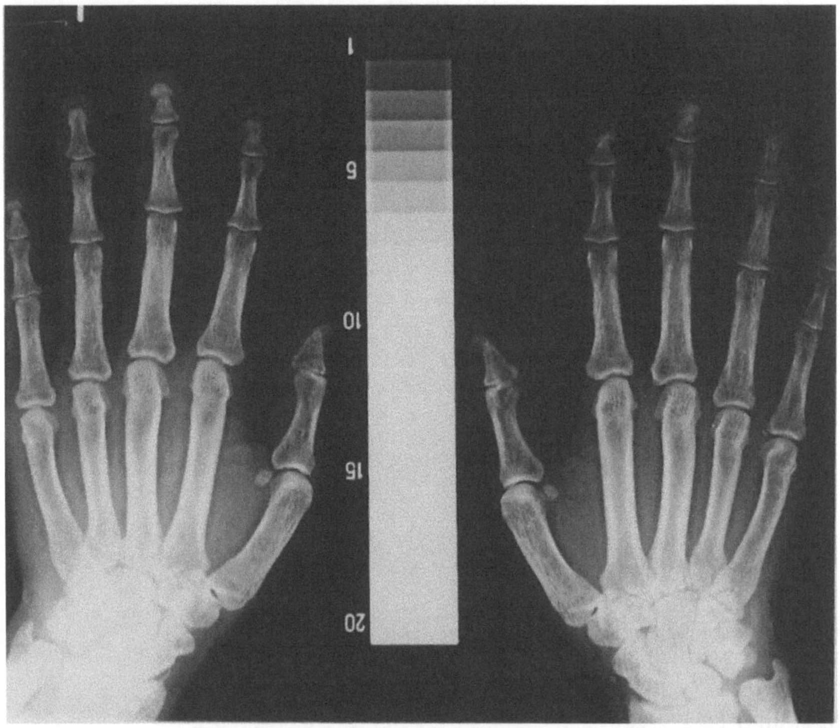

c

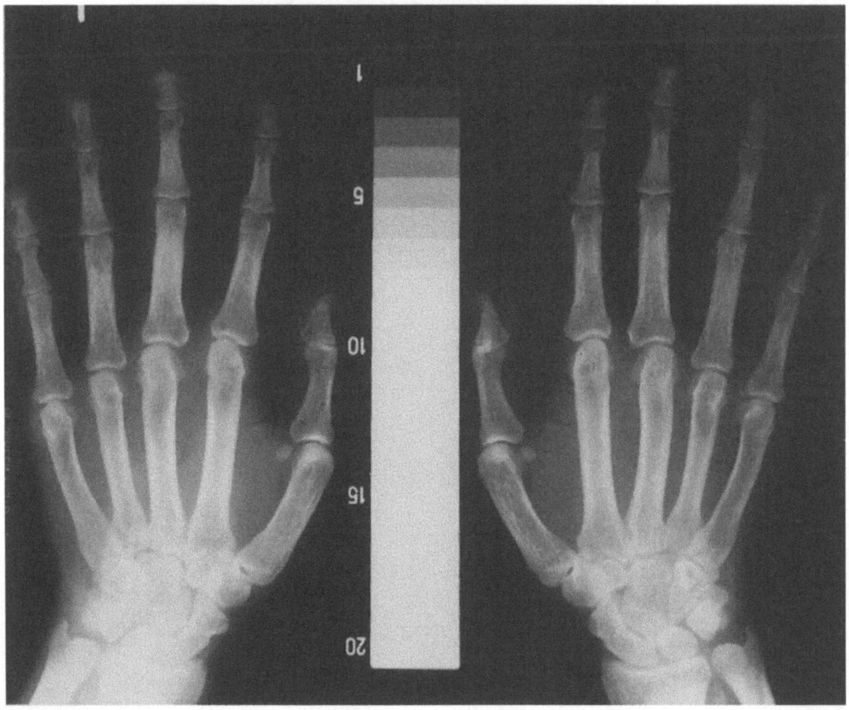

d

Fig. 17.9a–d. Secondary hyperparathyroidism (a 26-year-old male). **a** Image with weak frequency enhancement, mainly at around 0.5 lp/mm. A typical "salt-and-pepper" pattern is seen in the skull. **b** Image obtained by making the gradation flat and applying strong frequency enhancement. The "salt-and-pepper" pattern is emphasized. **c** CR image of hand bones with weak frequency enhancement, mainly at around 0.25 lp/mm. Slight subperiosteal bone absorption is seen in the fingers. **d** Image of hand bones taken by screen/film radiography. Contrast is not as great as in **c**, but slight subperiosteal bone absorption is observed in the fingers

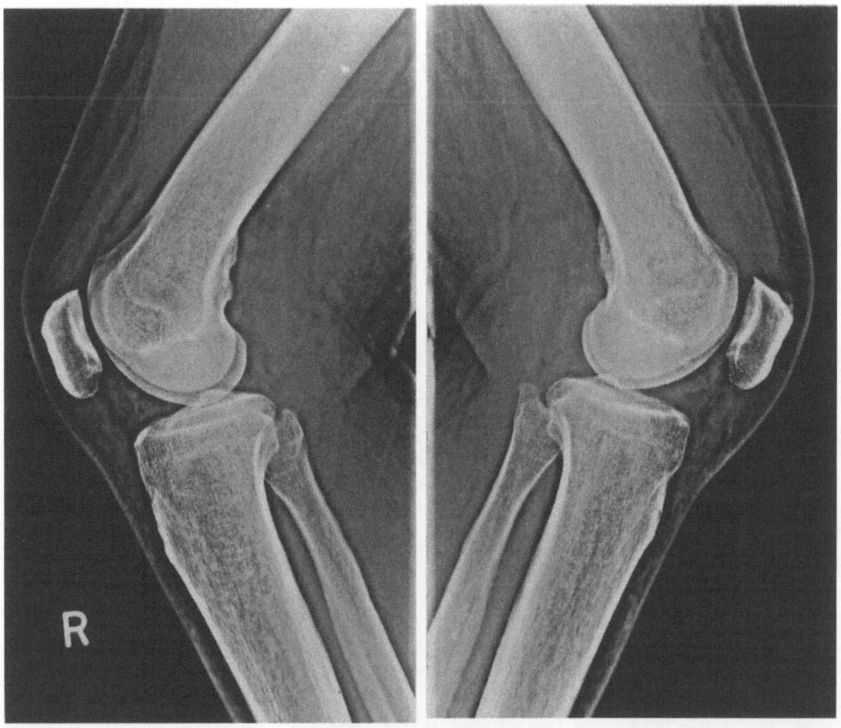

Fig. 17.10. Arthritis in the left knee (a 57-year-old female). A lateral view of both knee joints with strong frequency enhancement, mainly at around 0.25 lp/mm. Pooling of liquid is noted in the suprapatellar bursa of the left knee cap and also in the synovial bursa behind the knee joint

Both of these images show slight subperiosteal bone absorption in the fingers. The density of the output image of CR is automatically controlled within a certain range irrespective of distribution of the transmitted X-ray intensity. Therefore, even when radiograms are taken with the same X-ray equipment and under the same exposure settings, a temporal change in the X-ray attenuation of the object may not be displayed as a change in density on the film. Evaluation of osteopenia, which is important in reading bone radiograms, may present difficulties. A possible solution to this problem is the use of an aluminum step wedge. It would seem necessary to install a function displaying the X-ray absorption values, as in the case of computed tomography (CT).

Lesions in and around articulations

Arthritis of the left knee (57-year-old female)

This patient complained of pain in the left knee joint. The lateral view of the knee joint shown in Fig. 17.10 was obtained with strong frequency enhancement, mainly at 0.25 lp/mm. Collection of fluid is evident in the suprapatellar bursa of the left knee cap; it can also be recognized in the synovial bursa behind the left knee joint.

Synovial sarcoma in the superior part of the left patella (17-year-old female)

The patient visited our clinic complaining of pain in the left knee. The lateral view of the left knee joint shown in Fig. 17.11 was obtained by the same image-processing procedure as that used in Fig. 17.10. In the superior left patella, a smooth tumor of the soft tissues (1.5 × 1.8 cm) is seen. After total surgical removal, the tumor proved to be a synovial sarcoma measuring 1.5 × 1.5 cm.

Calcification of the knee joint cartilage (80-year-old female)

A bone scintigram taken to check metastasis of cervical cancer to the bones disclosed the accumulation of radionuclide (RN) in both knee joints. Following this finding, the knee joint was subjected to radiography. The patient was asymptomatic. The image shown in Fig. 17.12 was obtained by the same image-processing procedure as that used in Fig. 17.10. In Fig. 17.12, a calcified meniscus at both knee joints and osteophyte formation, each attributable to advanced age, are recognized. No swelling is seen around the knee joint. Frequency enhancement made calcification more evident.

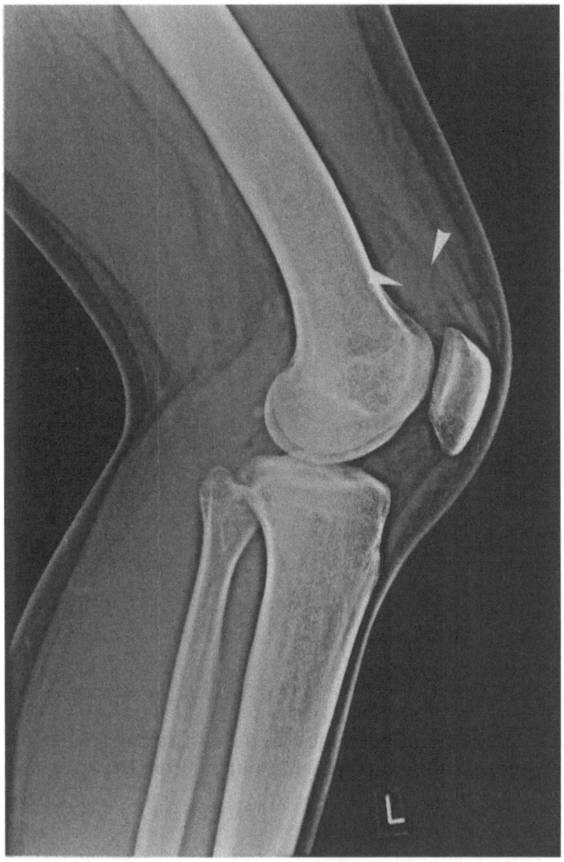

Fig. 17.11. Synovial sarcoma in the superior part of the left patella (a 17-year-old female). A lateral view of the left knee joint with the same image processing as in Fig. 17.10. Smooth tumor shadow (1.5 × 1.8 cm) is seen in the superior left patella (*arrowheads*)

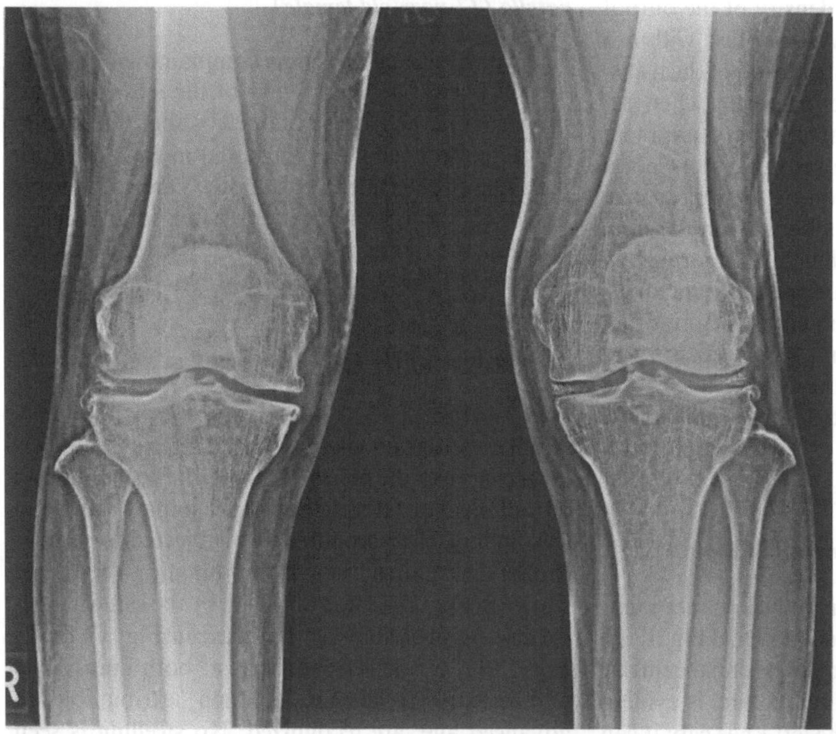

Fig. 17.12. Calcification of the knee joint cartilage (an 80-year-old female). A frontal view of both knee joints with the same image processing as in Fig. 17.10. Calcification of the meniscus in both knee joints and osteophyte formation are observed, possibly attributable to advanced age

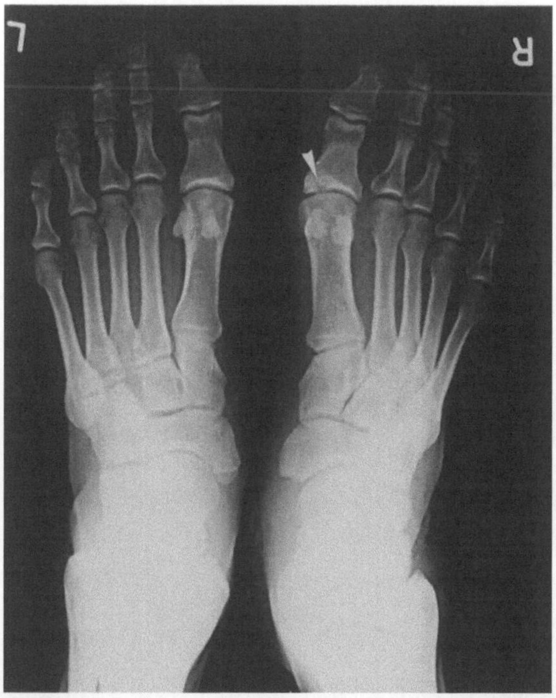

a

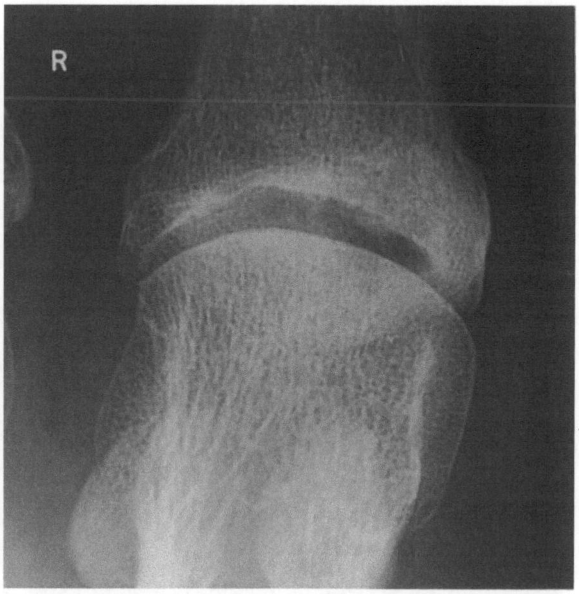

b

Fig. 17.13a, b. Gout (a 48-year-old male). **a** Image processed with weak frequency enhancement, mainly at around 0.38 lp/mm. Swelling of soft tissues is seen in the medial side of the first metatarsal joint of the right foot. Bone erosion (approximately circular) is seen in the proximal part of the first os compedale of the right foot (*arrowhead*). **b** A tenfold magnification. Bone erosion as well as the bone cristae and cortex in the surrounding regions are visualized in detail

Gout (48-year-old male)

Because the patient had hyperuricemia and pain in the hallux of the right foot, the foot was examined by radiography. The image shown in Fig. 17.13a was obtained by weak frequency enhancement (mainly at 0.38 lp/mm) at the exposure settings of 43 kV and 4 mAs. In this image, swelling of the soft tissues is seen at the medial side of the first metatarsal articulation of the right foot; bone erosion (approximately circular) is found in the proximal part of the right first proximal phalanx.

In Fig. 17.13b, an image at tenfold magnification taken at 150 cm FFD, 80 kV, 3mA, and 0.07 s shows the bone erosion and the surrounding bone cristae and cortex in detail following weak frequency enhancement (mainly at around 0.38 lp/mm).

Lesions in soft tissues

Lipoma (66-year-old female)

The patient visited our clinic following the detection of tumors in the soft tissues of the lateral side of the right brachium. The right brachium was X-rayed at 46 kV, 200 mA, and 0.064 s. A tumor with high X-ray transmission was seen in the lateral side of the right brachium. Details of this tumor are difficult to read on the normally processed image (Fig. 17.14a; after weak frequency enhancement mainly at around 0.25 lp/mm). In the enhanced image (Fig. 17.14b; obtained by making the gradation flat and applying strong frequency enhancement), blood vessels on the tumor surface and their relationship to the muscles are clear. The lipoma (11 × 12 cm, 390 g) was totally removed surgically. Adhesion of the lipoma to the fascia was not noted.

Angioma (5-year-old boy)

Frontal and lateral views of the knee were taken at 46 kV, 200 mA, and 0.05 s. The images shown in Fig. 17.15 underwent strong frequency enhancement mainly at 0.25 lp/mm. In these images, swelling of the soft tissues, dilated vessels, and calcification in the frontal part of the knee joint (particularly the superior part of the patella) are evident. No change can be noted in any bone.

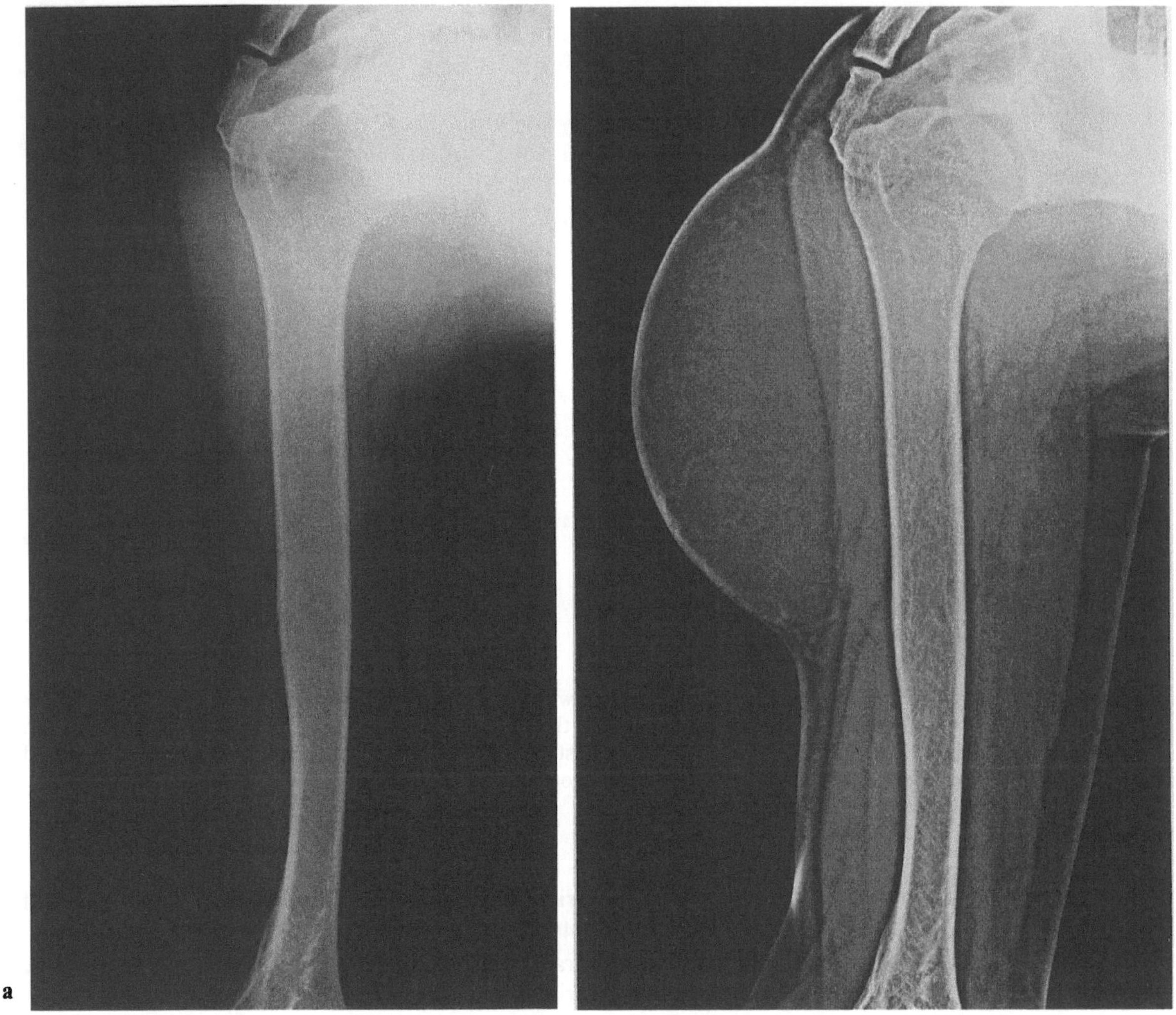

Fig. 17.14a, b. Lipoma (a 66-year-old female). **a** Image with weak frequency enhancement, mainly at around 0.25 lp/mm. A tumor with high X-ray transmission is seen on the lateral side of the right brachius (details obscure). **b** Image obtained by making the gradation flat and applying strong frequency enhancement. There is a clear relationship between the tumor and the blood vessels and muscles on its surface

Plantar ulcer (41-year-old male)

Lateral views of the calcaneus magnified fourfold were taken at the setting of 80 kV, 3.0 mA, 0.1 s, and 100 cm FFD. The normally processed image (Fig. 17.16a) was obtained by weak frequency enhancement mainly at around 0.18 lp/mm. In this image, bone prominence in the inferior calcaneus and bone sclerosis are evident, whereas changes in the soft part are obscure.

The enhanced image (Fig. 17.16b) was obtained by making the gradation flat and applying strong frequency enhancement. In this image, the relationship of the plantar ulcer to the bone lesions is clear.

Postoperative recurrence of malignant neurilemoma (41-year-old female)

Malignant neurilemoma in the posterior thigh had been removed 3 years previous to examination at our clinic. Later, the patient showed local recurrence of the neurilemoma and underwent angiography to diagnose the extent of the tumor. Intra-arterial digital subtraction angiography using CR was carried out by means of injecting 30 ml 65% Angiografin (diluted to one-third with physiological saline) at the rate of 8 ml/s via a catheter in the femoral artery (Fig. 17.17). Exposure settings were 72kV, 400 mA, and 0.04 s. Injection of the diluted contrast medium caused no pain. The femoral artery and its

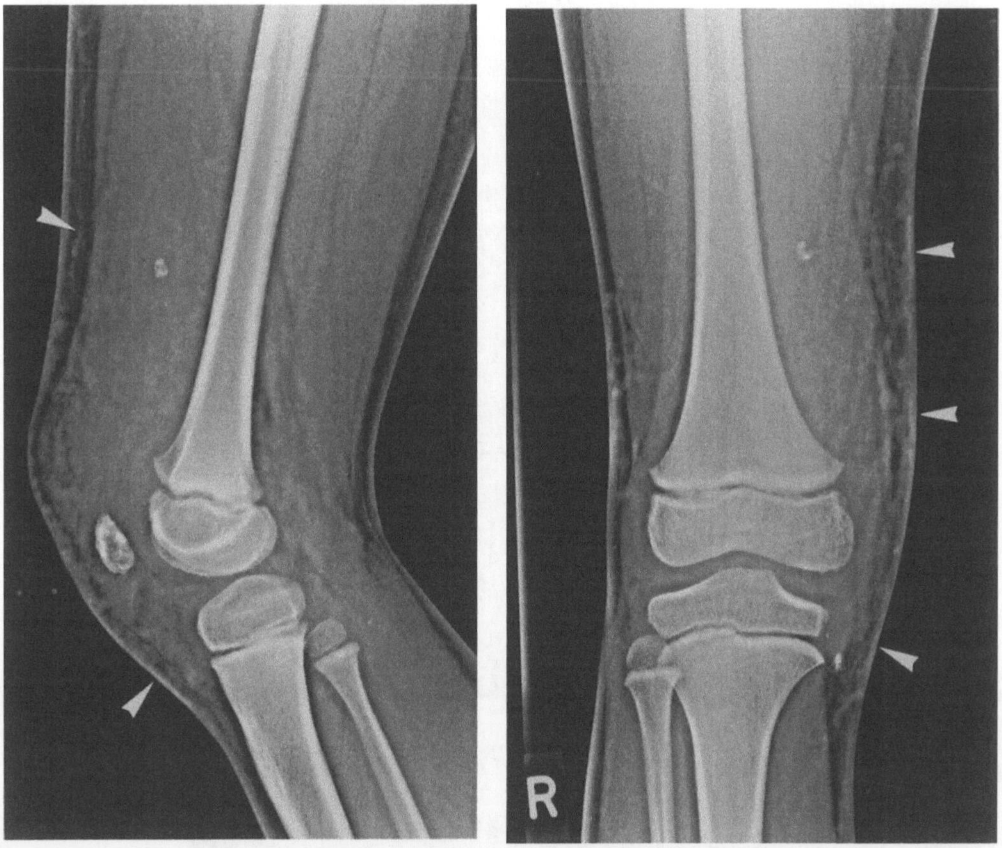

Fig. 17.15. Angioma (a 5-year-old boy). Image with strong frequency enhancement, mainly at around 0.25 lp/mm. Soft-tissue swelling centered at the superior patella and extending to the anterior part of the knee joint is seen. Dilated blood vessels (*arrowheads*) and calcification are observed

branches (even fairly thin ones) can be visualized clearly. At the site of tumor recurrence, an increase in blood vessels and tumor staining was evident.

Advantages and disadvantages of CR

The advantages and disadvantages of CR in its clinical application to the field of bones and articulations will be discussed in comparison with conventional screen/film techniques.

Wider latitude

The expanded latitude with CR allows the simultaneous observation of both the bones and the soft tissues, which is often necessary in the diagnosis of bones and articulations.

Easier application of magnification radiography

As stated above, the contrast of the image is reduced in magnification radiography due to X-ray dispersion. However, CR can make up for the reduced contrast by means of tonal conversion. Magnification radiography thus becomes easier with CR.

Spatial resolution

The best spatial resolution with CR is 2–2.5 lp/mm for an imaging plate size of 12 × 10 in. This limited spatial resolution may cause problems when CR is applied to the imaging of bones and articulations, where evaluation of minute bone cristae is often required. In such cases, the application of magnification radiography may be preferable.

We believe that further intensive research is necessary to optimize the diagnostic capability of the CR system as follows.

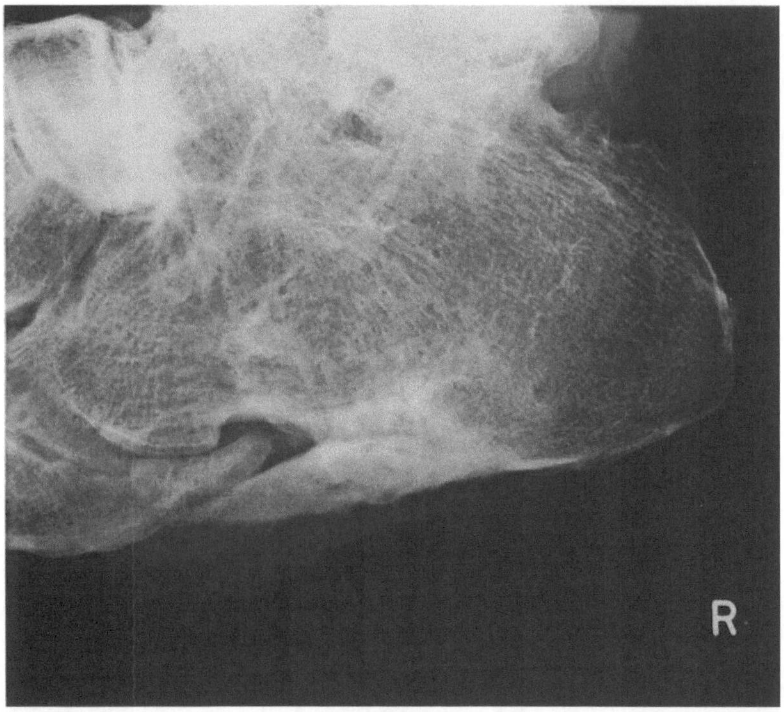

a

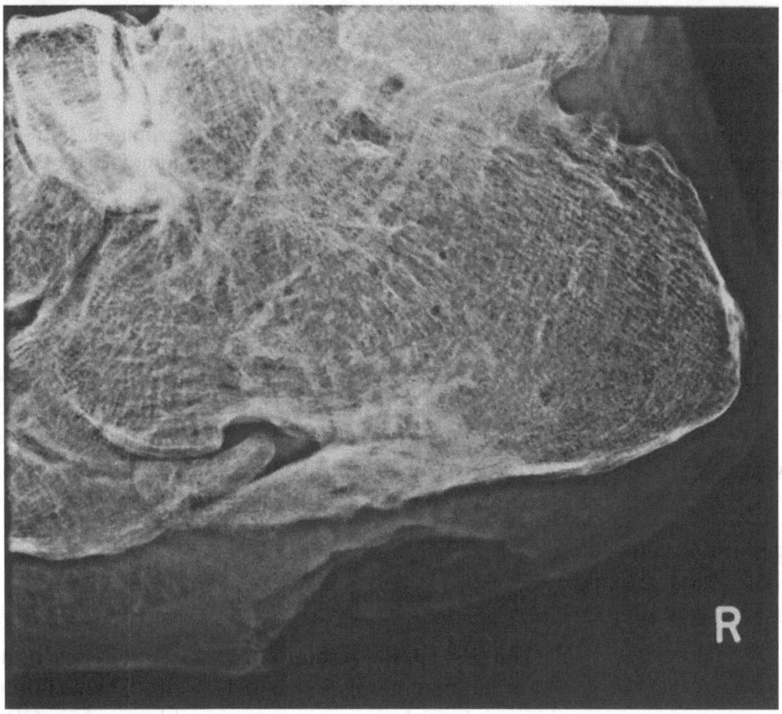

b

Fig. 17.16a, b. Plantar ulcer (a 41-year-old male). A lateral view of the calcaneus magnified fourfold. **a** Image with weak frequency enhancement, mainly at around 0.18 lp/mm. **b** Image obtained by making the gradation flat and applying strong frequency enhancement. Calcaneal lesions are evident in **a**, and the relationship between the plantar ulcer and calcaneal lesions is clear in **b**

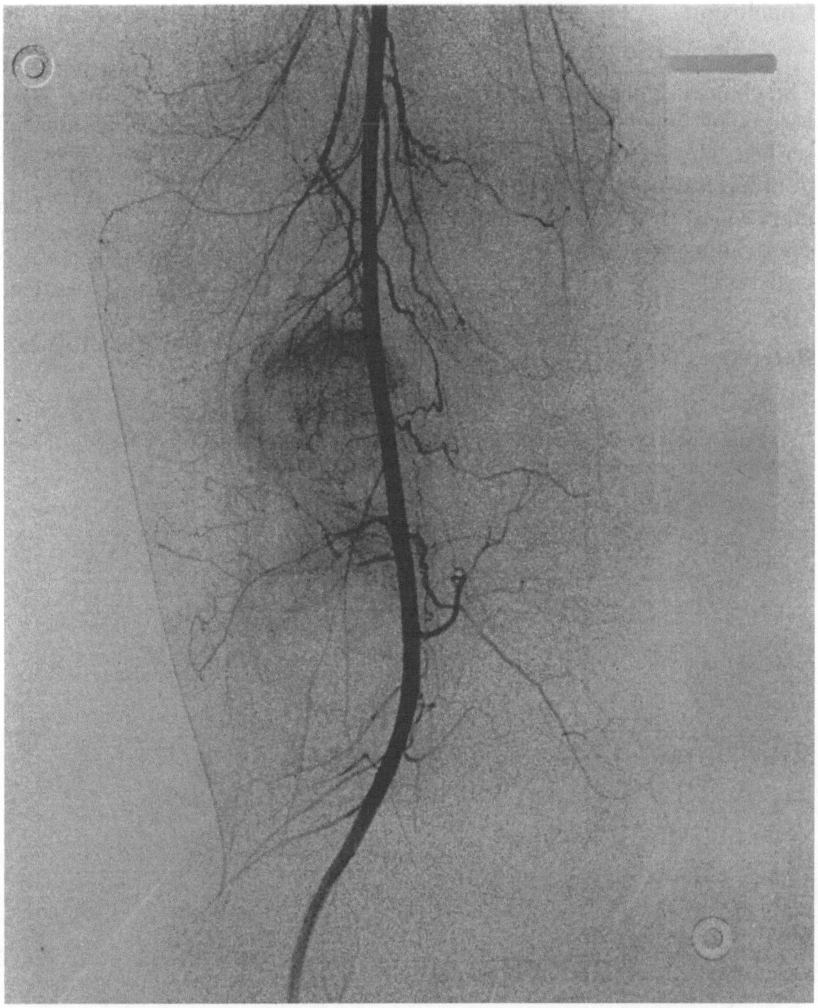

Fig. 17.17. Postoperative recurrence of malignant neurilemoma (a 41-year-old female). Intra-arterial digital subtraction angiography using CR with the injection of 30 ml 65% Angiografin (diluted to one-third) into the femoral artery at the rate of 8 ml/s. Hypervascularity and tumor stain are noted

Establishment of an appropriate type of image processing

The present CR system can be programmed for the image processing of up to 24 types of examination (programmed types). Once the program is optimized for each examination, it can be routinely used. However, our clinical experience in using CR is still not sufficient; the optimal type of image processing for bones and articulations has yet to be established. The present study indicate- that a more appropriate way to observe the bone structure involves the techniques corresponding to relative X-ray absorption without strong frequency enhancement. However, in the visualization of a soft tissue it is better to make the gradation flatter and to apply strong frequency enhancement. Therefore, when observing both bones and soft tissues, displaying the normally processed and the enhanced image on an output film seems to be the best approach.

Display of X-ray absorption values

In the determination of the absolute amount of X-ray absorption by objects, it is necessary to take a radiogram using an appropriate instrument for calibration. A system which can easily display the absolute values of X-ray absorption would be particularly useful in the diagnosis of lesions in bones and articulations.

Conclusion

The clinical application of CR in the diagnosis of lesions in bones and articulations has been described. The usefulness of CR in this field is beyond doubt; it is expected that this system will be utilized more widely from now on, though it has some minor disadvantages.

References

1. Takano M (1983) New computed radiography using stimulable phosphor. Jpn J Clin Med 41: 1395–1403
2. Matsubara K, Otou K, Sakuma S (1982) Intelligent diagnostic X-ray system—its clinical application. J Med Imagings 2: 208–213
3. Matsubara K, Sakuma S, Okae S, Yamaguchi H, Matunaga T, Otou K, Matuura H (1983) Clinical application and problems of computed raiography. Jpn J Clin Med 41: 124–133
4. Sakuma S (1972) Small field macroradiography in high magnification. Nipp Act Radiol 31: 1115–1121
5. Sakuma S (1970) Macroroentgenography in twenty-fold magnification taken by means of 50 μ focal spot X-ray tube and evaluation of its reduced image. Nipp Act Radiol 30: 205–209

Chapter 18

Pelvimetry

Katsuhiko Aoki,[1] Tokuroh Nobechi,[2] Osamu Doi,[2] Etsuko Fujimaki,[2] and Tomikazu Mizuno[2]

X-ray pelvimetry was performed on 19 subjects using computed radiography (CR) and on 100 subjects using a conventional screen/film (SF) system. CR enables the accurate identification of measuring points through image processing at a low dose; the accuracy obtained with CR in predicting the relationship between cephalopelvic disproportion and subsequent cesarean section was better than that with the conventional system. CR is an important new radiological examination modality whereby better images can be obtained at a lower dose, thereby drastically reducing the radiological exposure to both mother and fetus.

Introduction

Developments in ultrasonography have resolved the problems related to radiation exposure to the mother and fetus in the field of obsterics. However, ultrasonography is not yet capable of measuring the size or shape of the pelvic inlet. Pinard and Varnier [1] introduced radiological pelvimetry in 1897, only 2 years after Röntgen discovered X-rays. Since Albert [2] established the examination modality in 1899, the method has been in daily use as a result of its efficiency, despite the potential problems of radiation exposure. The problems encountered in radiation pelvimetry are: (1) maternal and fetal exposure to radiation and (2) measurement inaccuracy due to inferior image quality arising from the difficult exposure conditions related to the highly thickened abdominal wall in the last stage of pregnancy. CR is advantageous in that it solves these problems by reducing the exposure dose and providing an image of sufficient quality for accurate measurement through low dose radiography and image processing, employing, e.g., tonal conversion and spatial frequency enhancement.

[1] Department of Radiology, Shizuoka Children's Hospital, 860, Urushiyama, Shizuoka, 420 Japan

[2] Department of Radiology, St. Luke's International Hospital in Tokyo, 10-1, Akashi-cho, Chuo-ku, Tokyo, 104 Japan

Subjects

Since October 1983, we have been able to use CR in less urgent cases 1 day a week; Fuji Medical System processed the images. Nineteen subjects with cephalic presentation were examined by CR and one hundred by SF from October 1983 to February 1985. There were 1736 deliveries at St. Luke's International Hospital during this period, 126 (7.3%) of which were given radiological pelvimetry and 163 (9.4%) underwent cesarian section (CS).

Method

Both CR and SF were employed using the same type of X-ray equipment—GE, DXS 3560. Stationary and moving grids were employed, respectively, for the Guthmann and Martius views, and both had the same specifications of 40 lines/cm, grid ratio of 4 : 1, and FFD of 100 cm. LT-II screen and RX film were used in the SF system. With both the Guthmann and Martius views, we employed exposure conditions of 120 kV_p and 40–60 mAs for an abdominal thickness of 30–34 cm.

Three imaging plates were placed together in a cassette and exposed simultaneously as a single exposure for each CR examination. From these three imaging plates, four images were made—three directly from the respective plates and an additional one by digitally summing the three images in a computer (superimposed image). Image processing, e.g., tonal conversion and spatial frequency enhancement, was conducted on all four images. The conditions employed for image processing are shown in Fig. 18.1a, b. Four subjects were exposed at the same dose as in the conventional SF system. From the images thus obtained, the exposure time was gradually shortened in two cases at the same tube current and voltage; the doses were progressively decreased to 60% (30 mAs), 50% (24 mAs), 25% (15 mAs), 17% (10 mAs), and 10% (6 mAs) that of the SF system in order to obtain a minimum exposure dose for CR at which measurable images could be obtained. Seven cases were exposed using CR at

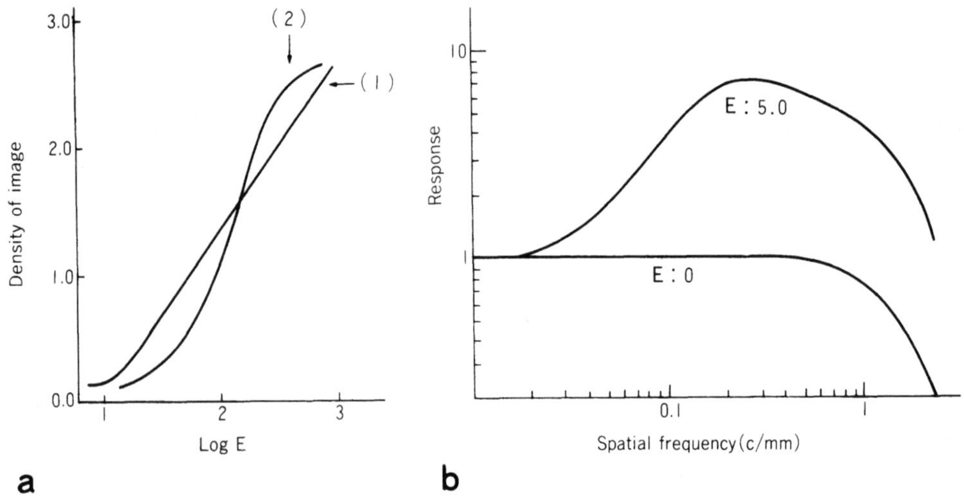

Fig. 18.1. **a** Characteristic curve of pelvimetry. *(1)* Characteristic curve of CR. This curve had low γ and good linearity for wide range. *(2)* Characteristic curve of the conventional SF system. **b** Response curve of spatial frequency enhancement; the degree of enhancement (E) was 5.0

10% the exposure dose of the SF system.

Using the same X-ray pelvimetric equipment, thermoluminescence dosimeters (TLD) were placed at 6-cm intervals in the water phantom at a depth of 34 cm, and doses at different levels were measured. The radiation conditions were based on those of the SF system, combining Kyokko LT-II and Fuji RX to achieve 120 kV_p, 60 mAs, and half, one-quarter, one-sixth, and one-tenth the mAs value for TLD measurement was carried out using a TLD reader (Kyokko TLD reader 1200).

Positive predictive values for CS based on the results obtained with SF and CR pelvimetry were compared. The recommendation for CS was made by assessing the cephalopelvic disproportion CPD index [3]: the greatest anteroposterior (AP) diameter of the fetal head/transverse diameter of the pelvic inlet in relation to the greatest AP diameter of the fetal head. If the CPD index was 7.1 or less, vaginal delivery was recommended; in the range 7.1–9.8, cases were considered borderline; CS was recommended for cases above 9.8. Since it was necessary in practice to prepare for both vaginal delivery and CS in borderline cases, they were included in the CS group for statistical analysis.

Results

Low dose radiography

As shown in Fig. 18.2b, images obtained with CR at the same dose as the SF system were far better in displaying and measuring the AP and transverse diameters of the pelvic inlet, the greatest AP diameter of the fetal head, obstetrical true conjugate diameter, and the position of the fetal head. Figure 18.3 shows the Martius views at 100% (Fig. 18.3a), 60% (b), 50% (c), 25% (d), 17% (e), and 10% (*f*) the dose of the conventional SF system. Even at one-tenth the dose, the pubic symphysis, sacrum, and ischial spine were clearly demonstrated and thus presented no problems in measurement.

Figure 18.4 shows the Martius views of the first, second, and third plates as well as the superimposed image at 100% and 10% the SF system dose. As the energy becomes attenuated passing through the plates, the image quality worsens. However, even at the dose equivalent to 10% that of the SF system, excellent measurable images could be obtained with the superimposed image method by CR.

Dose measurement using the water phantom

Figure 18.5 shows the dose exposed on the incident surface of the film and at 0, 8, 16, and 24 cm below the water surface when the tube voltage and current were maintained at a constant level and the exposure time varied. A decrease in exposure dose on the water surface was reflected by the decrease in the dose at each level of the water phantom at substantially the same ratio. For instance, the dose at 16 cm, which is considered to be the center of the body, is 135×10^{-2} mGy at the same mAs as that of the SF system (200 mA, 0.25 s), while it is 16×10^{-2} mGy at one-tenth the mAs (100 mA, 0.05 s). This indicates that the exposure dose becomes about one-tenth that of the SF system.

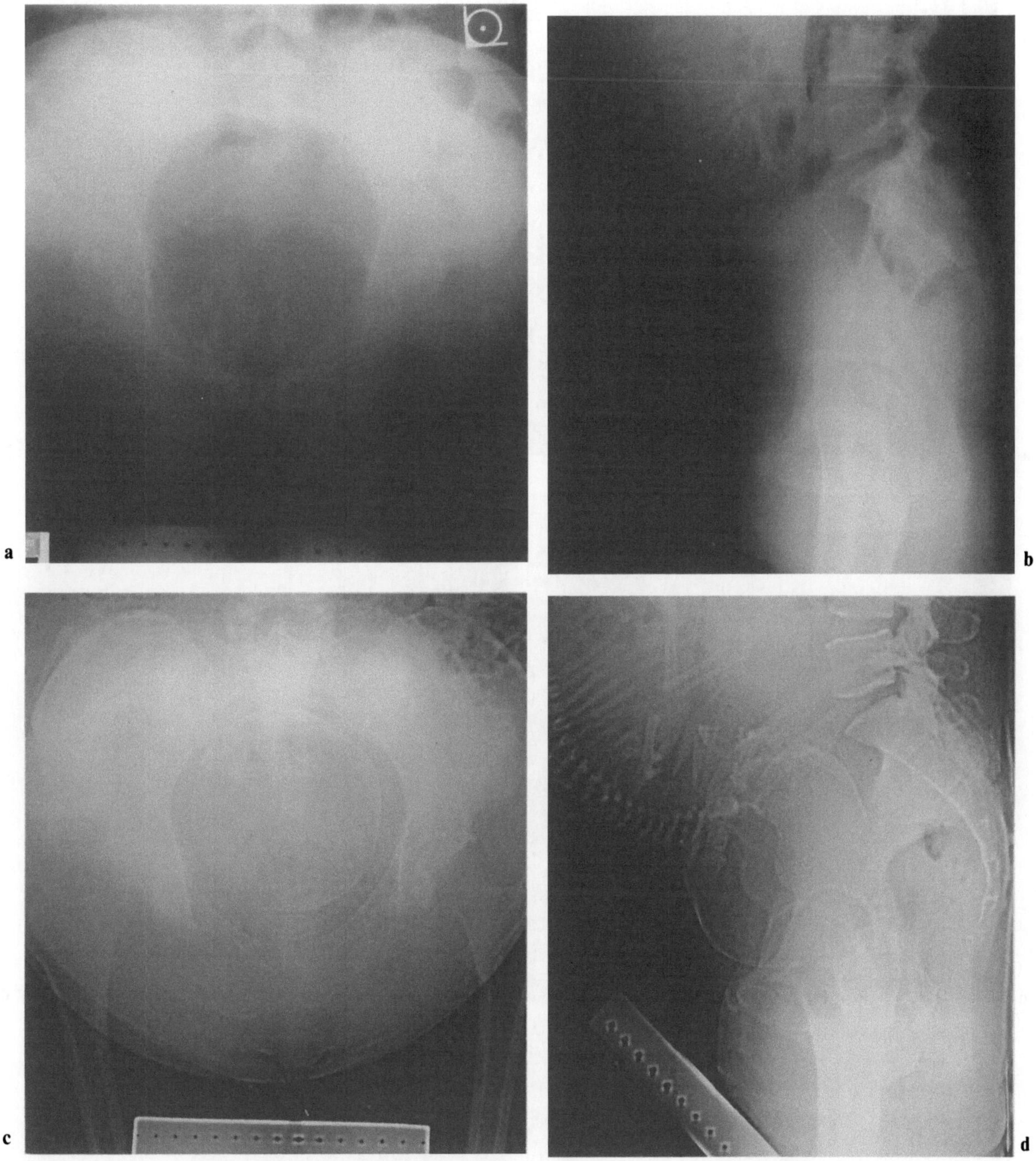

Fig. 18.2a–d. X-ray pelvimetry. **a**, **b** Conventional SF system. **c**, **d** CR

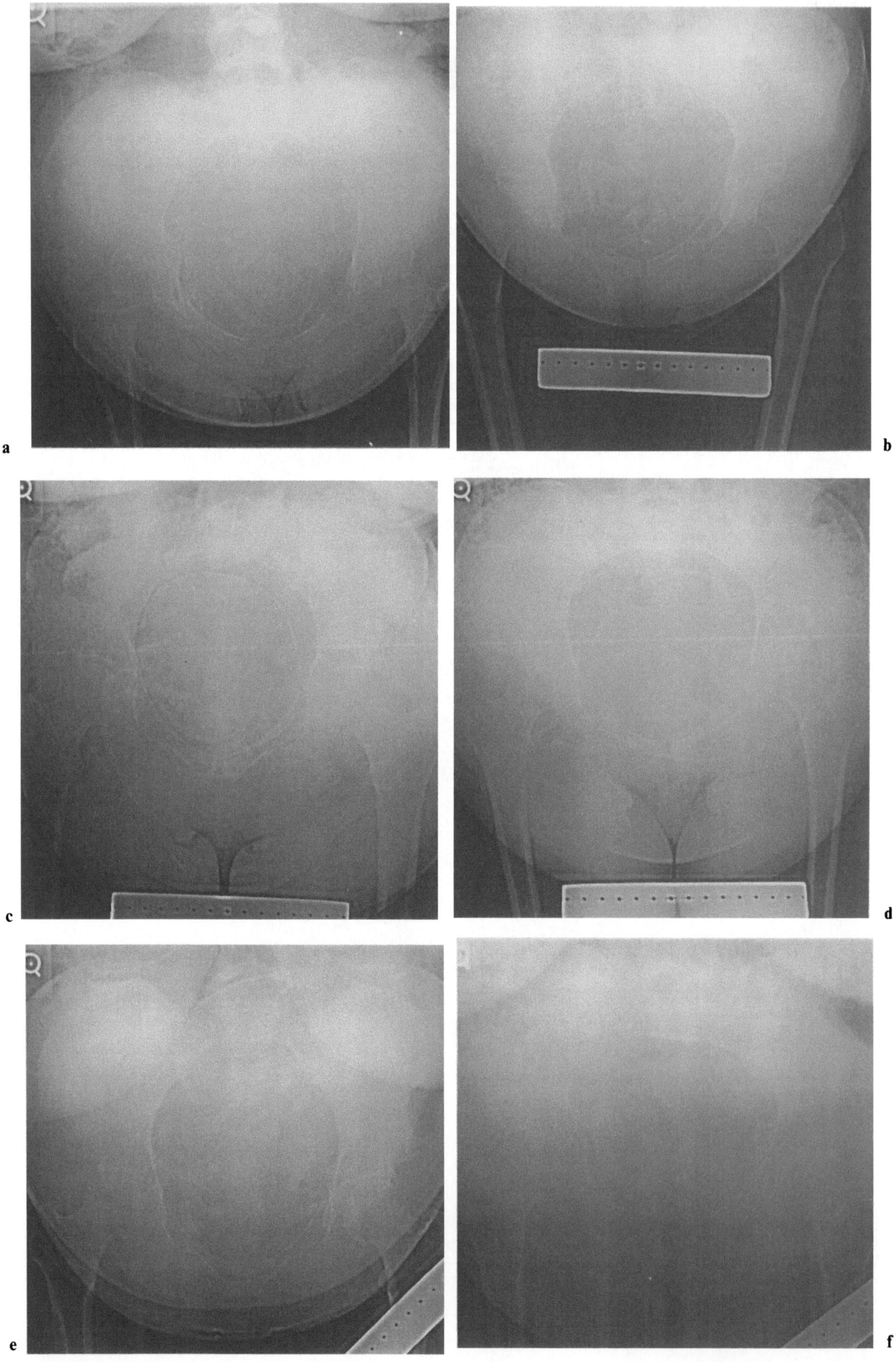

Fig. 18.3a–f.

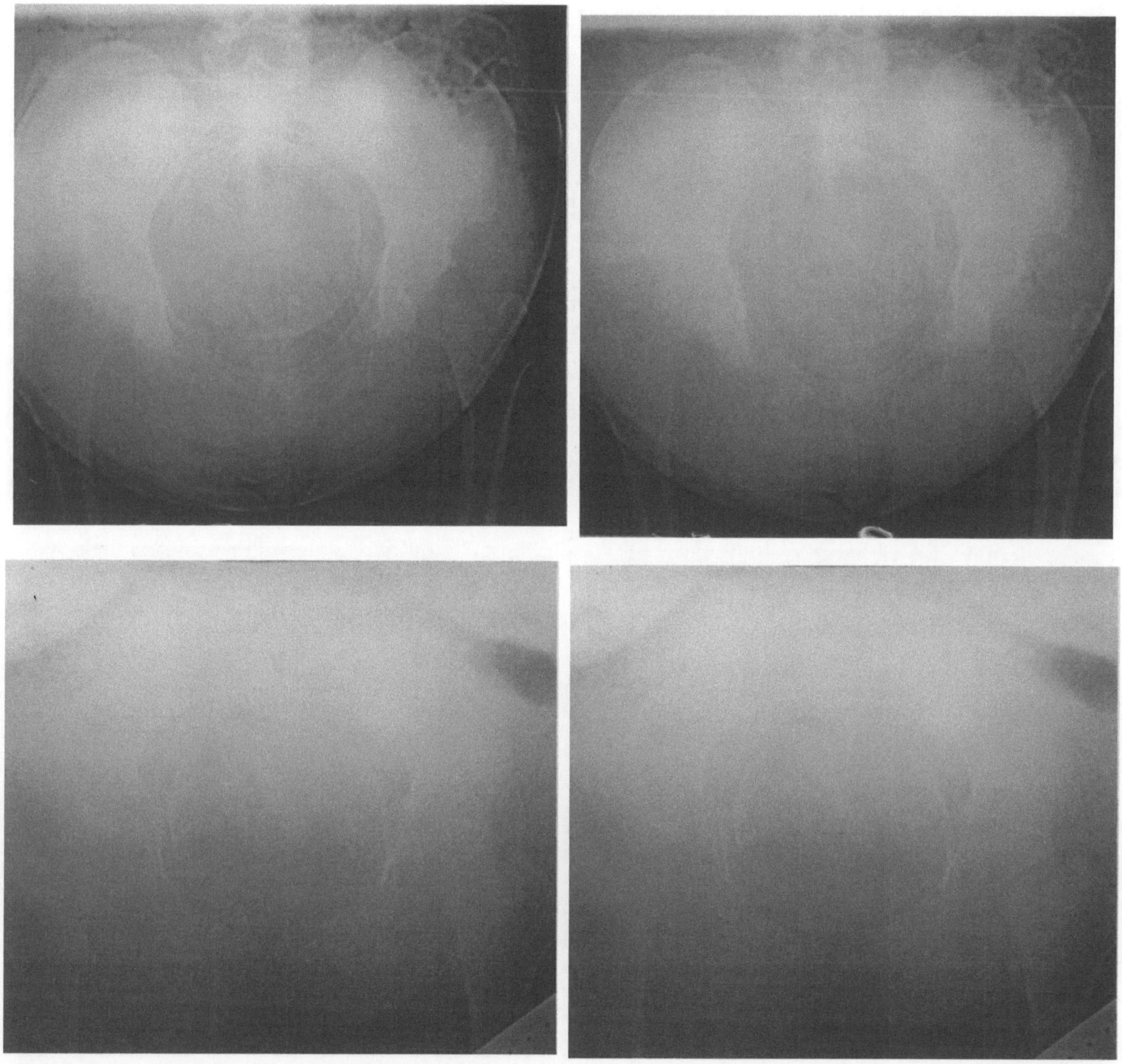

a b c d

Fig. 18.4a–h. Martius view by CR. **a** Superimposed image using three plates (the first, second, and third plates) 100% dose. **b** Image of the first plate, 100% dose. **c** Image of the second plate, 100% dose. **d** Image of the third plate, 100% dose. **e** (*see over*) Superimposed image using three plates (the first, second, and third plates) 10% dose. **f** Image of the first plate, 10% dose. **g** Image of the second plate, 10% dose. **h** Image of the third plate, 10% dose

Accuracy in predicting indication for cesarian section

Measurement was possible for all 100 cases examined with the SF system except for those with a breech presentation; however, some positions were inadequate and a few were underexposed. All 19 cases examined by CR showed a satisfactory measurement result. With the Martius view, the lower the station (STN) of the fetal head, the more enlarged was the head shown on the film. Therefore, the results were classified into two groups in order to evaluate predictive accuracy; STN 0 to −3 and STN −4 or less (floating). Table 18.1 shows the cephalopelvic disproportion (CPD) index. While the CPD obtained in radiography does not necessarily reflect all the factors predicting the necessity for CS, we shall provisionally assume that there exists a direct correspondence between a positive CPD and the indication for CS and vice versa. Then, if the STN is 0 to −3, the sensitivities of the SF system and CR are 50% and 100%, the specificities 83.3% and 100%, and the accuracies 75% and 100%, respectively. With floating, the sensitivities of the SF sys-

◁ **Fig. 18.3a–f.** Martius views by CR. **a** 100% dose of the convention SF system; **b** 60%; **c** 50%; **d** 25%; **e** 17%; **f** 10%

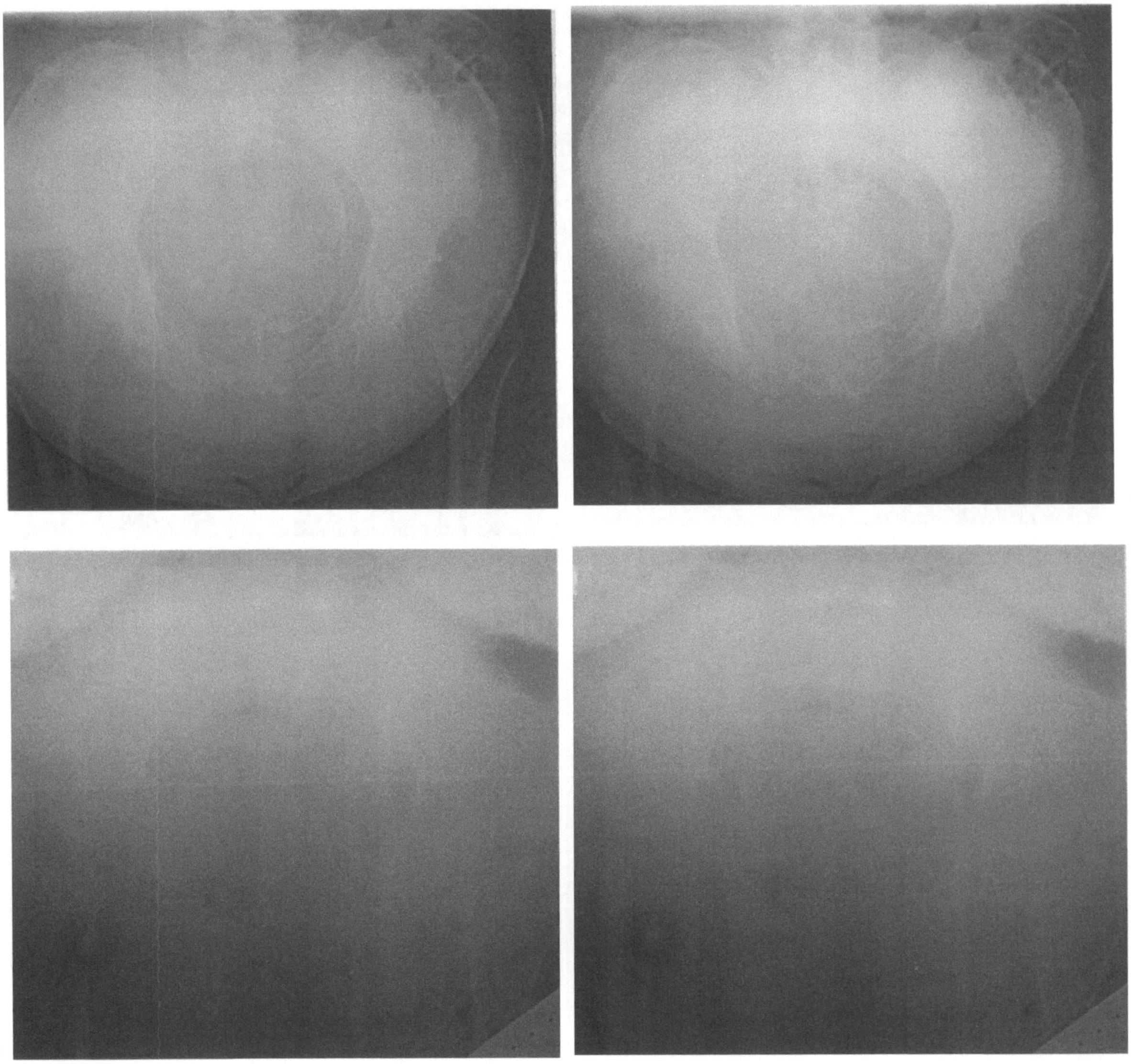

Fig. 18.4e–h.

tem and CR are 59.3% and 60%, the specificities 47.6% and 62.5%, and the accuracies 51.4% and 61.5%, respectively. While accuracy in the CPD on which the recommendation for CS is based seemed to be better in CR, there was no statistically significant difference between the SF system and CR irrespective of STN.

Discussion

There are problems of radiographic exposure to the maternal gonads and whole fetal body in X-ray pelvimetry, and therefore it should only be done when there is a definite indication. The incidence of infantile leukemia and cancer increases when pregnant women are exposed to X-rays [4, 5]. Various efforts such as radiography without grids [6] and the use of highly sensitive SF combinations [5, 7] are being made in order to obtain pelvic radiographs at low doses. The conventional SF system often produces an underexposed film and consequent measurement inaccuracy because it is not expedient to reduce the radiation dosage with the maximally thick abdominal wall of a pregnant subject.

The cases examined with the SF system in the present study included emergency cases and cases with high STN classified by the CPD. Accuracy in the CPD for predicting subsequent CS was somewhat inferior in the cases examined with the SF system to that with CR. Besides, there was no statistically significant difference between the two systems, and therefore the modality seems to be irrelevant so long as the measurement points are precisely identified.

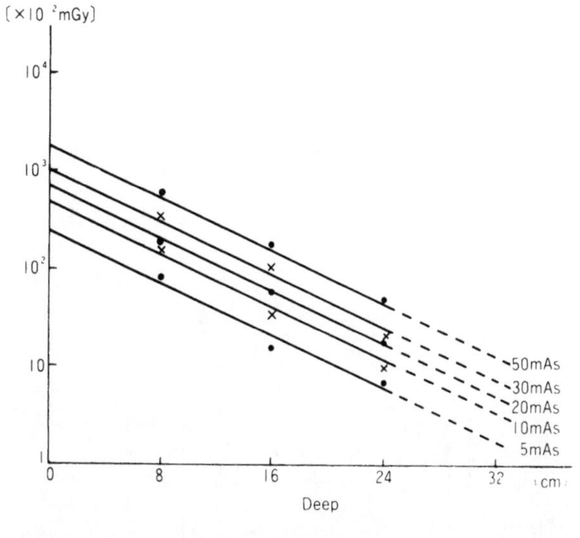

Fig. 18.5. Dose measurement using the water phontom

Table 18.1. CPD index.

	STN: 0 to 3		STN: floating	
	CR (%)	SF system (%)	CR (%)	SF system (%)
Sensitivity	100	50	60	59.3
Specificity	100	83.3	62.5	47.6
Accuracy	100	75	61.5	51.4

Pelvimetry requires clear images of the pelvic bone structure and cortical bone margin of the fetal head; the other soft tissues are not considered important in the diagnosis. Even at the low exposure dose, CR can give a high contrast to the bone margins by means of image-processing techniques, such as tonal conversion and spatial frequency enhancement; therefore, it becomes easy to make the measurements required for diagnosis. In the case of a pregnant woman whose abdominal thickness is 37 cm, her pelvic bone structure and the cortical bone margin of the fetal head are not so clearly visible even when the image is processed under normal conditions which are not optimized for pelvimetry in CR (Fig. 18.6a). By enhancing the spatial frequency and increasing the gradient of the characteristic curve (Fig. 18.6b), both the maternal bone structure and fetal bone margin became distinctly visible, thereby rendering easy and accurate measurement. Although enhancement of spatial frequency deteriorates granularity, this presents no problems if measurement is the sole object. A superimposed image, which utilizes the sum of the X-ray energy deposited on the three imaging plates, improves granularity.

A superimposed image is a requisite in reducing dosage. If the plates are not neatly piled inside the cassette, the contours become blurred after summation Figure 18.7a shows double cortical bone margins because of a slight misalignment of the third plate. This caused difficulties in measurement. In this case, the third plate was removed and the remaining two plates were used for summation and subsequently films with precise contours were obtained (Fig. 18.7b).

Plates that are other than 8 × 10 in. in size cannot be used to take life-size pictures as in the SF system, and measurement errors may become multiplied when obtaining actual values. While pelvimetry does not require a precise error margin, such precision is necessary in transplantation and prosthetic implantation.

Watanabe et al. [8] obtained deviations in pelvimetry noise variation from the images obtained through shortened exposure time of the pelvic phantom in CR and sought the best image quality obtainable. They reported a reduction of the exposure dose to one-eighth that of the SF system with the one-plate method and 1/19 in the superimposed method using three plates. Since the cases studied were actual clinical cases, the authors were not in a position to perform repeated exposures. Therefore, we conducted radiography by sequentially reducing the dose from 100% to 10% that with the SF system, but not below 10%. Considering the simplicity in actual operation and the minimum image quality required for measurement, the optimal dose for pelvimetry in CR is considered to be one-tenth that of the conventional SF system.

Digital radiography [9] using computed tomography (CT) is a means of reducing the exposure dose. The Martius view is difficult with this system because of the positioning. Reports have also been made on observations on the bone structure of pregnant women and the fetus using magnetic resonance imaging (MRI), which is entirely free of exposure to ionizing radiation [10, 11]. While MRI still needs higher spatial resolution to display the contours more sharply, this imaging diagnosis method should receive further attention as a technique which eliminates the problem of exposure.

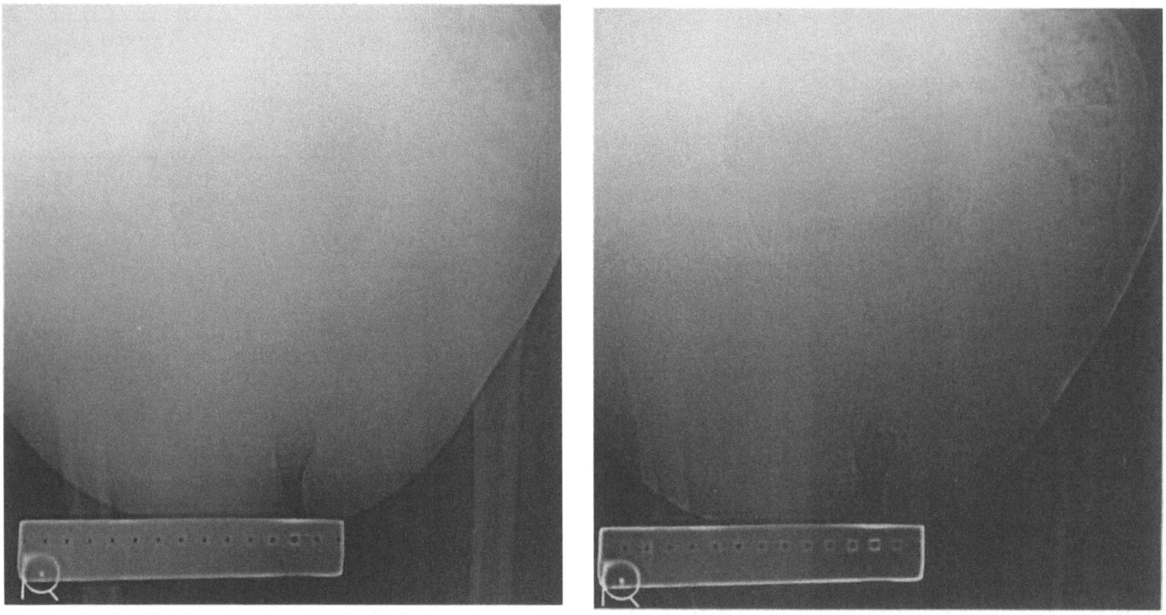

a b

Fig. 18.6a, b. CR images (the case of a pregnant woman whose abdominal thickness was 37 cm). **a** Image processed under normal conditions (low γ and E of 5.0, see Fig. 18.1). **b** Image processed under other conditions (high γ and E of 9.0)

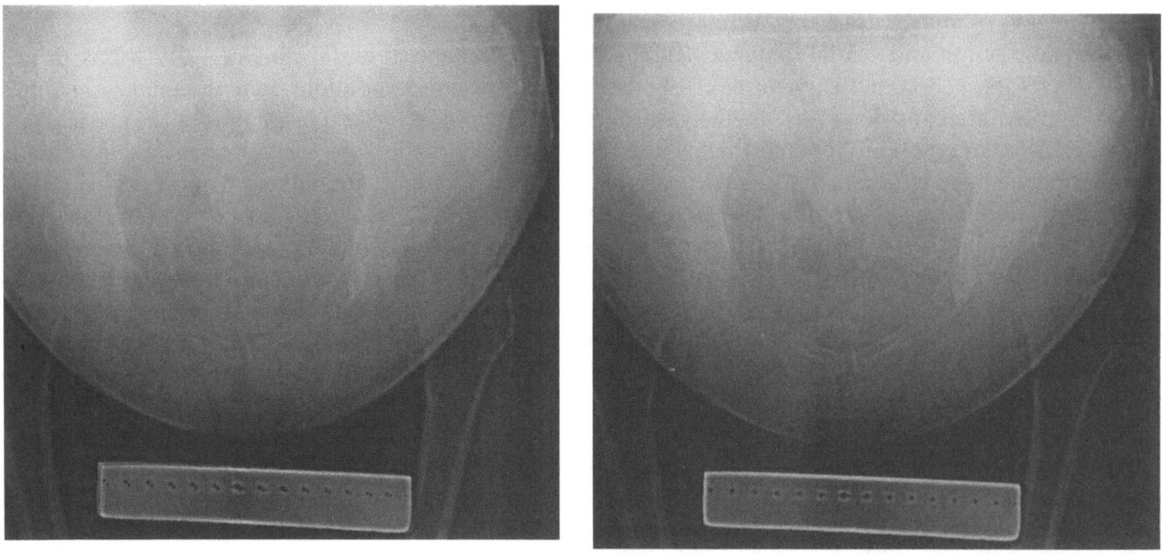

a b

Fig. 18.7. Superimposed images by CR. **a** Superimposed image using three plates. **b** Superimposed image using two plates. Sharpness of the superimposed image using three plates was reduced because third plate was misregistered

Conclusion

Advantages of pelvimetry with CR

1) Tonal conversion and spatial frequency enhancement enable the identification of distinct points in the bone structure and fetal head necessary for measurment.

2) Reduction in exposure dose radically reduces the dose on the maternal gonads and the whole fetal body, which are considered to be most sensitive to radiation.

3) Deterioration of image quality accompanying the decrease in exposure dose is sufficiently compensated for by the superimposed image using three plates.

Disadvantages of pelvimetry with CR

1) Precise alignment of the plates is an essential requirement in radiography in order to prevent blurring of the contours in a superimposed image.

2) The small size of the output film involves contraction of the images and increases measurement errors. These errors, however, are negligible in pelvimetry.

References

1. Pinard, Varnier (1897) Beckenphotographie und Beckenmessung mittels X-Strahlen. Zentralbl Gynäk 21:1145
2. Albert (1899) Verwertung von Röntgenstrahlen für die praktische Geburtshilfe. Berlin Klin Wchnschr 36:535
3. Sawasaki S (1968) Cephalo-pelvic disproportion index. Sanka to Fujinka (Obstetrics and Gynecology) 43:833–840
4. Stewart Am, Kueale GW (1970) Age-distribution of cancer caused by obstetric X-rays and their relevance to cancer latent periods. Lancet II:4–8
5. Iba S, Satoh K, (1983) Estimation of foetus risk from X-ray pelvimetric examinations. Japanese Journal of Radiological Technology 39(6):862–870
6. Russell JGB, Hufton A, Pritchard C (1980) Technical notes, gridless (low radiation dose) pelvimetry. Br J Radiol. 53:233–236
7. Hatanaka M, Koyama Y, Akashi K, Kusakai K, Sumita K (1983) A new method of X-ray pelvimetry. Fuji Med Forum 142:24–26
8. Watanabe Y, Itoh E, Kitamura M, Ikawa M, Miyasaka Y, Hachiya J, (1983) Dose reduction in X-ray pelvimetry (Martius method) using Fuji computed radiography. Fuji Med Forum 143:15–18
9. Federle MP, Cohen HA, Rosenwein MF, Brant-Zawadzki MN, Cann CE (1982) Pelvimetry by digital radiology: A low-dose examination. Radiology 143: 733–735
10. McCarthy SM, Stark DD, Filly RA, Callen PW, Hricak H, Higgins CB (1985) Obstetrical magnetic resonance imaging: Maternal anatomy 1. Radiology 154(2):421–425
11. McCarthy SM, Filly RA, Stark DD, Hricak H, Brant-Zawadzki MN, Callen PW, Higgins CB (1985) Obstetrical magnetic resonance imaging: Fetal anatomy 1. Radiology 154(2):427–432

Chapter 19

Pneumoconiosis

Hisao Shida[1]

The International Labor Organization (ILO) 1980 international classification of radiographs of pneumoconioses is used widely throughout the world [1]. It has long been recognized that exposure due to radiography of the chest has a marked influence on the radiographic appearance of lesions of pneumoconiosis. Consequently, in radiographs showing evidence of pneumoconiosis, readers have difficulty in applying the ILO classification in the diagnosis of the disease unless the exposure when making the radiographs is maintained within an optimum range. There are some problems with the ILO standard films: These selected films have various types of tonal conversion, inadequate optimum density, the effect of a characteristic curve of image-intensifying screens, and faults such as unevenness in processing. These ILO standard radiographs for pneumoconiosis do not then always have a standardized film quality, even though there was general consensus about them.

The system of computed radiography (CR) is without any of the faults of ordinary films and is probably the most advanced imaging system available at present [2, 3]. It should thus be capable of making an excellent set of standard radiographs for pneumoconiosis that would exactly suit the needs of readers. CR has the merits of improvement of diagnostic accuracy, provides much more diagnostic information, and the dosage is far lower [4]. The application of CR will, however, only be satisfactory if the radiographic quality is reasonable for the diagnosis of pneumoconiosis. CR images are highly compatible with human vision and permit a more accurate diagnosis by the use of tonal conversion.

The spatial frequency modification entails that images of fine, small, rounded, or irregular opacities can be selectively accentuated to the desired degree to eliminate contrast deficiencies and blurring, and this insures sharp images that can be easily read. However, there is one shortcoming with this system in the film size, which is half the length of a routine chest X-ray film. This means that fine, small, rounded opacities (diameter up to 1.5 mm) are difficult to discern on the CR film. By the use of appropriate spatial frequency modification, it is considered that the small size of the film is not an impediment.

Figure 19.1 shows an appropriate type of tonal conversion. The E curve is usually taken for the diagnosis of pneumoconiosis or other pulmonary disease. Figure 19.2 shows the application of variable spatial frequency modification in the diagnosis of pneumoconiosis. It is supposed that a frequency of 2–4 and enhancement of 0.5–2.0 are adequate for reading.

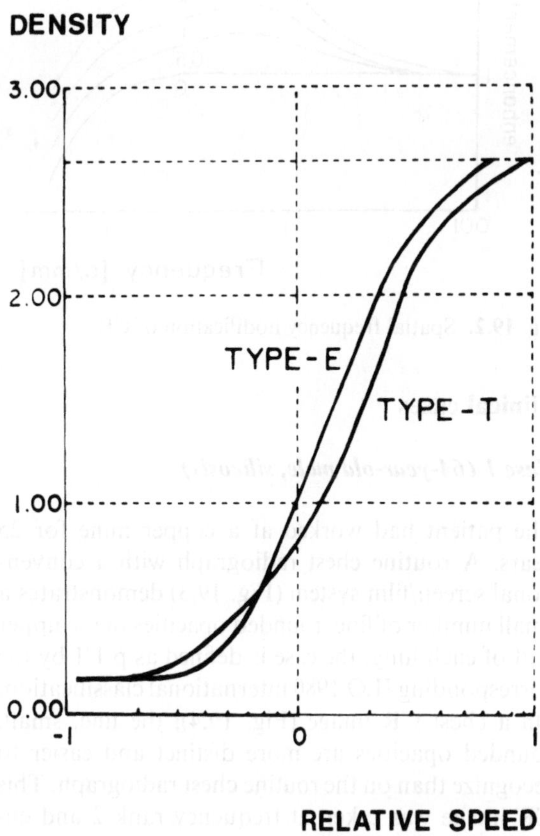

Fig. 19.1. Total conversion. The E curve is usually taken for diagnosis of pneumonconiosis

[1] Chief of Department of Radiology, Rosai Hospital for Silicosis. 632 Takatoku, Fujiwaramachi, Shioya-gun, Tochigi, 321-25 Japan

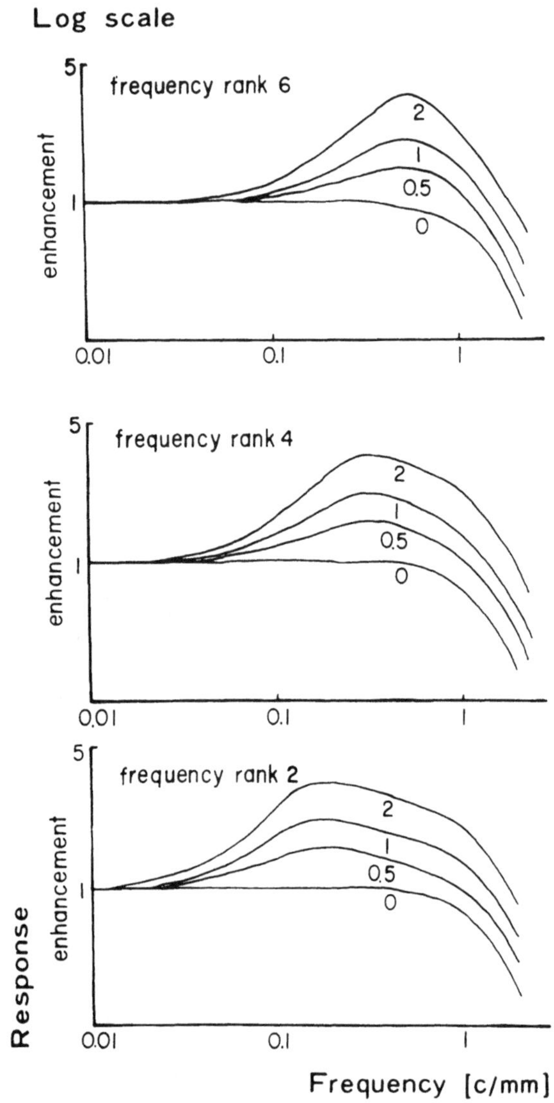

Fig. 19.2. Spatial frequency nodification of CR

Clinical cases

Case 1 (64-year-old male, silicosis)

The patient had worked at a copper mine for 25 years. A routine chest radiograph with a conventional screen/film system (Fig. 19.3) demonstrates a small number of fine, rounded opacities in the upper half of each lung; the case is defined as p 1/1 by the corresponding ILO 1980 international classification. On a chest CR image (Fig. 19.4), the fine, small, rounded opacities are more distinct and easier to recognize than on the routine chest radiograph. This CR image was taken at frequency rank 2 and enhancement 0.5. At the level of frequency rank 2 and enhancement 2.0 (Fig. 19.5), the small, rounded opacities are much more clearly visualized than in Fig. 19.4. The chest image of Fig. 19.6 shows a different type of tonal conversion, high-frequency modification, and strong enhancement. The bronchial system, hilar region, and aorta are sharply demonstrated, but fine vascular shadows are diminished due to increased noise.

Case 2 (64-year-old male, silicosis)

The patient had worked at a copper mine for 25 years. A plain chest roentgenogram (Fig. 19.7) reveals coarse, disseminated, small, rounded opacities, and eggshell calcification of the hilar lymph nodes associated with pleural thickening of both apices. The case is defined as q 2/2, es, pl by the corresponding ILO 1980 classification. Figures 19.8 and 19.9 are close-up views of the right lung and demonstrate various types of tonal conversion and spatial frequency modification. Figure 19.8a was taken at frequency rank 2, enhancement 0.5, and Fig. 19.8b at frequency rank 2, enhancement 2.0. The small, rounded opacities are more clearly visible than on the routine chest roentgenogram.

Case 3 (68-year-old male, foundry worker's lung)

The patient worked in a foundry for 13 years. A plain chest radiograph (Fig. 19.10) reveals densely distributed irregular opacities throughout the lungs and honeycombing of both lower lungs. The left lung appears hyperlucent compared with the right; bullae are also evident in the upper lateral margin of the left lung (Fig. 19.10, arrow). This radiograph is defined as s 2/2, bu, ho by the corresponding ILO 1980 classification. A chest digital radiograph with frequency rank 2 and enhancement 0.5 (Fig. 19.11) demonstrates clearly a pattern of honeycombing, bullae, and an especially accentuated reticular pattern. In the chest image with strong enhancement of 2.0 (Fig. 19.12), edge sharpness is generated, but due to increased noise it is difficult to distinguish irregular opacities and the vascular pattern. The image of Fig. 19.13 taken with moderate frequency rank 4 and enhancement 2 reveals a sharply visualized fine reticular formation. It is noticeable that there is a change from a reticular to a nodular pattern with high-frequency modification.

CT images (Fig. 19.14) show numerous bullae (arrows), fibrotic strands, or membranous opacities (arrowheads) associated with various forms of pleural thickening. These findings are compatible with those of the digital chest images, and the hyperlucency of the left lung is caused by the presence of the large bullae in the posterior lung.

Fig. 19.3. Silicosis pl/1 in a 64-year-old male. Routine chest film

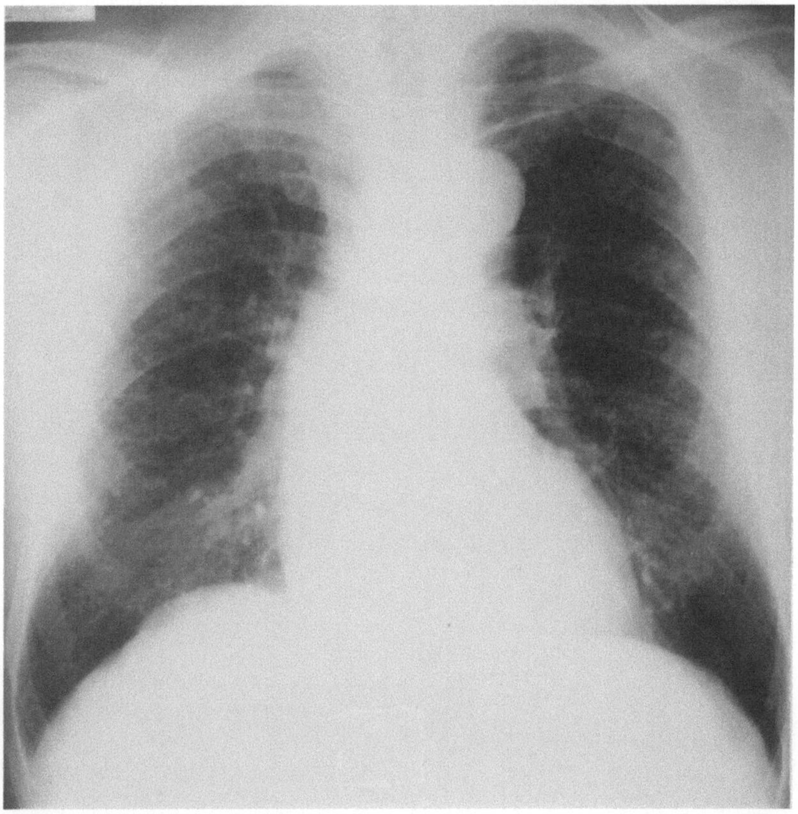

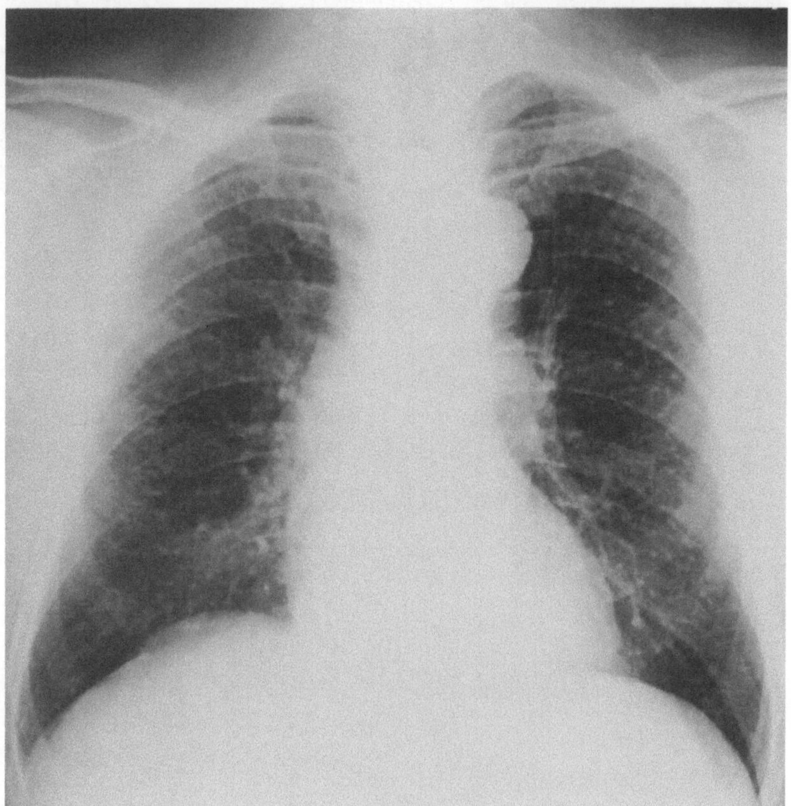

Fig. 19.4. Digital image. Tonal conversion E, frequency rank 2 (low), enhancement 0.5

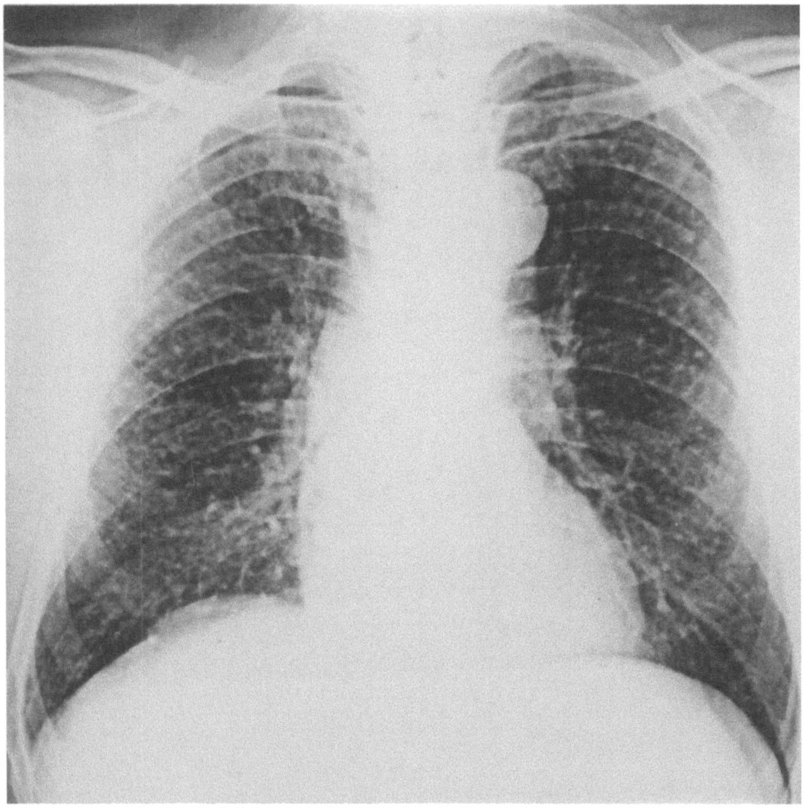

Fig. 19.5. Digital image. Tonal conversion E, frequency rank 2 (low), enhancement 0.5

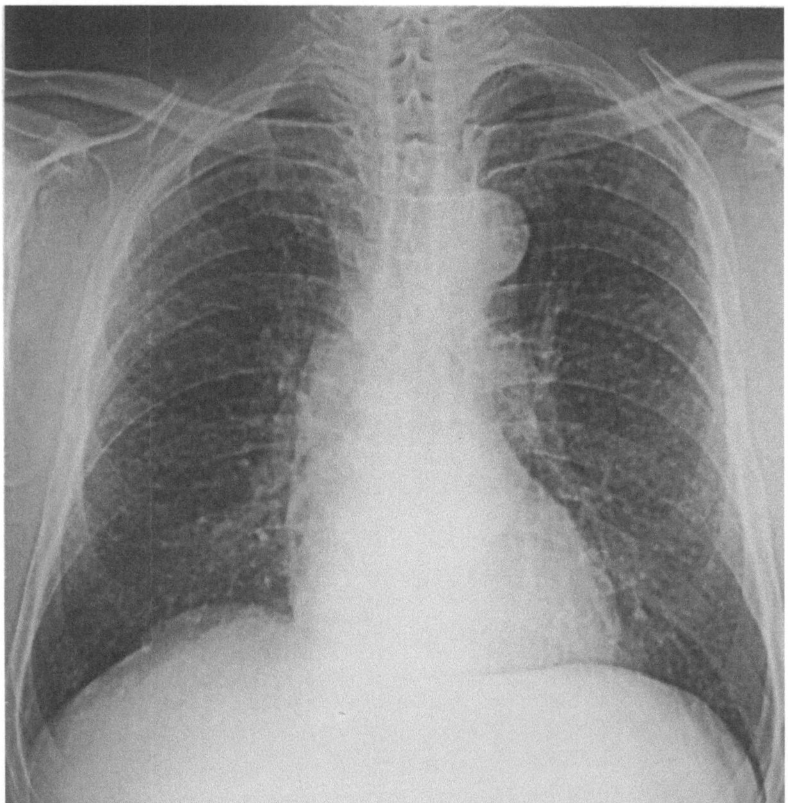

Fig. 19.6. Digital image. Tonal conversion S, frequency rank 3 (moderate), enhancement 3.0

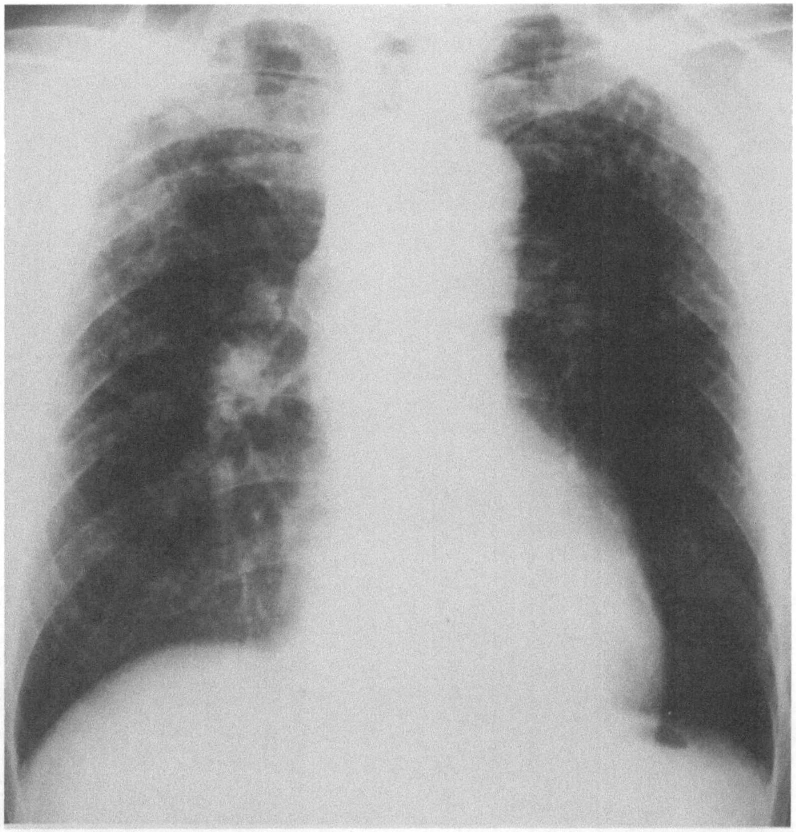

Fig. 19.7. Silicosis q2/2, es, pl in a 64-year-old male. Routine chest film

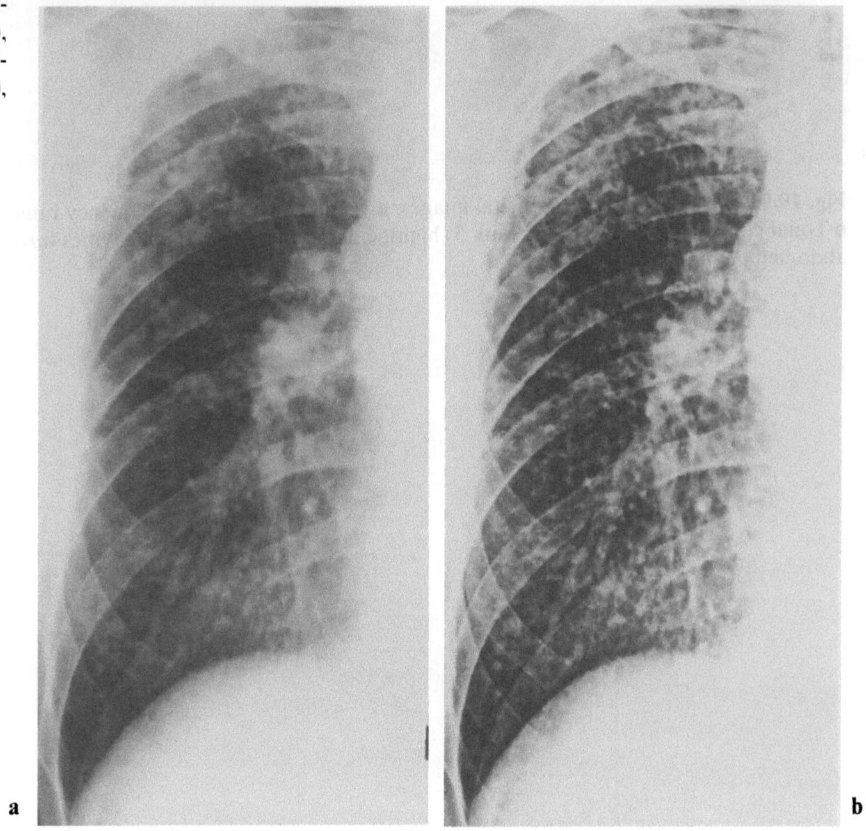

Fig. 19.8a, b. Close-up view of digital images. **a** Tonal conversion E, frequency rank 2 (low), enhancement 0.5. **b** Tonal conversion E, frequency rank 2 (low), enhancement 2.0

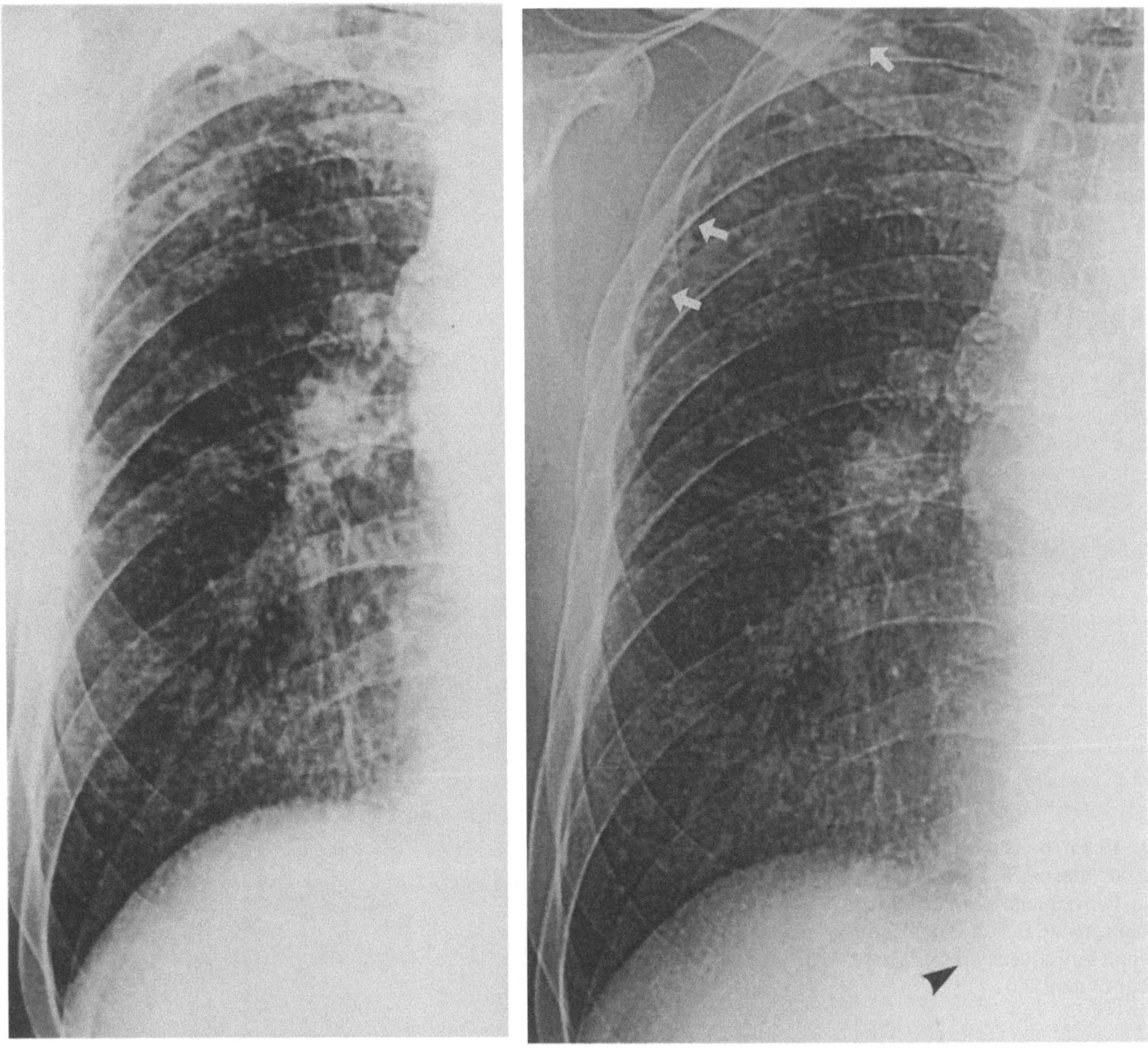

a b

Fig. 19.9a, b. Close-up view of digital images. **a** Tonal conversion E, frequency rank 4 (moderate), enhancement 2.0. **b** Tonal conversion G, frequency rank 5 (high), enhancement 5.0. Small cavities (*arrows*) and eggshell calcification of abdomen (*arrowhead*) are present

Fig. 19.10. Foundry worker's lung s2/2, bu, ho in a 68-year-old male. A bulla is present (*arrow*)

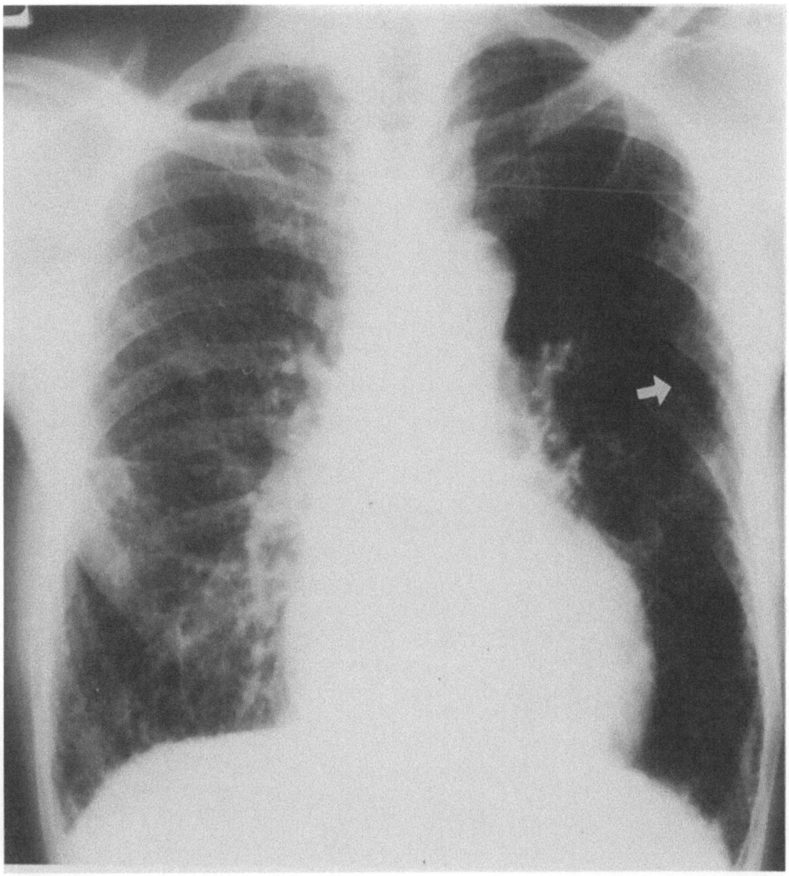

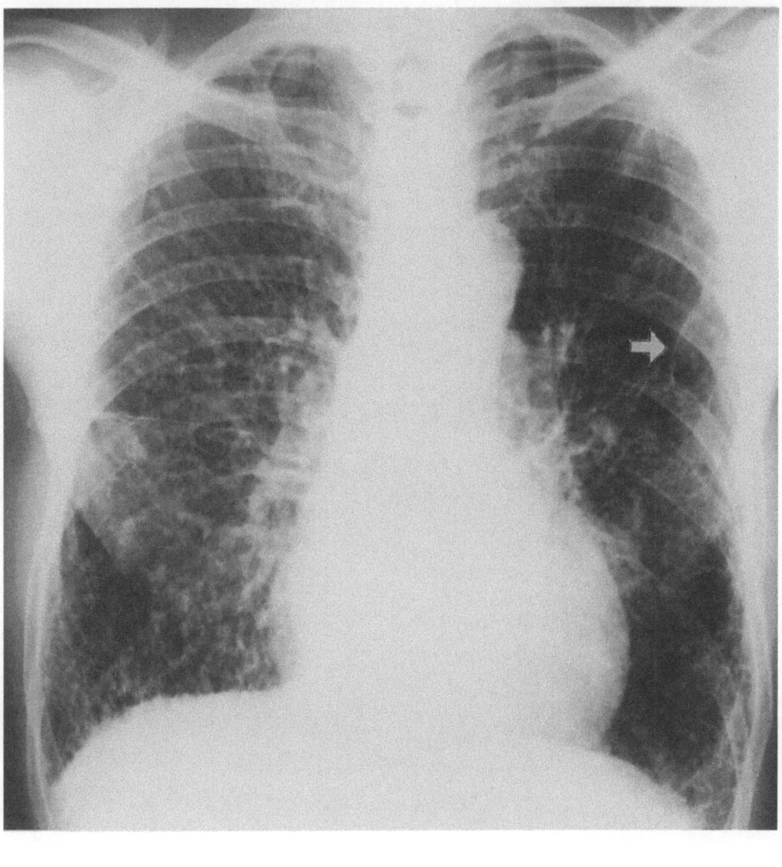

Fig. 19.11. Digital image. Tonal conversion E, frequency rank 2 (low), enhancement 0.5. A bulla is present (*arrow*)

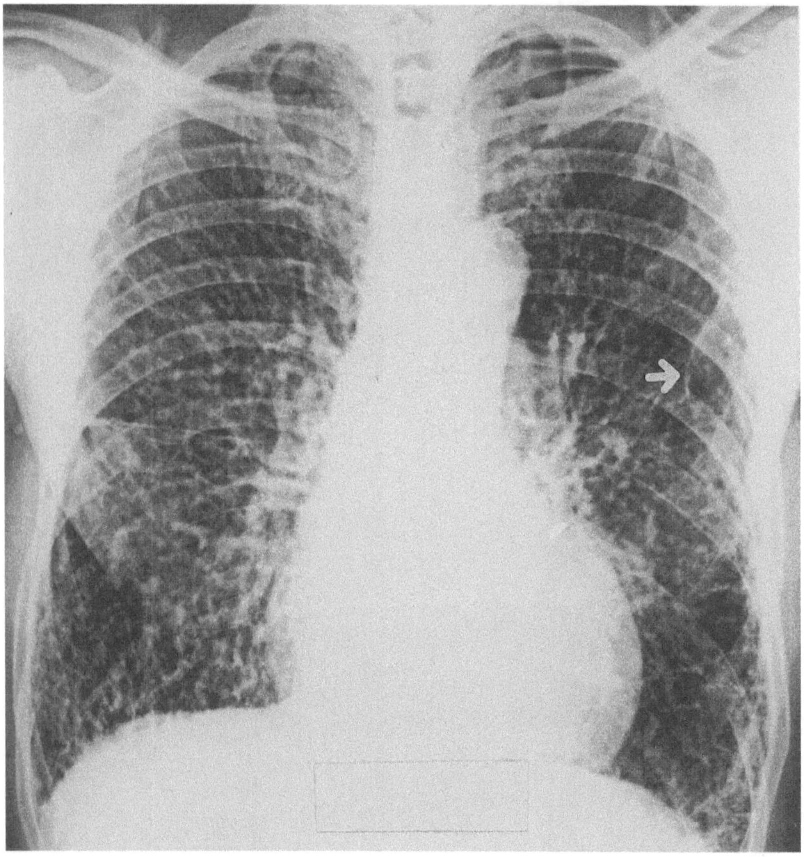

Fig. 19.12. Digital image. Tonal conversion E, frequency rank 2 (low), enhancement 2.0. A bulla is present (*arrow*)

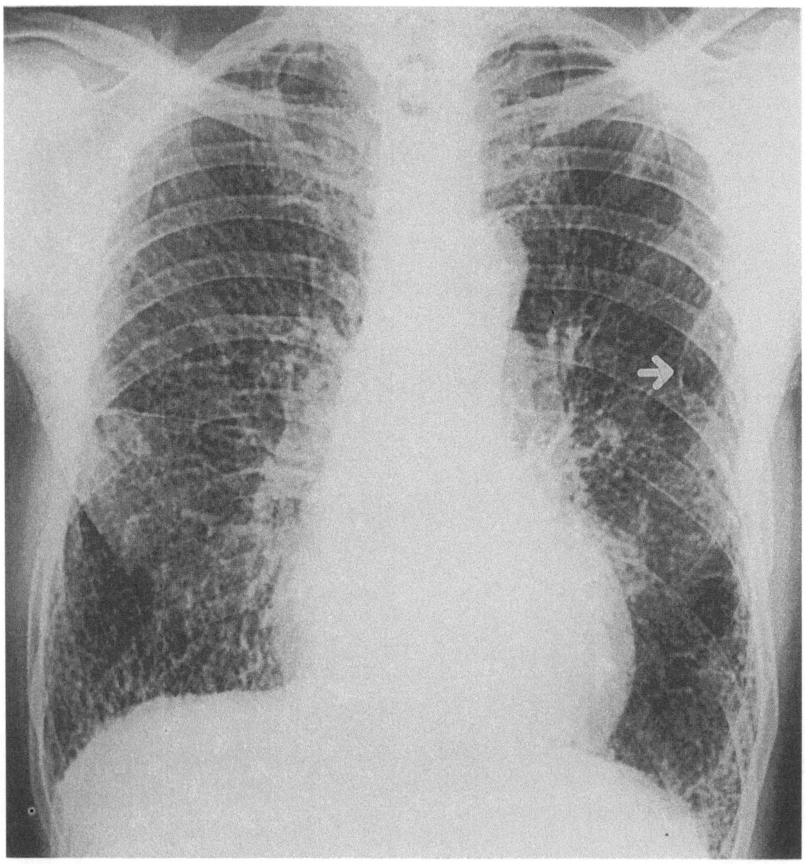

Fig. 19.13. Digital image. Tonal conversion E, frequency rank 4 (moderate), enhancement 2.0. A bulla is present (*arrow*)

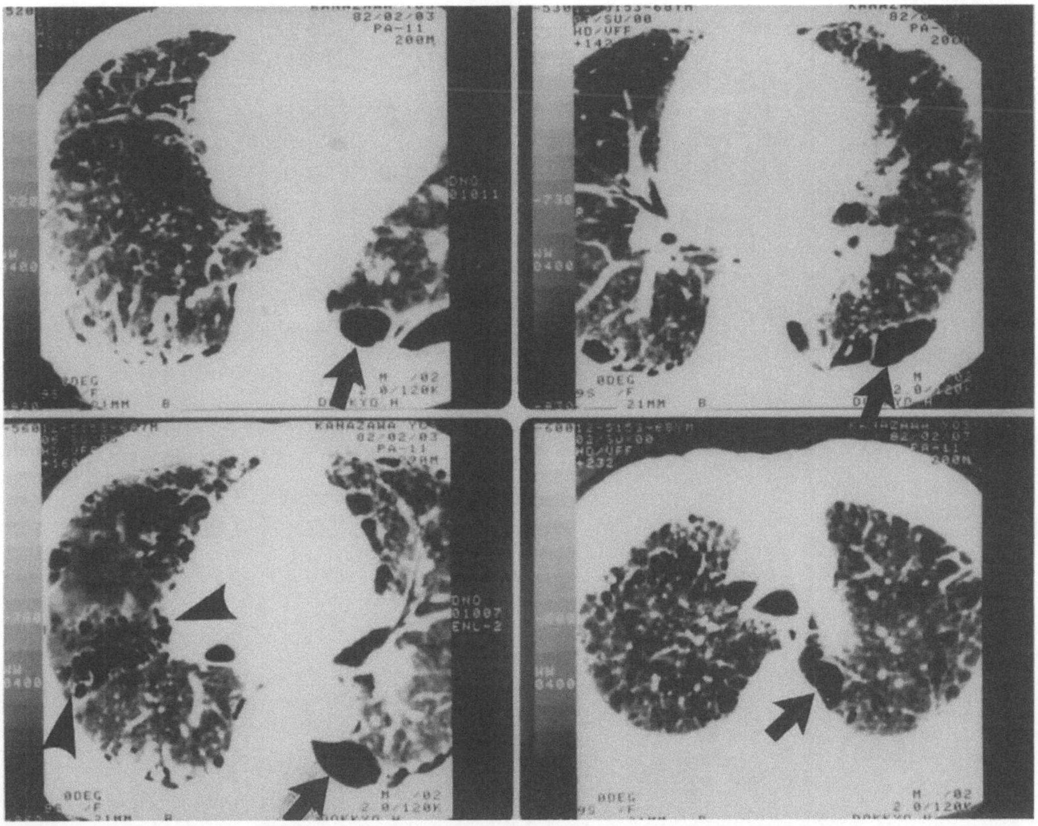

Fig. 19.14. Computed tomograms. Bullae (*arrow*), fibrosis, and emphysema are present (*arrow heads*)

Discussion

The CR system is at present under investigation for use in the radiological diagnosis of pneumoconiosis. From the reduced X-ray exposure, clear visualization of opacities, and wide dynamic range, it appears that it will make a large contribution to X-ray film interpretation. Furthermore, it can be asserted that CR will be an excellent device for mass surveys and its use of digital signals will facilitate automatic diagnosis.

References

1. ILO International Classification of Radiographs of Pneumoconisoses (1980) ILO, Geneva
2. Sonoda M, Takano M, Miyahara J, Kato H (1983) Computed Radiography utilizing scanning laser stimulated luminescence. Radiology 148:833–838
3. Ishida M, Kato H (1984) Hardware and software for image data processing-digital radiography Japan. J. of M. E. 22, 1:53–60
4. Eguchi K, Anan M, Yamada T, Ikeda S (1983) Evaluation of diagnostic digital radiography of the chest in comparison with conventional chest X-ray films. Japan Chest Dis. 21, 12:1139–1144

References

Chapter 20

Digital X-ray images of pneumoconiosis and their evaluation

Kenjiro Fukuhisa[1], Takeshi A. Iinuma[1], Toru Matsumoto[1], Yukio Tateno[1], Tokuro Nobechi[2], Hisao Shida[3], Yutaka Hosoda[4], and Masamitsu Ishida[5]

Introduction

The method of radiographic diagnosis of pneumoconiosis has been internationally standardized; namely, each film is evaluated according to the profusion, shape, and size of small opacities with reference to preselected standard X-ray films of typical pneumoconiosis [1, 2]. Recently, new diagnostic technologies such as computed tomography (CT) have been introduced, although they are still in the experimental stage [3–5]. In this chapter, we report the results of an objective study performed by many specialists on the applicability of the computed radiography (CR) system in the diagnosis of pneumoconiosis, a field which has a long history.

Purpose

The radiographic diagnosis of pneumoconiosis is carried out according to a world-wide diagnostic system standardized by the International Labor Organization (ILO) [6–10]. By mutual agreement among specialists in various countries, ILO designated standard X-ray films for the diagnosis of this condition. Currently, each X-ray film for diagnosis is evaluated with reference to such standard films. In this diagnostic system, the profusion of opacities on a film is classified as 0, 1, 2, or 3 (or more precisely according to a 12-grade scale: 0/–, 0/0, 0/1, 1/0, 1/1, 1/2, 2/1, 2/2, 2/3, 3/2, 3/3, or 3/+); the shape and size of small opacities are classified and indicated by the six letters p, q, r, s, t, and u. Reference X-ray films consist of typical X-ray films corresponding to this classification. However, with this diagnostic system, different exposure conditions (intensifying screen, kilovoltage, development processing, etc.) between the test and reference radiographs may induce a bias in the quality of films, particularly the contrast, thus strongly affecting the criteria of film reading (profusion, shape, and size of small opacities). This is the largest obstacle to achieving accurate diagnosis with this system [11].

On the other hand, CR, a computer-controlled digital system, is reported to be capable of providing final output images of high stability (i.e., eliminating the fluctuation of film density); furthermore, this system is said to require a lower skin X-ray dose than that with conventional screen/film systems. From these features of CR we would thus anticipate that the shortcomings of conventional chest radiography may be surmounted by the use of this system [1, 2, 11].

However, two problems remain unsettled. One of them is the difference in the size of output films between conventional radiography (which has long used 354 × 354-mm films for the diagnosis of pneumoconiosis) and CR (which uses films half the length, i.e., one-quarter the area of conventional films). This difference requires us to determine in advance what effects can be expected from such size reduction. The other problem is related to the capability of CR to process the images freely with respect to tonal conversion and spatial frequency enhancement. This feature of CR requires the determination of the optimum processing conditions for the diagnosis of pneumoconiosis. The purpose of this investigation is to study the applicability of CR by answering these questions.

Materials and methods

Materials for image processing

Materials for the study had to be selected with the view of achieving the purpose stated above. However, the diagnosis of pneumoconiosis is made by

[1] National Institute of Radiological Sciences, 9-1 Anagawa 4-chome, Chiba, 260 Japan
[2] St. Luke's International Hospital, 10-1 Akashi-cho, Chuo-ku, Tokyo, 104 Japan
[3] Rosai Hospital For Silicosis, National Labor Welfare Corporation, 632 Takatoku, Fujiwaramachi, Shioya, Tochigi, 321-25 Japan
[4] Radiation Effects Research Foundation, 5-2 Hijiyama Park, Minami-ku, Hiroshima, 732 Japan
[5] Miyanodai Development Center, Fuji Photo Film, Co. Ltd., 798 Miyanodai, Kaiseimachi, Ashigarakami-gun, Kanagawa, 258 Japan

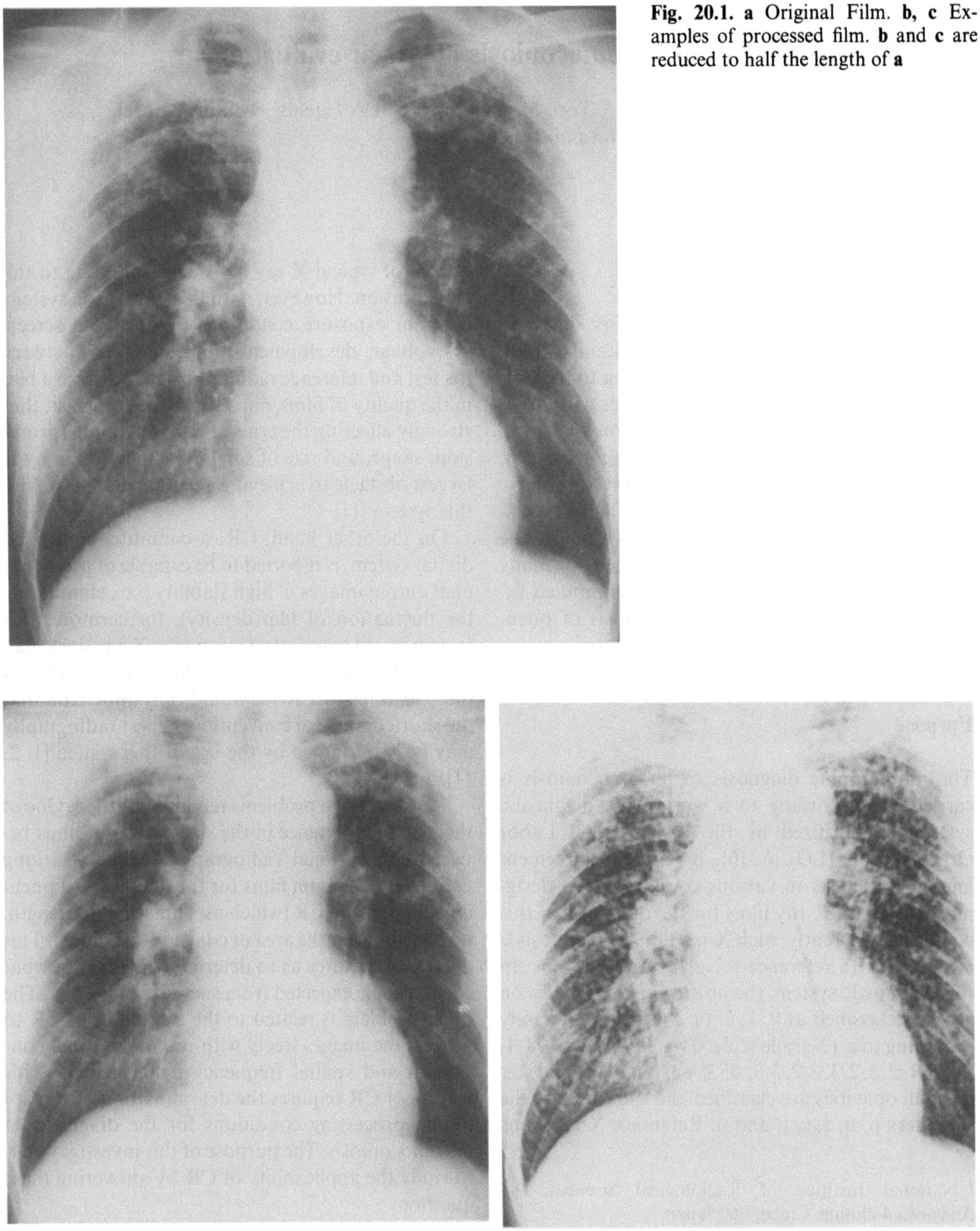
a b c

Fig. 20.1. **a** Original Film. **b, c** Examples of processed film. **b** and **c** are reduced to half the length of **a**

several physicians, and the diagnosis once made in this way is not always confirmed by surgery or autopsy. Therefore, the diagnostic value of CR cannot be correctly evaluated even when films of patients with pneumoconiosis are directly taken with CR. For this reason, US films for training in the diagnosis of pneumoconiosis and German films, which headed the list of unadopted candidates for ILO reference films, were selected for the study; the information derived from these films with a drum scanner was digitized and processed with respect to tonal conversion and spatial frequency enhancement by a computer in the same way as in CR. ILO reference and test films (US and German films) were equal in size

(354 × 354 mm), while the size of the processed films was reduced to half the length to resemble CR output images. Figure 20.1a–c shows an example of CR images (Japanese patient). Figure 20.1a is an original film which was classified as 2/2 (profusion) and q (shape and size), and Fig. 20.1b, c shows reduced films processed differently.

To investigate the effects of size reduction, we subjected ILO reference films to similar types of image processing by using a radiographic image-processing system, which included a drum scanner with both reading and writing units as well as a minicomputer and its subsystems. We selected the processed films that most resembled the original full-size ILO films.

Terminology

We here define the major terms used in this chapter.

Full-size ILO films. Fifteen sheets of the 1980 ILO reference films (three profusion grades—1/1, 2/2, 3/3; five grades of shape and size—p/p, q/q, r/r, s/t, t/t) were used as well as two films taken from normal Japanese volunteers; thus, we had 17 films in total.

Reduced ILO films. These were full-size ILO films processed in the same way as in CR. Of the reduced films obtained by this type of processing, the ones which were regarded from a preliminary experiment as the best representatives of full-size ILO images in a reduced form were selected for study.

Full-size test films. Twenty-three films were selected from the US films for training specialists in pneumoconiosis and the German films which headed the list of unadopted candidates for ILO reference films. Three films of normal Japanese volunteers, each of a different age and with a different thickness of chest, were also used.

Reduced test films. These were full-size test films processed in the same way as in CR; namely, the length of each side was reduced to half.

Specialists. These included six internationally recognized Japanese specialists on pneumoconiosis (members of the ILO pneumoconiosis committee or central pneumoconiosis examiners of Japan). They participated in a preliminary experiment.

Film readers. Eight physicians specializing in the diagnosis of pneumoconiosis participated in the main experiments and evaluated the full-size and reduced test films.

Table 20.1. Processing modes used in preliminary experiment 1

Processing mode	γ	Rank	E
I	1.0		
II	1.0	4	0.25
III	1.0	4	0.50
IV	1.0	4	0.80
V	1.0	2	0.25
VI	1.0	2	0.50
VII	1.0	2	0.80
VIII	0.85		
IX	0.85	4	0.25
X	0.85	4	0.50
XI	0.85	4	0.80
XII	0.85	2	0.25
XIII	0.85	2	0.50
XIV	0.85	2	0.85

E degree of spatial frequency enhancement

Methods of study

The study consisted of five steps, two of which are referred to as preliminary experiments and the other three as main experiments. They are described in outline below.

Preliminary experiment 1

To investigate the applicability of reduced films (processed in the same way as in CR) in the diagnosis of pneumoconiosis as well as to study the features of reduced ILO reference films, we processed some reference ILO films (profusion of 1/1 or 2/2) in 14 different ways as shown in Table 20.1. The films processed were presented for evaluation to the six specialists, who unanimously confirmed that diagnosis of pneumoconiosis was possible from the processed films and that the films processed by some of the 14 modes were very close to full-size ILO films in terms of image characteristics.

Preliminary experiment 2

On the basis of the results of preliminary experiment 1, six modes of processing were selected as shown in Table 20.2. Each of the full-size ILO films was processed by these six modes to yield 102 processed films in total. Dividing these 102 films into groups of six films of the same origin (i.e., processed from the

Table 20.2. Processing modes selected by preliminary experiment 1 and used in preliminary experiment 2

Processing mode	γ	Rank	E
1	1.0		
2	1.0	4	0.50
3	1.0	2	0.25
4	0.85		
5	0.85	4	0.50
6	0.85	2	0.25

E degree of spatial frequency enhancement

same ILO film), the components of each group were numbered, whereby the sequence of numbering was changed among the groups. Thereafter, films were evaluated by the six specialists, who recorded their assessment on the work sheets, which were later analyzed by computer.

Main experiment 1

The main experiments were performed by eight film readers (six film readers in some tests) who had been routinely engaged in the diagnosis of pneumoconiosis but had no knowledge of CR. Figure 20.2 presents a flow chart of the main experiments, and Fig. 20.3 shows the work sheet for film reading.

In main experiment 1, full-size test films were evaluated with reference to the full-size ILO films. This study was performed in view of the opinion that the profusion, shape, and size of small opacities seen on full-size test films were in some parts not consistent with the criteria of evaluation used in Japan. Another purpose of this study was to obtain one set of basic data through the evaluation of two kinds of film, each 354 × 354 mm in size.

Main experiments 2 and 3

Full-size test films were processed by each of the four modes selected by preliminary experiment 2. The processed films were given random numbers and evaluated in the sequence of the numbers by the same film readers as those for main experiment 2. The films in main experiment 2 were evaluated with reference to the reduced ILO films selected by preliminary experiment 2, while those in main experiment 3 were evaluated with reference to full-size ILO films.

Statistical analysis

Analysis of data from preliminary experiment 2

In preliminary experiment 2, the specialists evaluated only the profusion of small opacities on the films which were obtained by processing full-size ILO films by six different modes. This evaluation was made according to the ILO classification (1980); namely, the profusion was assessed according to a 12-grade scale (0/−, 0/0, 0/1, 1/0, 1/1, 1/2, 2/1, 2/2, 2/3, 3/2, 3/3, or 3/+). Each grade was given a weight ranging from 1 to 12. The weight for each grade was multiplied by the number of film readers who assessed the grade of the film. The total of this value for all grades was divided by the total number of film readers to yield the average profusion for a given film. For example, if a film is assessed as grade 0/1 (weight 3) by three readers, grade 1/1 (weight 4) by two readers, and grade 1/2 (weight 5) by one reader, the average density for this film can be obtained as follows:

$$(3 \times 3 \text{ readers} + 4 \times 2 \text{ readers} + 5 \times 1 \text{ reader})/$$
$$6 \text{ readers} = 3.67 = 4 - 0.33$$

This value is equivalent to grade 1/1 (weight 4). If this film is originally graded as 1/1, the expectation is 4; therefore, −0.33 can be regarded as the mean assessment value for this film calculated from the expectation. In a similar way, its unbiased variance is obtained as follows:

$$[(3-4)^2 \times 3 \text{ readers} + (4-4)^2 \times 2 \text{ readers} +$$
$$(5-4)^2 \times 1 \text{ reader}]/(6-1) \text{ readers} = 0.8$$

Assuming that it has a normal distribution, its standard deviation is obtained as follows:

$$\sqrt{0.8} = 0.894$$

If these calculations are made for all films, the total tendency will become evident. All values related to the profusion of small opacities were obtained on the basis of the above calculations.

Table 20.3 shows the means, standard deviations, and Student's t-test results (when the population variance was unknown and the population mean was 0.0) for each processing mode. The values were shifted to the minus side for processing modes 1, 3, and 4 and to the plus side for 2 and 5. Both the mean and standard deviation were smallest in processing mode 6. The t-values for modes 1, 2, 4, and 5 showed a significant difference at the level of 5%. Furthermore, film readers were requested to select one film out of the six processed films which they thought

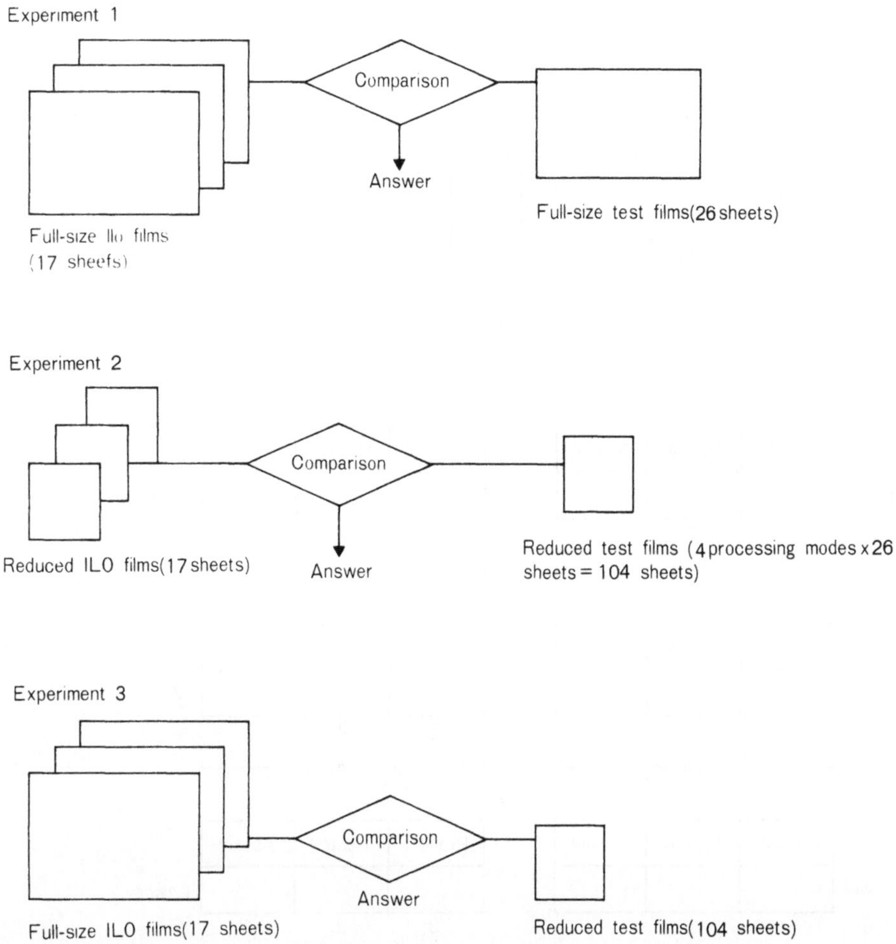

Fig. 20.2. Flow chart for profusion assessments. Three experiments are to be conducted in numerical order. Films are to be evaluated by independent reading. Each test film (particularly in experiment 2) is to be evaluated sufficiently with reference to the reference films. Results of film reading are to be entered on the work sheet

most accurately represented the information on the original full-size ILO films as well as one film which they considered to be the most appropriate for the diagnosis of pneumoconiosis. Table 20.3 includes the responses to these requests, although responses were not given by all film readers. From this table, the tendency can be seen that processing modes 4 and 6 produced images close to ILO films and that mode 5 was the most appropriate for the diagnosis of pneumoconiosis. On the basis of these and other analyses, we adopted four processing modes—4, 5, and 6 and a modification of 5, i.e., enhancement at a slightly lower frequency range for processing full-size test films. Thus, 104 reduced test films were produced. On the basis of the data shown in Table 20.3, mode 6 was selected for use in reducing ILO films in main experiment 2. The above four modes selected for the processing of films will now be renamed A, B, C, and D.

Mode A: $\gamma = 0.85$
B: $\gamma = 0.85$, rank 4, $E = 0.50$
C: $\gamma = 0.85$, rank 2, $E = 0.25$
D: $\gamma = 0.85$, rank 4, $E = 0.30$

Where E signifies the degree of spatial frequency enhancement.

Results of main experiment 1

From the data on profusion assessment for full-size test films, we calculated the mean values and standard deviations by using the above formulae. Since the data contained some evident errors, we eliminated the data exceeding two standard deviations from such calculations, and the mean and standard deviation were calculated from the remaining data; the results of this calculation were treated as correct answers. The shape and size of the small

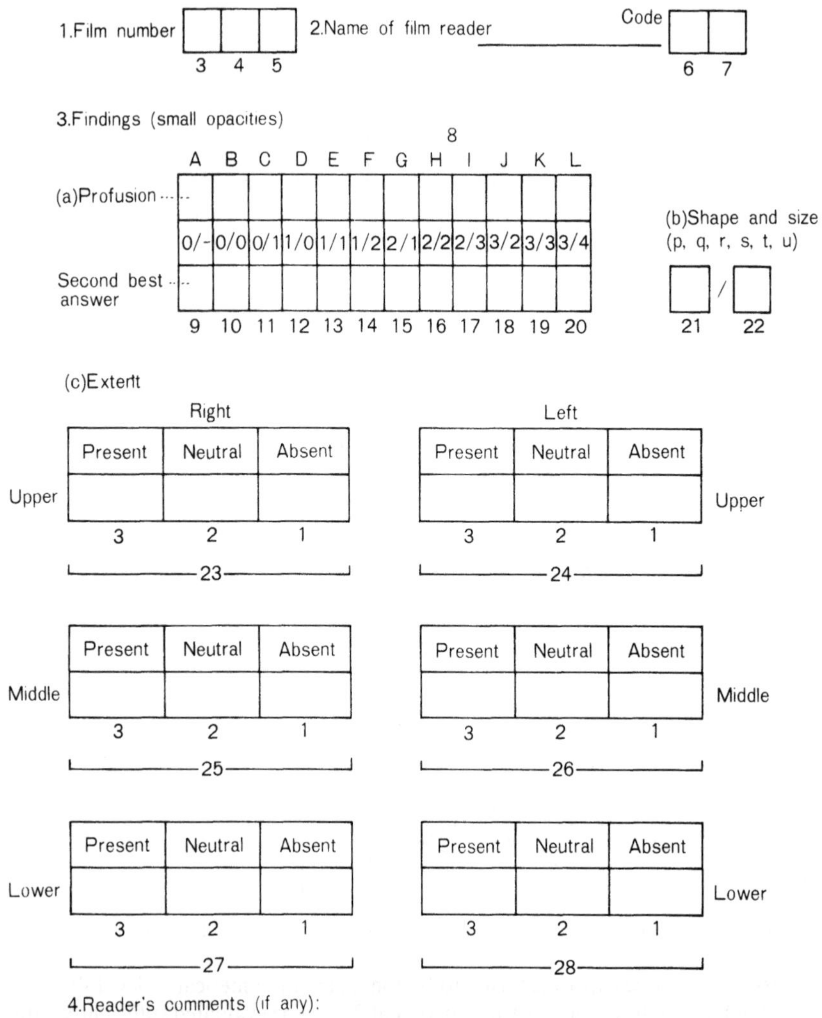

Fig. 20.3. Work sheet for evaluation of pneumoconiosis on X-ray films (prepared by the Committee for Research and Investigation of Pneumoconiosis Check Systems). This work sheet was used for comparison between full-size and reduced test films by committee members and comparison among full-size test films, reduced test films (referring to reduced ILO films), and reduced test films (referring to full-size ILO films). In analysis of data, the second best answer for profusion and the second answer for shape and size (for example, answer "S" in cases of "P/S") were eliminated

opacities were assessed according to the ILO classification (p, q, r, s, t, or u), whereby each film reader was requested to give two choices. The grade given by the majority of film readers as the first choice was adopted for the film concerned. If two or more grades were given by an equal number of film readers, the second choices were taken into consideration in deciding the grade for the particular film. In the evaluation of profusion, film readers were allowed to give two choices if they wished. In the analysis, second answers were not utilized because we felt that such dual grading had no practical meaning. Table 20.4 shows the results of the assessment of profusion and shape/size in 26 full-size test films. The data from main experiments 2 and 3 were analyzed by treating the values shown in Table 20.4 as correct answers. Table 20.4 also presents the original assessments of the profusion, shape, and size of the small opacities. This table shows that Japanese film readers tended to grade the films at higher levels of profusion than did the original US/European readers.

Analysis of profusion data obtained from main experiments

Table 20.5 presents the statistical analysis of the data obtained from main experiments 1–3. When full-size test films were evaluated with reference to full-size

Table 20.3. Results of preliminary experiment 2

Processing mode	Assessment 1: density			Assessment 2: resembling full-size ILO films	Assessment 3: appropriate for diagnosis of pneumoconiosis
	Mean	SD	Student's *t*-test		
1	−0.765	1.021	−7.646	7	5
2	0.216	0.709	2.591	4	19
3	−0.137	0.642	−1.426	6	6
4	−0.531	1.030	−5.646	43	5
5	0.382	0.897	4.074	4	36
6	−0.088	0.526	−1.224	33	9

films, both the mean and standard deviation were small, as expected. The test films processed by mode C provided data very similar to those for full-size test films irrespective of the reference films (full-size ILO or reduced ILO films). These data from the first evaluation of size-reduced films (half the length on each side and one-quarter the area), involving minute processing of tonal conversion and spatial frequency enhancement, suggest a high reliability for this processing technique. When Student's *t*-test was conducted in the same way as in preliminary experiment 2 (i.e., population mean at 0.0 and population variance unknown), only the films processed by mode C showed no significant difference at the level of 5%. Test films processed by any other mode provided quite similar results within each mode irrespective of the reference film (full-size ILO or reduced ILO films); namely, the films processed by mode A were rated as low profusion when either of the two reference films was used, the film processed by mode B as high profusion, and the films processed by mode D as slightly high profusion compared with the films processed by mode C.

Table 20.4 Comparison of assessments (profusion, shape, size) between film readers for the present study and original diagnosis

Film no.	Present study		Original diagnosis	
	Profusion	Shape/size	Profusion	Shape/size
1	0/0	—	0/0	—
2	0/0	—	0/0	—
3	0/0	—	0/0	—
4	0/1	s	0/1	s
5	0/1	p	1/0	s
6	0/1	s	0/1	p
7	0/1	p	0/1	q
8	0/1	p	0/1	t
9	1/0	p	0/1	t
10	1/0	p	1/0	t
11	1/0	q	1/0	q
12	1/0	p	0/1	q
13	1/0	s	1/0	p
14	1/1	s	1/1	p/s
15	1/1	p	1/0	q
16	1/2	q	1/1	q/s
17	2/1	q	2/2	p/q
18	2/1	t	1/1	s
19	2/2	q	2/1	q/s
20	2/2	q	2/2	q
21	2/2	q	2/2	q
22	2/3	q	2/1	p
23	2/3	q	2/1	q
24	3/2	r	3/2	q
25	3/2	q	3/2	q
26	3/3	q	3/2	q

Data on shape and size of small opacities

Tables 20.6 and 20.7 show the data on the shape and size of small opacities. These tables present the percentages of variable assessment on the shape and size of a given type of small opacity. In accordance with the ILO classification (1980), p, q, and r stand for rounded opacities and s, t, and u for irregular opacities. The diameter of the opacities is less than 1.5 mm for p, 1.5–3.0 mm for q, and 3.0–10.0 mm for r. The width is less than 1.5 mm for s, 1.5–3.0 mm for t, and 3.0–10.0 mm for u. There is no linear relationship between p, q, and r on the one hand and s, t, and u on the other. In the Japanese criteria for evaluation of small opacities, the classes of s, t, and u are not used. For this reason, in Table 20.7 we summarize the opacities as round and irregular, while in Table 20.6 we give the data for each ILO class of small opacity (p, q, r, s, t, u). The top of the left column shows the evaluation of full-size test films with reference to full-size ILO films and indicates that recognition by film readers may vary even between two films of the conventional full size (354 × 354 mm). The right column presents the evaluation of reduced test films

Table 20.5 Summarized results of porfusion evaluation by film readers

Reference	Test	Processing mode	Mean	SD	Student's *t*-test
Full-size ILO film	Full-size test film	—	−0.034	1.041	−0.475
Reduced ILO film	Reduced test film	A	−0.657	1.132	−8.820
		B	0.847	1.442	10.050
		C	0.096	1.151	1.318
		D	0.221	1.168	2.921
Full-size ILO film	Reduced test film	A	−0.662	1.401	−6.961
		B	0.675	1.885	6.101
		C	0.007	1.471	0.071
		D	0.129	1.681	1.239

with reference to full-size ILO films, and the left column (excluding the top) shows the evaluation of reduced test films with reference to reduced ILO films. Furthermore, for better comparison, we conducted a chi-square test (including Yates' correction) by using a 2 × 2 contingency table based on the assessment of full-size test films with reference to full-size ILO films; we marked the figures with * and ** if they had a significant difference at the level of 5% and 10%, respectively. The chi-square test was also conducted for each processing mode between the assessment referring to full-size ILO films and the assessment referring to reduced ILO films by using a 2 × 2 contingency table; we marked the figures with + if they had a significant difference at the level of 5%.

In Table 20.6, the evaluation of reduced test films with reference to full-size ILO films shows a larger difference in assessment among film readers than with reference to reduced ILO films. On the films processed with mode A, p and s tended to be evaluated as normal shadows, and q tended to be assessed as r or p, thus indicating that small opacities were difficult to read with this processing mode. On the films processed by modes B and D, p was evaluated correctly at a higher rate, but s tended to be evaluated as p, and q (when evaluated in reference to full-size ILO films) tended to be read as p, thus indicating that the evaluation of processed films becomes more accurate when reduced ILO films are used as reference films. Between the assessments referring to reduced and full-size ILO films, a difference was noted for class q when the processing was mode B or D; and the difference was significant for mode B at the level of 5%. As a general judgment, including the data for classes r and t whose evaluation showed no difference between the two reference films, we can see that modes C and D gave excellent results. Of these two modes, mode D was regarded as superior to mode C when judged on the basis of the accurate evaluation rate for classes p and q; the assessment results for the films processed by this mode were comparable with those obtained from the conventional evaluation of full-size test films using full-size reference films.

Table 20.7 can be considered a summary of Table 20.6. Table 20.7 also indicates that the difference in results of evaluation by different film readers is larger when full-size ILO films are used as reference films. On the films processed by mode A, rounded opacities and even irregular opacities were assessed as normal at a high rate. On the films processed by modes B or D, irregular opacities were read as rounded opacities at a high rate, although rounded opacities were correctly evaluated at a high rate.

General analysis of the accurate evaluation rates for profusion, shape, and size of small opacities revealed that the films processed by mode C provided data close to those obtained from the conventional method of film evaluation (i.e., evaluation of full-size test films using full-size reference films) and that the film-size reduction process in CR causes no difficulty in film evaluation.

Discussion

The chest has complicated structures composed of various organs with different absorption coefficients of X-rays. Therefore, chest radiographs contain much information which may vary minutely according to the chest thickness, age, sex, and other variables of the subject. For this reason, standardization of the optimum exposure and evaluation settings for the target organ is indispensable for accurate radio-

Table 20.6. Percentage of correct answers for each subcategory of shape and size

Processing mode	Interpretation	Full-size ILO films							Reduced ILO films						
		Normal	p	q	r	s	t	u	Normal	p	q	r	s	t	u
Full-size test films	Normal	96	4												
	p	18	45	5		20	13								
	q	1	4	71	15	3	6								
	r			25	75										
	s	19	34	3		41	3								
	t			25			75								
	u														
Mode A	Normal	89	6			6			88	8			4		
	p	43*	33	5		7	12		32	32	7		16	13	
	q	2	22*	53*	22		2		4	8	64	18	3	5	
	r			33	67						13	88			
	s	29	46		4	8*	13		41**	34			19**	6	
	t			50			50			13	13		25	50	
	u														
Mode B	Normal	89	11						75	17			4	4	
	p	12	71*	5		2*	10		7	68*	2		13	11	
	q	2	35**a	55*a	5*	2	2		1	9a	80**a	3*		8	
	r			67	33						38	63			
	s	13	63**			25			19	53	3		25		
	t		17	50			33				38		13	50	
	u														
Mode C	Normal	89	11						83	4			8	4	
	p	24	48	7		10	12		23	34	7		25	11	
	q	2	15**	60	22		2		3	6	58**	31*		1	1
	r			17	83						13	88			
	s	21	42			25	13		28	31			34	6	
	t			50			50						13	88	
	u														
Mode D	Normal	94	6						88	8			4		
	p	17	67*	5		5**	7		18	50	5		16	11	
	q	2	25*	68	3*		2		1	8	81	9	1	0**	
	r			50	50						38	63			
	s	17	50			29	4		25	53			22		
	t		33	17			50				25		25	50	
	u														

* Significantly different from full-size test film at the 95% confidence level
** Significantly different from full-size test film at the 90% confidence level
[a] Significant difference was noted between the assessment referring to full-size ILO films and the assessment referring to reduced ILO films at the 95% confidence level

graphic diagnosis in a given organ. Pneumoconiosis is recognized worldwide as an occupational disease for which a standardized method of radiographic diagnosis exists. However, radiographic diagnosis does not always agree with the symptoms, and this disease (particularly in mild cases) cannot be confirmed by surgery or biopsy as practiced in cancer cases. Therefore, diagnosis of pneumoconiosis often involves difficulties. The largest factor in such difficulties is, as stated above, the variable criteria of diagnosis arising from the lack of uniformity in the quality of chest X-ray films (i.e., differences in contrast, etc.). Efforts to obtain better films and to make a better diagnosis have been made for many years. However, considering the nature of radiography, we cannot expect much improvement in its diagnostic capability in the present situation.

With the CR system, on the other hand, radiographs are taken using an imaging plate instead of a conventional X-ray film; with this system, any part

Table 20.7. Percentage of correct answers for round and irregular opacities

Processing mode	Interpretation	Full-size ILO films			Reduced ILO films		
		Normal	Round	Irregular	Normal	Round	Irregular
Full-size test film	Normal	96	4	0			
	Round	8	75	17			
	Irregular	15	35	50			
Mode A	Normal	89	6	6	88	8	4
	Round	18*	74	8**	15**	70	15
	Irregular	23	50	27**	33	33	35
Mode B	Normal	89	11	0	75	17	8
	Round	6	88**	7*	4	83**	13
	Irregular	10	63*	27**	15	53	33
Mode C	Normal	89	11	0	83	4	13
	Round	10	81	9	10	74	15
	Irregular	17	43	40	23	25	53
Mode D	Normal	94	6	0	88	8	4
	Round	7	87*	6*	8	81	11
	Irregular	13	50	37	20	48	33

* Significantly different from full-size test films at the 95% confidence level
** Significantly different from full-size test films at the 90% confidence level

of the information contained on the imaging plate can be freely enhanced or weakened and images added or subtracted by means of converting the image into digital signals and processing the graduation or spatial frequency of an image with a computer. Therefore, this system is expected to be very useful in radiography of the chest, which involves complicated information. Among other uses, CR appears suitable for use in the diagnosis of pneumoconiosis, which requires evaluation of the profusion, size, and shape of opacities on X-ray films. However, as mentioned above, we have to explore the possible effects of different film sizes and processing modes on film evaluation before CR is applied to this field.

The present study revealed that the films processed by modes C or D could be utilized for diagnosis of pneumoconiosis without inducing a large difference from the diagnosis on full-size films. The reduced films used in the present study were prepared by means of the radiographic image-processing system. A similar experiment of film evaluation should be conducted using films which are taken and processed by CR itself so as to find a method of processing that enables much more accurate diagnosis.

In addition, it may be necessary to prepare reference films taken with CR so that diagnosis of pneumoconiosis may be performed with reference to them. If this is done, it may be possible to establish a criterion of diagnosis which can be accepted by everybody.

A possible theme for a future study on the diagnosis of pneumoconiosis is automated diagnosis with a computer. The logistics for this automated diagnosis have been established by Hall et al. [12] and Kobatake and Takatani [13] to such an extent that a fairly high level of accuracy can be expected from such a means of diagnosis. However, the irregular quality of X-ray films has been pointed out as an obstacle to automated diagnosis. The incorporation of the logistics of this automated diagnosis into CR may lead to the development of high-accuracy automated diagnostic systems. However, even when such systems are developed, the final diagnosis of pneumoconiosis and assessment of its improvement should be made by specialists using CR films. The methods of statistical analysis shown in this chapter should be useful as a basic tool for assessment of the efficacy of pneumoconiosis diagnosis.

Conclusion

This report can be summarized as follows.

1) We studied the possibility of using the CR system for making the quality of X-ray films uni-

form, an issue of great concern in radiographic diagnosis of pneumoconiosis.

2) Conventional diagnosis was made by referring to standard films, 354 × 354 mm in size, while the films for the CR system are a quarter the size of conventional ones. For this reason, we conducted an objective study to determine whether some errors in evaluation might arise from film-size reduction.

3) Fourteen pneumoconiosis specialists participated in the experiments on film reading. The films used in the experiments were ILO reference films, films used in diagnostic training for pneumoconiosis in the United States (full-size and reduced ones), and films which were not adopted as ILO reference films but which headed the list of unadoped candidates (full-size and reduced ones).

4) The experiments revealed that films which were reduced by an appropriate method of processing could be utilized for the diagnosis of pneumoconiosis without damaging the information contained on the original films.

5) The above results indicate that the technique studied has a great possibility of inducing a very significant technological innovation in the long history of pneumoconiosis diagnosis.

References

1. Sonoda M, Takano M, Miyahara J, Kato H (1983) Computed radiography utilizing scanning laser stimulated luminescence. Radiology 148:833–838
2. Ishida M, Kato H (1984) Seminar on digital radiography. Jpn J of Med. Electronics and Biomed Eng. 22:53–60
3. Shida T (1983) CT for pneumoconiosis. Kokyu 2: 201–20
4. Shida T, Hyodo H, Iwasaki N, Mishiro T, Kuramae S, Matsumoto M (1983) Suggestions on criteria of chest CT reading. J Tomography Research Soc 10:100–106
5. Saito N, Hosoda H, Chikauchi Y, Hachiya J, Horie S, Chiba Y (1983) Computed tomography of asbestos-caused pleural hypertrophy. J Jpn Assoc Chest Dis 21:1038–1048
6. Hosoda H (1979) Radiographic diagnosis of pneumoconiosis. Jpn Indust Sci J 2:1–6
7. Shima S (1983) Introduction to occupational diseases-Pneumoconiosis. Rodo Eisei 24:46–48
8. Ciyotani K (1980) Diagnostic criteria, typing and severity degrees for pneumoconiosis. Jpn J of Int Med 45:961–965
9. Labor Safety Section of Ministry of Labor (1979) Handbook for diagnosis of pneumoconiosis. Central Association for Prevention of Labor Accidents
10. Hosoda H, Nobutomo K, Saito N, Chiba Y (1981) Guideline for use of ILO 1980 classification of chest X-ray films of pneumoconiosis (Summary). Kokyuki Shikkan 32, 4:1–23 (Data)
11. Central Association for Prevention of Labor Accidents (1985) '83 Report of the committee for research and investigation of pneumoconiosis check systems (Image processing subcommittee)
12. Hall EL, Crawford W Jr, Roberts EL (1975) Computer classification of pneumoconiosis for radiographs of coal workers. IEEE (Trans Biomed Eng BME) 22, 0:518–527
13. Kobatake H, Takatani O (1982) Automated diagnosis of pneumoconiosis with chest radiographs. Jpn J of Med Electronics and Biomed Eng 20:307–313

Subject index

MIX
Papier aus verantwortungsvollen Quellen
Paper from responsible sources
FSC® C105338

If you have any concerns about our products,
you can contact us on
ProductSafety@springernature.com

In case Publisher is established outside the EU,
the EU authorized representative is:
Springer Nature Customer Service Center GmbH
Europaplatz 3, 69115 Heidelberg, Germany

Printed by Libri Plureos GmbH
in Hamburg, Germany